CW01067062

Gases in Medicine
Anaesthesia

Gases in Medicine
Anaesthesia

Edited by

E.B. Smith and S. Daniels
University of Wales, Cardiff, UK

The Proceedings of the 8th BOC Priestley Conference organized by The Royal Society of Chemistry, held at the Scientific Societies Lecture Theatre, London and Burlington House, London on 22–24 April 1997.

Special Publication No. 220

ISBN 0-85404-718-2

A catalogue record for this book is available from the British Library

Published by The Royal Society of Chemistry,
Thomas Graham House, Science Park, Milton Road,
Cambridge CB4 0WF, UK

For further information see our web site at www.rsc.org

Typeset by Computape (Pickering) Limited, North Yorkshire

Preface

The Priestley Conferences were established in 1977 by the Royal Society of Chemistry and are sponsored by BOC Gases. The triennial conferences recognise Joseph Priestley's contribution to science and the subject of this, the 8th Priestley Conference, anaesthesia is particularly appropriate given Priestley's work on gases.

The conference was organised by a committee under the chairmanship of Professor E.B. Smith around three scientific sessions covering new anaesthetics, the molecular and cellular mechanisms of anaesthesia and non-hypnotic effects of anaesthetics and other medical gases. In addition, there was an historical session that addressed the origins and introduction of anaesthesia organised by Professor C.A. Russell.

Professor Cedric Prys-Roberts, who gave the BOC Centenary Lecture, opened the conference, addressing the question of anaesthesia in the 21st century. Professor Prys-Roberts looked forward to the day when our understanding of anaesthesia will allow the automated administration of anaesthetics together with the other drugs that might be needed and that might interact with the anaesthetic. At this point, he argued, anaesthesia would truly become a 'science'.

The first paper in the historical session highlighted the contribution that Priestley made to the introduction of anaesthesia by his discovery of nitrous oxide. Priestley's work on simple gases had stimulated Thomas Beddoes who established a 'pneumatic' institute for the investigation of the physiological effects of breathing exotic gases. It was at this institution that Humphry Davy investigated the effects of breathing nitrous oxide. Amongst the many famous people who tried breathing nitrous oxide was Joseph Priestley who apparently did not find it as pleasant as did others. Thus Priestley gave impetus to the scientific investigation of anaesthesia and, in the great tradition of early physiology, was also an experimental subject.

Professor R.M. Jones, who gave the closing Plenary Lecture 'Is there a need for new anaesthetics', brought the technical sessions to a close. In the course of a highly entertaining lecture Professor Jones answered this question in a most positive fashion stressing the need to consider the clinical aspects of anaesthetics as well as the molecular biology.

Professor Salvadore Moncada FRS gave the 1997 Priestley Lecture. He reviewed the biological role of another simple gas discovered by Joseph Priestley, nitric oxide. Professor Moncada reviewed the remarkable discoveries over the past 10 years of the role nitric oxide plays in many biological systems, including chemical neuro-transmission. He finished with the thought that it was over 200 years from the discovery of nitric oxide to the elucidation its biological role and reminded us of Bacon's statement that:

'*For it appears at first incredible that any such discovery should be made, and when it has been made, it appears incredible that it should so long have escaped men's research*'.

As has been the practice at previous Priestley conferences there was a Schools Competition open to children between the ages of 13 and 18 at schools in the UK. A number of colourful and inventive submissions were received and the late Lord Dainton, on behalf of the Royal Society of Chemistry, presented prizes to those judged most successful.

It is hoped that this collection of papers will illustrate to those in academia and medicine the diversity of research into anaesthesia, the long history to this endeavour and a flavour of its fascination to the many distinguished scientists and clinicians who have contributed over the years.

E.B. Smith and S. Daniels.
Editors.

Contents

These proceedings are dedicated to three members of the BOC Priestley Conferences Organising Committee, all of whom died early in 1998: Dr Jack Barrett, the founding Chairman of the Organising Committee, Lord Dainton of Hallam Moors and Professor Max McGlashan.

BOC Centenary Lecture 1997
Anaesthesia in the 21st Century

Cedric Prys-Roberts

SIR HUMPHRY DAVY DEPARTMENT OF ANAESTHESIA,
BRISTOL ROYAL INFIRMARY, BRISTOL, UK

Applying the premise that an oarsman steers by his wake, I intend to tackle my impossible task of predicting what might happen to anaesthesia in the whole of the next century, on the basis of forward extrapolation from our present state of knowledge. Given that extrapolation is potentially a statistically weaker process than interpolation, I shall limit that degree of extrapolation to no more than ten years into the 21st century. To extend my prediction beyond that limit would be to court ridicule let alone downright rejection. Our ability to interpret phenomena is no greater than the sum of our knowledge of the natural sciences which pertain to anaesthesia, surgery and other branches of medicine which we practice. Lord Kelvin expressed it perfectly [1]:

'... when you can measure what you are speaking about, and express it in numbers, you know something about it; but when you cannot measure it, when you cannot express it in numbers, your knowledge is of a meagre and unsatisfactory kind; it may be the beginning of knowledge, but you have scarcely, in your thoughts, advanced to the stage of Science whatever the matter may be.'

Thus restricting my crystal ball gazing, I shall attempt to use numbers such as we know them at present to justify any planned leaps into the dark. A reliable paradigm for such an extrapolation is what is known in control theory as 'two step ahead adaptive control with exponential forgetting' [2]. This wonderful piece of jargon gave me the key to presenting a scientifically valid prediction of where the science and chemistry of anaesthesia may be by the year 2010.

The broad mathematical principle of this method of adaptive control is that the variable to be predicted is measured at a given sample interval, determined by the precision of control required, and sequential values are submitted to a standard least squares regression, using the same sample interval, the next two values in the sequence are determined by extrapolation, and the accuracy of the prediction is governed by the residuals in the least squares calculation. As this prediction is made on-line, usually at intervals of about 30 seconds for human cardiovascular variables, in order to minimise the effects of distant historical values, these are 'forgotten' in an exponential manner, so that only relatively recent values influence the prediction [3,4]. We applied this process to the closed loop control of blood pressure using a stepping motor connected to the concentration control bezel of an isoflurane vaporiser This allowed the computer to determine the required concentration of isoflurane to maintain a predetermined blood pressure level, and allowed us to determine, in a completely unbiased way, the interaction of pre-dosing with clonidine, an α_2-adrenoceptor agonist, in decreasing the required concentration of isoflurane [5]. This is a wonderful tool which could be applied to investigating the interaction of other drugs in decreasing the dose of an inhalational or intravenous anaesthetic required to maintain a given level of anaesthesia. As we can apply such control theory to the regulation of a physiological variable such as blood pressure or heart rate, both of which can be measured with a remarkable degree of precision (direct blood pressure to ± 1 mm Hg during quiet spontaneous breathing; ± 2 mm Hg during IPPV) it is reasonable to ask whether we could control the delivery of other drugs, to produce a given state of anaesthesia. The existing control algorithm is certainly robust enough, and new ones controlled by artificial neural networks may solve any non-linearity problems. The problem does not lie in the control algorithms but, precisely as Lord Kelvin suggested, in that we cannot adequately measure anaesthesia; we find it difficult to express it in numbers. Will that be the major breakthrough by or at the start of the next century? As I look at the programme for this meeting I see considerable evidence of progress towards understanding how the gases and vapours which produce clinical anaesthesia act at membrane and synaptic sites, and in hippocampal slices, but I detect no attempt to express clinical anaesthesia in numbers.

General anaesthesia is an unnatural but reversible state of depression of the central nervous system of such a degree that consciousness is lost and that on recovery nothing is recalled relating to the period of anaesthesia [6]. In humans there is also depression of spontaneous ventilation, motor hypotonia [7], and suppression of physiological responses to noxious stimulation. But which of these functions are effects of anaesthetics on the brain, as opposed to the spinal cord or the motor neurons? Most of the evidence now suggests that the suppression of somatic motor responses to noxious stimulation is a function which is suppressed at a motor neuron level, in that increasing concentrations of anaesthetics hyperpolarize the motor neurones at a spinal level. The index of equipotency of inhalational anaesthetics, the minimum alveolar concentration (MAC) is independent of cerebral function [7,8]. A new set of correlations between the human MAC and various (lecithin, olive oil, benzene) solvent/gas partition coefficients [9] emphasise that the site of action may be related to neuronal cell membranes. It is now clear that to explain cerebral actions of anaesthetics (loss of consciousness and new learning) in such terms we need new MAC values for suppression of consciousness rather than for suppression of somatic motor action. Such data are difficult to come by and will require studies of large numbers of patients to provide complete concentration/effect curves, rather than relying on the rather small numbers on which traditional

MAC values have been based [10]. It is fortuitous that the concentrations of both inhalational or intravenous anaesthetics required to suppress movement in response to noxious stimuli are considerably higher than those required to suppress consciousness and short term memory. Nevertheless, if we are to make any progress in the next century in the direction of automatic control of anaesthesia we must be able to correlate some measurable index of the state or level of anaesthesia with varying degrees of wakefulness, through sedation, to varying degrees of the anaesthetic state. Such correlations were achieved to a remarkable degree during induction of anaesthesia with five intravenous anaesthetics in volunteers by Schüttler and colleagues [11,12]. They infused the relevant drug at a constantly increasing rate until the subject lost consciousness and certain reflexes, at which point the infusion was turned off and the subject allowed to awaken. This process was repeated twice. They then correlated an index of neurophysiological activity, the median power frequency (MPF) of the electroencephalogram, with blood concentrations of the relevant anaesthetic. For propofol, for instance, the Cp_{50} (the blood concentration at which the MPF is decreased by 50%) was found to be 2.3 $\mu g\ ml^{-1}$. Those studies were based on transient states of anaesthesia, with the inevitable hysteresis between the pharmacokinetic and dynamic responses of the subjects, both during increasing and decreasing drug concentration. What happens during quasi-steady state anaesthesia, in which the drug is infused according to a pharmacokinetic model to try and reach a stable state of anaesthesia?

Forrest and colleagues [13] maintained 72 patients in states varying from awake, through light to heavy sedation, to light and deep anaesthesia, by predetermined infusion schemes designed to achieve and maintain steady blood propofol concentrations over a range from 1 to 9 $\mu g\ ml^{-1}$. At 30 minutes into such infusions there was a clear and strong correlation between the blood propofol concentration and the patient's state of consciousness, and the eye-lash reflex; and also with the suppression of the EEG median power frequency. The minimum MPF during stable anaesthesia was 6 Hz, substantially higher than that recorded by Schüttler and his colleagues [12] although the Cp_{50} for EEG suppression (2.3 $\mu g\ ml^{-1}$) was identical in both studies. A subsequent study [14] using a different approach in volunteers found a Cp_{50} (awake) for propofol of $2.69 \pm 0.56\ \mu g\ ml^{-1}$.

Based on their foregoing studies the Bonn group [15] chose an MPF of 3 Hz as the set point for closed-loop of a propofol infusion to maintain surgical anaesthesia. Based on the stable infusion studies [13] it would seem that this would ensure a more than adequate level of anaesthesia, indeed they would be overdosing the patients by at least 30%. You simply do not need a computer controlled infusion system to overdose patients by that margin. It can be done just as effectively and much cheaper with a simple manually controlled step infusion scheme [16].

This is a good example of progress being held up by the application of information derived from inappropriate studies. One cannot fault the original studies [11,12], indeed they were innovative and brought considerable new insights into the relationships between the pharmacokinetics and dynamics of intravenous anaesthetics. The many subsequent studies of closed-loop control by the same group [15 is a good example] were also innovative, and the control programme was properly applied, but the chosen set point resulted in systematic overdosing and a potentially excessive level (depth) of anaesthesia. One has to ask why the market has not been inundated with commercially available devices to control anaesthesia.

Another example of this problem concerns the application of mid-latency auditory evoked potentials (MLAEP) as an indicator of so-called depth of anaesthesia [17-19] and as a controlled variable for closed-loop control of anaesthesia. There is no doubt that the latency of the MLAEP increases and the amplitude decreases, but there have been few studies which adopt critical statistical estimates of the selectivity and sensitivity of the various parameters derived from the MLAEP as a means of classifying correlations between such parameters and the state of the patient's consciousness or, for that matter, indices of suppression of the physiological response to noxious stimulation during anaesthesia. There has also been a good deal of publicity, but regrettably few published data, about the control of anaesthesia using very simple computers to control drug infusions on some derived index from the auditory evoked potential. Based on figures in published articles one might imagine that it should be easy to derive a value for the shift of the latency, but therein lies the fallacy. It is very easy to publish a figure showing an almost perfect evoked potential, whatever that may be, but it is a quite different order of difficulty to persuade such a potential to repeat itself, from one period to the next, and one minute to the next with any consistency. It can be done by very complex digital filtering of the original signal – the main problem is that we are dealing with a microvolt signal in a forest of millivolt bushes and larger trees.

Tooley and colleagues [20] applying such digital filtering techniques to evoked potentials derived from groups of patients similar to those in the Forrest study [13], showed a 'best correlation' between the latency of the Nb wave and blood propofol concentrations during stable anaesthesia, with a threshold for loss of consciousness of 53 ms, a specificity of 96% and a sensitivity of 100%. Clearly there is a correlation, but how much confidence can we have in the values that we measure? In other words we have to modify Kelvin's dictum and introduce the concept that it simply isn't good enough to express anaesthesia in numbers – we must be able to do so with confidence, and that is the beginning of statistical science.

When we come to correlations between neurophysiological signals and conscious or anaesthetised states, the problem which arises is with boundaries. When we define good or adequate anaesthesia, we must be able to differentiate such a state from bad anaesthesia, during which the patient may be aware. Can we use current neurophysiological signals to aid such differentiation? Is there further information that can be derived from signals such as the electroencephalogram which can enable us to better classify the patient's descriptive state?

The EEG is basically chaotic. It is a stochastic process, a set of random variables which unfold in time, strictly stationary only if all its statistical properties are invariable with time. Fourier analysis, widely used to analyse the complex frequency spectrum, is not strictly appropriate for a stochastic or truly random variable. There are three popular models for analysis of stochastic data in the time domain, autoregressive, or moving average, or a combination of both. Unlike Thomsen and colleagues [21] who also used autoregressive modelling, we allowed the EEG data from [13] to be analysed by a multi-layer neural network [22]. Bayes's theorem enables the classification of events according to conditional probabilities. Such classifications can be determined by computer cells called perceptrons which are programmed to classify events by forming decision boundaries based on statistical probabilities [23]. Thus a modern computer perceptron can be considered as a discriminant function for a two class problem. Multi-dimensional or non-linear problems can be classified by using multiple layers of perceptrons – a so called neural network [24].

The EEGs from our digitised database [13] were analysed by a Multi Layer Perceptron developed by Holt and Tarassenko in the Department of Engineering Science in Oxford. Their multi-layered perceptron was trained on examples of the awake EEG, and was subsequently able to classify three stages of natural sleep in another database of sleep EEGs [22]. Based on the same training samples, the perceptron was able to classify the EEG of our sedated or anaesthetised patients, with a high degree of probability on to quite different mappings to those of natural sleep. Thus our ability to use statistical processes to better predict the state of anaesthesia in our patients may lead to safer, more efficacious, and economical anaesthetic practice in the early years of the next century.

Do these predicitive methods allow us any glimpses into the future of new anaesthetics? Do we need any new anaesthetics or adjuvant drugs? We could manage quite effectively with what we have at present. We can give almost perfectly satisfactory anaesthesia and post-operative pain relief with the drugs that we have – the main adverse outcome for the majority of patients is the prevalence of nauusea and vomiting in the post-operative period. It must be clear from what I have said up to now that I believe that any major developments in anaesthesia during the next ten to twenty years will be the result of new computer technology. This will not only improve the standards by better controlling the delivery of drugs to achieve improved efficacy, greater safety, and above all greater economy; but it will also give us all better insights into the mechanisms of anaesthesia, which in turn may also improve our practice of anaesthesia.

References

1. McDonald DA. *Blood flow in arteries 2nd Ed.* London: Edward Arnold, 1974; p3.
2. Kulhavy R. Restricted exponential forgetting in real time identification, IN: *7th IFAC Symposium on Identification and Systems Parameter Estimation*, edited by Barker HA, Young PC. Oxford: Pergamon, 1985;pp 1143-1148.
3. Millard RK, Hutton P, Pereira E, Prys-Roberts C. On using a self-tuning

controller for blood pressure regulation during surgery in man. *Comput. Biol. Med.* 1987;**17**: 1-18.

4. Millard RK, Monk CR, Woodcock T, Prys-Roberts C. Controlled hypotension during ENT surgery using self-tuners. *Biomed. Meas. Infor. Contr.* 1988;**2**:59-72.
5. Woodcock TE, Millard RK, Dixon J, Prys-Roberts C. Clonidine premedication for isoflurane-induced hypotension. *Br. J. Anaesth.* 1988; **60**: 388-394.
6. White DC. Anaesthesia: a privation of the senses. In: *Consciousness, awareness and pain in general anesthesia.* Edited by Rosen M, Lunn JN. London: Butterworth, 1987.
7. King BS, Rampil IJ. Anesthetic depression of spinal motor neurons may contribute to lack of movement in response to noxious stimuli. *Anesthesiology* 1994;**81**:1484-1492.
8. Rampil IJ, Mason P, Singh H. Anesthetic potency (MAC) is independent of forebrain structures in the rat. *Anesthesiology* 1993;**78**:707-712.
9. Taheri S, Halsey MJ, Liu J, Eger EI, Koblin DD, Laster MJ. What solvent best represents the site of action of inhaled anesthetics in humans, rats and dogs? *Anesth. Analg.*1991; **72**:627-634.
10. Prys-Roberts C. A philosophy of anaesthesia: some definitions and a working hypothesis. In : *International Practice of Anaesthesia.* edited by Prys-Roberts C, Brown BR. Oxford: Butterworth-Heinemann, 1996; pp **1**/4/1-14.
11. Schwilden H, Schüttler J, Stoeckel H. Quantitation of the EEG and pharmacodynamic modelling of hypnotic drugs: etomidate as an example. *Eur.J.Anaesth.* 1985;**2**:121-131.
12. Schüttler J. Pharmakokinetik und Dynamik des intravenosen Anästhetikums Propofol. *Anästhesiologie und Intensive- medizin Bd 202.* Stuttgart: Springer Verlag, 1990.
13. Forrest FC, Tooley MA, Saunders PR, Prys-Roberts C. Propofol infusion and the suppression of consciousness: the EEG and dose requirements. *Br.J.Anaesth* 1994;**72**:35-41.
14. Chortkoff BS, Eger EI, Crankshaw DP, Gonsowski L, Dutton RC, Ionescu P. Concentrations of desflurane and propofol that suppress response to command in humans. *Anesth. Analg.*1995;**81**:737-743.
15. Schwilden H, Stoeckel H, Schüttler J. Closed-loop feedback control of propofol anaesthesia by quantitive EEG analysis in Humans. *Br.J.Anaesth.* 1989; **62**: 290-296.
16. Roberts FL, Dixon J, Lewis GTR, Prys-Roberts C, Harvey JT. Induction and maintenance of propofol anaesthesia: a manual infusion scheme. *Anaesthesia* 1988; **43** (Suppl):14-17.
17. Thornton C, Konieczko KM, Knight AB, *et al.* Effect of propofol on the auditory evoked response and oesophageal contractility. *Br.J.Anaesth.* 1989;**63**:411-417.
18. Heneghan CPH, Thornton C, Navaratnarajah M, Jones JG. Effect of isoflurane on the auditory evoked responses in man. *Br. J. Anaesth.* 1987;**59**:277-282
19. Thornton C, Jones JG. Evaluating depth of anaesthesia. Review of methods. In: *Depth of Anaesthesia* (Jones JG. ed). *Int. Anesthesiol. Clin.* 1993, No 4:67-88.
20. Tooley MA, Greenslade GL, Prys-Roberts C. Concentration-related effects of propofol on the auditory evoked response. *Br. J. Anaesth.* 1996; **77**: 720-726.
21. Thomsen CE, Rosenfalck A, Christensen KN. Assessment of anaesthetic depth by clustering analysis and autoregressive modelling of the electroencephalogram. *Comp. Method. Prog. Biomed* 1991;**34**:125-138.

22. Roberts S, Tarassenko L. Analysis of the sleep eeg using a multilayer network with spatial organisation. *IEE Proc-F.* 1992;**139**:420-425.
23. Rosenblatt F. the perceptron: a probabilistic model for information storage and organization in the brain. *Psychol. Rev.*1958;**65**:386-408.
24. Richard MD, Lippman RP. Neural network classifiers estimate Bayesian a posteriori probabilities. *Neural Comp.* 1991;**3**: 461-483.

New Anaesthetics

Intravenous Anaesthetics: the Alternative to Gases

John W. Sear

JOHN RADCLIFFE HOSPITAL, OXFORD, UK

The provision of clinically adequate general anaesthesia can be attained by either inhaled agents or intravenous hypnotic agents given alone or in combination with nitrous oxide.

The onset of action of the inhaled agents depends on their inspired concentration, alveolar ventilation and cardiac output; while offset depends on the alveolar concentration and the latter two factors as well as the extent of metabolism (particularly so in the case of halothane). One of the problems with all anaesthetics is that when used alone those concentrations needed to maintain hypnosis may cause significant cardiorespiratory depression; as well as other organ and drug specific effects such as altered cerebral blood flow regulation; altered hepatocyte integrity; fluoride-induced nephrotoxicity; uterine relaxation; risk of seizures; malignant hyperpyrexia; and impaired cellular immunity. Because of these concerns, there has been a trend towards the development and evaluation of new agents having improved pharmacological profiles. One obvious possibility is intravenously administered drugs.

For i.v. agents, onset of effect depends on dose administered, the speed of administration, cardiac output, and the blood-brain (=biophase) equilibration time. Offset is complex, and is the resultant of drug redistribution and metabolism, and the presence or absence of active metabolites (as is seen with thiopentone and ketamine, and some of the benzodiazepines). The side-effects of i.v. agents are similarly numerous. These can range from minor sequelae (e.g. pain on injection, hiccoughs, excess salivation, thrombophlebitis and other venous sequelae; excitatory involuntary muscular movements) to the more major (e.g. cardiorespiratory depression; laryngospasm and bronchospasm; rash, urticaria and true allergic reactions; and convulsions).

For the volatile agents, delivery needs a 'carrier gas' – usually nitrous oxide, which also provides analgesia as well as reducing the inspired concentration needed to maintain anaesthesia. In the case of i.v. agents, many are hydrophobic and need to be formulated as salts or in lipid emulsions. The choice of the carrier does not seem to affect potency, although it may alter the side-effect profile.

Over the past five years, there have been several new chemical entities and formulations evaluated as intravenous anaesthetics.

I. New Formulation of Intravenous Anaesthetic Agents

Lipid emulsions are normally used to formulate drugs that are water-insoluble; however they may also be used as carriers of hydrophilic compounds that cause significant side-effects such as pain on injection (methohexitone) or pain on injection and venous sequelae (etomidate).

a. Methohexitone: In volunteers, Westrin et al compared aqueous methohexitone with an emulsion formulation [1]. The incidence of spontaneously reported 'pain on injection' was reduced from 83% to zero, and visual analogue pain scores from 38 mm to 5 mm. In a further evaluation in 42 unpremedicated patients undergoing a variety of surgical procedures, the ED_{50} (hypnotic dose causing loss of consciousness in 50% subjects) for the emulsion formulation was 1.2 (0.1) mg/kg and 1.1 (0.1) mg/kg for aqueous methohexitone – showing that change in formulation did not appear to affect potency.
b. Etomidate: Studies with an emulsion formulation of etomidate have shown no change in the pharmacodynamic properties, but lower incidences of pain on injection, myoclonus and local thrombophlebitis [2, 3]. An additional advantage of the emulsion formulation is its lower osmolality and higher pH (400 mosmol/kg and pH 7.6 compared with 4965 mosmol/kg and pH 5.1 for the propylene glycol formation); the former results in less red cell haemolysis [4]. More recently, another etomidate formulation solvented in hydroxypropyl-β-cyclodextrin has been evaluated [5]. Its use was associated with less myoclonia and pain (17% vs 92%; and 8% vs 58%), less thrombophlebitis (0% vs 42%), and no haemolysis. There were no alterations in the kinetics or dynamics of etomidate. Again the reformulation has a lower osmolality. However none of these reformulations overcome the action of etomidate in inhibiting adrenal steroidogenesis.
c. Propanidid: Originally introduced into clinical practice in the UK in 1964, the allergenic potential of the solvent micellophor led to the eugenol being withdrawn from clinical practice in 1984. However the drug had a number of important advantages – principally an ultra-short duration of effect due to rapid hydrolysis of the ester linkage by tissue and liver esterases. Complete recovery was seen within 1 hour of a bolus dose of 5-7 mg/kg. Other non-hypnotic effects included excitatory movements in about 10% of patients, and cardiovascular depression due to negative inotropy, afterload reduction and a delayed response to histamine release. Habazettl et al have recently evaluated a liposomal preparation of propanidid in rats [6]. At the highest infusion rate (120 mg/100 g/hour), this formulation (in comparison with the original preparation) caused less cardiorespiratory depression, a decreased inci-

dence of clonic seizures and reduced mortality. Discontinuation of propanidid again resulted in rapid awakening. However further evaluation in pigs showed this liposomal preparation to have cardiovascular stimulant properties, and to cause increased serum concentrations of histamine, adrenaline and noradrenaline [7]. Thus it seems unlikely that this reformulation is going to offer much to the clinician.

d. Propofol: This sterically hindered phenol is formulated in an oil emulsion, and has properties of the ideal i.v. anaesthetic agent – rapid redistribution from blood to tissues, and a fast clearance from the blood with biotransformation to inactive metabolites; anti-emetic properties at both subhypnotic and hypnotic doses; and suitable recovery profile for day case surgery after bolus dosing and infusions [8]. However propofol causes cardiovascular depression (hypotension and bradycardia) which are potentiated by vagotonic opioids (such as fentanyl or alfentanil) and hypovolaemia. The other main side-effects are pain associated with i.v. injection, involuntary movements, hypotension and bradycardia, and rash – each of which occurs with a frequency of greater than 1 in 100 administrations. There have also been reports of epileptiform movements and true convulsions.

Attempts at decreasing the pain (which varies in incidence between 30 and 70%) include pre-treatment with lignocaine, aseptically mixing with lignocaine immediately before dosing, or pre-treatment with fentanyl or alfentanil. More recently it has been shown that addition of LCT (long chain triglyceride) to propofol (as Diprivan) will reduce the incidence of severe pain from about 70% to zero [9]. The mechanism behind this is thought to be the result of a decrease in the propofol concentration in the aqueous phase secondary to the increase in the fat content. However this change in aqueous drug concentration does not seem to influence the kinetics or dynamics of propofol [10].

Other attempts to reformulate propofol (to reduce the pain) in 2-hydroxypropyl β-cyclodextrin have proven unsuccessful as the drug has caused severe bradycardia and hypotension when administered to rats [11]. Hence again this does not appear to be a viable alternative to the emulsion formulation.

II Ketamine and its Enantiomers

Ketamine is a phencyclidine derivative, with the advantages of being water-soluble, and producing profound analgesia at subanaesthetic doses. While lacking the cardiorespiratory depressive effects of other i.v. anaesthetic agents, its usefulness has been limited by a high incidence of disturbing emergence reactions (in up to 30% of patients). However these side-effects can be reduced in both incidence and severity by the co-treatment of patients with other supplementary drugs (e.g. midazolam and propofol).

TIVA (total intravenous anaesthesia) techniques using ketamine and propofol or midazolam have been evaluated by a number of authors [12]. Use of propofol both reduces anaesthetic requirements of ketamine, and acts to obtund the incidence of side-effects.

Current interest extends to investigation of the relevant potency of the two stereoisomers [R(−) and S(+)]. The potency ratio for anaesthesia and analgesia is approximately 4:2:1 for S(+) ketamine: racemic ketamine: R(−) ketamine. S(+) ketamine produces longer hypnosis than the R(−) isomer at equipotent doses, with the racemate being intermediate. In a cross-over study, the dynamics of the racemate and S(+) ketamine were comparable with regard to haemodynamic and metabolic responses. However there was improved recovery with the latter formulation [13]. S(+) ketamine also causes a lower median EEG power spectrum; decreased locomotor activity; but equipotent analgesia [14].

Hering and colleagues have examined the effects of racemic ketamine and the S(+) isomer on the EEG spectrum after either i.m. or i.v. midazolam (0.1 mg/kg) [15]. After i.m. midazolam given 45 minutes before induction of hypnosis with either racemic ketamine 2 mg/kg or S-(+) ketamine 1 mg/kg, there was a reduction in EEG α power and an increase in the β frequencies; i.v. midazolam similarly activated the low β range (13-18 Hz). Ketamine (both racemate and S-(+) enantiomer) increased fast β activity (21-30 Hz) with an accompanying reduction in delta power. The median frequency increased from 6 to 10 Hz. The S-(+) isomer therefore appears to have the same central actions as the racemate without the psychotomimetic emergence sequelae or cardiovascular stimulant properties. The IC_{50} [drug concentration of ketamine necessary to achieve a 50% depression of the maximal EEG median frequency reduction] for S(+) ketamine was less than that for the R(−) and the racemate preparations (0.8 μg/ml compared with 1.8 and 2.0 μg/ml respectively) [16].

On the basis of an assumed equipotency ratio of S(+) ketamine to racemate of 1:2, Geisslinger and colleagues compared the kinetics and dynamics of the enantiomers of ketamine in 50 surgical patients. There were no significant kinetic differences between S(+) ketamine alone and the enantiomer present in the racemic mixture [17]. However the R(−) enantiomer showed a lower clearance and smaller apparent volumes of distribution when compared with the S(+) enantiomer when administered as one component of the racemate. These data correspond with the *in vitro* observations showing a faster metabolic degradation of the S(+) isomer; and where the addition of the R(−) isomer to S(+) ketamine inhibits the latter's metabolism [18]. However *in vivo* there were no differences in the clearance rates of the S(+) and R(−) isomers when given as the racemate. The concentration-effect relationship for S(+) ketamine therefore lies to the left of that for the racemate, and it also has a steeper curve.

Clinical evaluation of S(+) ketamine. When used as part of the anaesthetic technique in surgical patients, the incidence of emergence reactions to be about 37% after the R(−) enantiomer, 15% after the racemate and 5% after the S(+) isomer with comparable incidences of dreaming with all three treatment groups [19].

One feature of ketamine anaesthesia is the significant increase in blood cortisol, catecholamines and glucose concentrations [20]; similar increases occur with the S(+) enantiomer in surgical patients when used for elective lower limb orthopaedic surgery, although the increases in circulating plasma adrenaline concentrations was greater after use of the racemic mixture [21]. There are no apparent differences between the enantiomers and racemate in their haemodynamic effects. However recovery is faster with the S(+) isomer.

The endocrine effects can be obtunded if S(+) ketamine (1 mg/ kg/hour) is given as supplement to propofol; although even then the hormonal responses are greater than during infusion of alfentanil and propofol [22].

III New Steroid Induction Agents

The hypnotic properties of steroid molecules were first recognised in 1927 by Cashin and Moravek who induced anaesthesia in cats using a colloidal suspension of cholesterol [23]. However present day use of these molecules stems from the first systematic review of the hypnotic properties of steroids (mainly belonging to the pregnane and androstane groups) was conducted in rats by Selye [24]. Of the screened steroids, there was no apparent relationship between hypnotic (anaesthetic) and hormonal properties; the most potent anaesthetic steroid, pregnane-3,20-dione (pregnanedione) was virtually devoid of endocrinological activity. However one of the major problems with these steroidal agents was their poor water solubility, and little further work was conducted until Laubach and colleagues synthesized hydroxydione [25].

This was the 21-hydroxy derivative of pregnanedione, and was made water soluble by esterification at the C21 position as the sodium hemisuccinate. Hydroxydione had a high therapeutic index, and few adverse effects in cats and dogs [26]. In clinical practice hydroxydione produced minimal changes in cardiorespiratory function, good muscle relaxation, a low incidence of coughing and pleasant recovery, with a very low incidence of vomiting [27, 28]. However induction took several minutes. As there was early obtunding of the pharyngeal and laryngeal reflexes, it was possible to achieve airway intubation. The respiratory rate increased with an accompanying decrease in tidal volume – with a resulting increase in minute volume. Marked respiratory depression and apnoea were not usually seen. Cardiac output and arterial blood pressure also fell.

There were several unwanted side-effects; pain on injection and a high incidence of post-anaesthetic irritation at the site of i.v. administration and along the associated vein.

Because of these side-effects, chemists and pharmacologists at Glaxo, UK began to look for other steroids having the clinical advantages of hydroxydione but without the tendency to cause pain on injection and thrombophlebitis.

Chemistry of the steroid molecule. All steroids have at least 6 asymmetric carbon atoms, resulting in a number of possible isomers. However, for the naturally occurring steroids, isomerism only occurs around the AB ring junction (cis–trans: 'chair' and 'boat' forms). The BC and CD ring interfaces are always in the 'trans' arrangement. This compares with the 'cis' arrangement of the CD rings in most plant sterols. Both 5α- and 5β-steroids are found naturally occurring, but the 5β-isomers are the more predominant.

Anaesthetically active steroids show a number of important structure-activity features:

- anaesthetic activity requires the presence of an oxygen function (either hydroxy or ketone) at each end of the steroid molecule (in the C3 position, and the C20 position of pregnanes or C17 position of androstanes).
- substitutions into the steroid structure, such as hydroxy groups, reduced anaesthetic activity and occasionally introduced convulsant properties (e.g. 11β-hydroxy).
- highly active compounds were found among both 5α- and 5β-series.
- the C3 hydroxyl group could be either in the α or β position. In general, 3α-hydroxy-5α- or 3α-hydroxy-5β-pregnanes showed the greatest activity, followed by 3β-hydroxy-5β- and 3β-hydroxy-5α- compounds. 3-keto substituents had little or no hypnotic activity.
- esters of hydroxy compounds are in general less active and more slowly acting than the parent alcohols.
- a single double bond in the A or B rings does not significantly affect anaesthetic activity, but 2 or more double bonds in these rings or a single double bond in the D ring is associated with non-activity.
- presence of C5 hydrogen atom which is 'cis' to the C10 methyl group is associated with increased hypnotic potency.

On the basis of these observations, a number of other steroid hypnotic agents have been evaluated over the last 30 years (5β-pregnane-3α-ol-11,20-dione 3-phosphate disodium: GR 2/146; alphaxalone-alphadolone acetate: Althesin; and minaxolone citrate). However all have been associated with significant adverse side-effects:

GR 2/146: delayed onset of anaesthesia
paraesthesia in arm and neck after i.v. dosing

Althesin: allergic reactions to the solvent (Cremophor EL)
and/or the constituent steroids

Minaxolone: slow onset of action and delayed recovery
high incidence of excitatory movements and hypertonus
possible oncogenic effect in rats

However all of these induction agents showed high therapeutic indices, and hence further research led to the formulation and evaluation of 5β-pregnanolone (eltanolone).

Water-insoluble pregnane derivatives

It had been known for many years that ovarian steroids affect central nervous activity. Oestrogens increase brain excitability, while large doses of progesterone could produce sedation or deep sleep in man [29], although the onset was slow (as was later seen with hydroxydione).

3α-Hydroxy, A ring reduced C_{19} and C_{21} steroids bind to the $GABA_A$ complex. Figdor and colleagues showed in male albino Swiss mice that progesterone (PG), 5α-pregnanedione (5α-P) and 3α-OH, 5α-pregnan-20-one (5α- or allo pregnanolone) were all potent hypnotic agents [30], with the latter having a therapeutic index of 29 [31]. Moreover there is conversion of the first two steroids to 3α-OH, 5α-pregnan-20-one in the brain, and recently it has been shown that the metabolite is responsible for PG and 5α-P induced anaesthesia in the mouse [32, 33] and in the tadpole [34]. Both 5α-pregnanolone and its β-enantiomer are themselves anaesthetically active in animals and man [35, 36].

Because of their hydrophobicity, 5α- and 5β-pregnanolone need to be formulated in lipid or similar solvents. Initial animal studies with pregnanolone were conducted with a propylene glycol formulation [31]. Single doses gave a shorter duration of anaesthesia than with hydroxydione. However even in these early studies, there was evidence of increased reflex excitability and muscular twitching. Respiratory and cardiovascular depression was less, at equipotent doses, than after thiopentone in cats.

In male rats, both 3α-hydroxy, 5α- and 5β-pregnanolone (formulated as an emulsion in Intralipid) caused hypnosis, with some involuntary movements, when infused until EEG burst suppression occurred [35]. In further experiments in mice, the 5β-isomer was found to have an ED_{50} for loss of the righting reflex of 3.64 mg/kg, compared with 21.8 mg/kg for thiopentone. Induction of anaesthesia was rapid with both drugs, but recovery was faster after pregnanolone at doses of 1.25 × and 5 × the ED_{50}. There was minimal cardiovascular depression; but recovery was not as rapid as following propofol or Althesin [37, 38]. Studies of pregnanolone emulsion in ventilated dogs receiving fentanyl (0.2 μg/kg/min) showed doses of 0.5-4 mg/kg pregnanolone to produce anaesthesia lasting 10-15 minutes. Cardiac output, SAP and contractility only decreased after doses >2-4 mg/kg. SVR also fell, but pulmonary vascular resistance appeared to increase [39].

Further animal evaluation of the drug in instrumented dogs has shown pregnanolone to exert negative inotropic properties. At high doses (2.5-5 mg/kg), pregnanolone caused a dose-dependent decrease in hepatic arterial blood flow, but little effect on portal venous flow or renal arterial flow [40].

Studies with eltanolone in volunteers

In healthy male volunteers, eltanolone caused anaesthesia at doses of 0.4-0.6 mg/kg, with loss of verbal contact occurring before loss of the eyelash reflex (unlike thiopentone). In studies where there has been arterial sampling, the disposition of eltanolone in adults showed a high systemic clearance (1.49-2.29 l/min), an elimination half life of 136-260 minutes, and an apparent volume of distribution between 58.4 and 146.3 litres [41]. Plasma protein binding was high (>99%), and mainly to albumin. Balance studies recovered <1% unconjugated pregnanolone and 7.9-16.2% as conjugated pregnanolone in the urine over 24 hours post-anaesthesia. The main reduced pregnanolone metabolite in man was 5β-pregnan-3α,20α-diol. Total urinary excretion accounted for about 57% of the steroid, with 28% appearing in the faeces.

In more recent kinetic studies in children and the elderly, the mean values of the elimination half life were 3.1 and 3.36 hours, clearance 1.91 and 1.57 l/kg/hour, and apparent volume of distribution 2.29 and 2.31 l/kg [42, 43].

In all these studies, there was a similar dynamic profile. Haemodynamic effects were minimal and dose-related. there was only mild ventilatory depression. Significant side-effects included excitation of short duration during the induction of sleep, and minor involuntary movements. Following a single dose of 0.6 mg/kg, there were also decreases in cerebral blood flow (−34%), and a comparable fall in oxygen consumption; thereby maintaining a coupling between metabolism and blood flow [44].

Clinical studies with eltanolone

The ED_{50} induction dose in benzodiazepine and opioid premedicated patients varied between 0.33 mg/kg and 0.44 mg/kg, with higher doses needed in the unpremedicated patient [45, 46, 47]. There was a low incidence of pain on injection; the main adverse side-effects being involuntary movements, apnoea and hypertonus. The relative potency of propofol (compared to eltanolone) was 0.313 in benzodiazepine premedicated patients. In all comparative studies, the onset of loss of verbal contact with eltanolone was slower than its comparators, and this would seem to support the findings of Schüttler *et al* who demonstrated a blood-brain equilibration time ($t_{1/2}\ k_{eo}$) of 8 minutes [41].

In children, the ED_{50} to loss of verbal command was 0.68 mg/kg in unpremedicated children aged 6-10 years, and 0.53 mg/kg between 11 and 15 years [48]. In elderly patients, there was a significant age effect on induction dose requirements needed to produce loss of verbal contact within 120 secs of the start of drug administration [49]. The ED_{50} dose was 0.29 mg/kg in young patients (aged 18-40 years), compared with 0.16 mg/kg for patients aged over 65 years. However at these ED_{50} doses, some patients started to awaken within four minutes of the start of injection. If we compare induction doses needed to both induce and maintain anaesthesia for 4 minutes, the potency ratio for old:young was 0.28 (95% CI 0.12-0.52).

i. Haemodynamic effects of eltanolone. Cardiovascular stability has been one of the important features of those steroid anaesthetic agents studied to date [50, 51]. In benzodiazepine premedicated ASA I and II patients, equipotent induction doses of eltanolone and propofol caused similar haemodynamic responses to induction and intubation, although there was a significantly greater increase in heart rate post-laryngoscopy and intubation in patients receiving eltanolone [52]. In patients undergoing coronary artery bypass grafting, the combination of eltanolone (0.5-1.0 mg/kg) and fentanyl (3 μg/kg) caused a greater depression of arterial pressure when compared with fentanyl and thiopentone (3 mg/kg) [53]. However cardiac output was unaltered after eltanolone, suggesting it had a greater effect on SVR than on contractility. There was also a reduction in PCWP and LVSWI.

The cardiovascular changes associated with eltanolone have been studied further *in vitro* by two groups of workers. Using the isolated perfused rat heart, Riou et al found that perfusate concentrations of eltanolone of up to 10 μg/ml did not significantly alter contraction-relaxation coupling under both low and high loading conditions in the rat myocardium [54]. There was no significant negative inotropic effect, although higher concentrations caused a decrease in calcium release from the sarcoplasmic reticulum. Furthermore, in contrast to the *in vivo* effects seen in the chronic instrumented dog studies of Wouters *et al* [55], Riou and colleagues were unable to demonstrate any significant negatively inotropic solvent effect. Similar data have been demonstrated by Mouren et al using the isolated blood-perfused rabbit heart preparation [56]. Coronary blood flow increased significantly with eltanolone concentrations up to 10 μg/ml; but comparable amounts of the vehicle decreased flow by 12%. At concentrations less than 10 μg/ml, neither the steroid nor the vehicle affected myocardial contractility and relaxation; although eltanolone decreased both at higher perfusate concentrations. In the catecholamine-depleted heart, the haemodynamics were unaltered, indicating that indirect sympathetic activation was not responsible for the positive inotropy.

ii. Other dynamic properties of eltanolone. These include a low incidence of pain to injection, and post-operative nausea and vomiting (both <5%). The respiratory effects of eltanolone (in keeping with other steroid induction agents) were less depressive than those seen with other i.v. induction agents [57]. The overall incidences of apnoea were 57% for eltanolone, 74% for thiopentone and 100% for propofol. Apnoea of **greater than 30 seconds** occurred in 30% of patients receiving eltanolone, 39% with thiopentone and 74% with propofol. The duration of this apnoea was least with eltanolone and greatest with propofol. The changes in rib-cage and abdominal components of ventilation were similar for all three induction agents. Thus, eltanolone appeared to be less ventilatory depressant than propofol, and to be comparable with thiopentone. There are no data on the effects of eltanolone on central respiratory control.

iii. Infusions of eltanolone. There are few data on the use of eltanolone by incremental dosing or infusion to supplement either nitrous oxide or opioid anaesthesia. Using increments of eltanolone to supplement nitrous oxide, maintenance requirements ranged between 0.015 and 0.025 mg/kg/min [58].

The incidence of side-effects was low, but most patients showed delayed recovery – in agreement with the incremental dose studies of Korttila *et al* [59, 60]. However, because of the slow blood-brain equilibration time, those studies examining recovery characteristics after increments or infusions of eltanolone may be flawed by use of inappropriate dosing strategies.

Further study of eltanolone has discontinued because of higher than anticipated incidences of rash and urticaria and involuntary excitatory movements, four cases of epileptiform convulsions, and no improved cardiorespiratory efficacy compared with other i.v. anaesthetic agents.

Water-soluble amino-steroids

Because of the difficulty in finding suitable solvents for water-insoluble steroids, Figdor and colleagues examined the properties of both pregnane and androstane esters and found that amino-esters of 21-hydroxypregnanedione were both water-soluble and caused loss of the righting reflex in animals [61].

The first important water-soluble agent was minaxolone citrate. Its main undesirable features were the high incidence of excitatory side-effects during induction, increased muscle tone intra-operatively, and involuntary movements during recovery. The latter was often prolonged. Minaxolone was withdrawn from clinical studies in September 1979 because of these adverse features, concern over toxicological effects of large doses in rats, and an absence of any clear advantage over the other agents available in clinical practice at that time.

Two new water-soluble steroid hypnotic agents have recently been evaluated by Organon. Gemmell and colleagues described the anaesthetic properties in the mouse, rat and dog of a water-soluble 2-substituted aminosteroid (ORG 20599) [62]. This appeared to have an efficacy similar to that of Althesin, a high therapeutic index of 13 and to be associated with an hypnotic effect of short duration. However, this has not been evaluated in man because of problems relating to the stability and solubility of the methanesulphonate salt. Another 2-substituted aminosteroid has also been evaluated in animals and in man (ORG 21465: base; ORG 21256: citrate salt) [63]. Again, this steroid shows a high therapeutic index in mice (13.8) when compared with propofol or thiopentone (4-5). In the monkey, ORG 21465 has been compared with propofol in doses of 4 and 3 mg/kg respectively. Both showed rapid onset of hypnosis, but duration of sleep and of recovery was slower with the aminosteroid. In the dog, the effective hypnotic induction dose of ORG 21465 was 3 mg/kg, and maintenance doses were 2.8 μmol/kg/min; giving a potency ratio of 1:8 with respect of propofol. Preliminary data from human volunteers were less reassuring. Doses in excess of 1 mg/kg caused loss of consciousness within one minute in unpremedicated subjects, and the duration of effect was dose-dependent over the range 1.0-1.8 mg/kg. However, as with many of the other steroids investigated in man, there was a high incidence (70%) of excitatory side-effects – although no accompanying EEG spike activity. Again there was an absence of cardiovascular and respiratory depression.

IV Other New Hypnotic Agents

There has been a recent evaluation of two water-soluble imidazo-benzodiazepines (Ro 48-6791 and Ro 48-8684). In comparison with midazolam, both showed a higher mean clearance (1.48 and 1.68 l/min respectively) and larger volumes of distribution, with an elimination half life of 180-210 minutes [64]. The IC_{50} values for 50% maximal EEG depression were 62 (28) ng/ml and 277 (147) ng/ml for the newer agents, and 531 (221) ng/ml for midazolam. The $t_{1/2}k_{eO}$ values were smaller for the newer agents, and recovery was also correspondingly faster.

Further studies with Ro 48-6791 have assessed its efficacy in young and elderly volunteers [65]. The kinetics of this new benzodiazepine agonist did not appear to be affected by age (24-28 years vs. 67-81 years), with mean clearances of 1410 and 1180 ml/kg respectively; initial volumes of distribution of 20.5 and 19.5 litres; and terminal half lives of 225 and 222 minutes. However, the IC_{50} was lower in the elderly patients (44 vs. 72 ng/ml). Recovery times after an infusion to loss of verbal command *and* a median EEG frequency of <4Hz were 18 and 25 minutes respectively.

In both age groups, Ro 48-6791 was about 2.5 times as potent as midazolam; and its use was associated with a low frequency of apnoea and only minimal cardiovascular depression. Further development of these agents is presently uncertain.

V Remifentanil

Remifentanil is a μ-agonist opioid of the 4-anilidopiperidine series of drugs, and as such has all the expected analgesic properties. Its main advantages are the short elimination half life (8-10 minutes), and high clearance of about 2.9 l/min which is independent of liver and renal function – metabolism being by non-specific tissue and plasma esterases to give an acid metabolite (GI-90291) which has 1/2000-1/4000 of the activity of the parent compound. The context-sensitive half time of the drug is also short (3.7 minutes) and independent of dosing duration [66], compared with 44 minutes for infusions of alfentanil of longer than 200 minutes duration.

Like alfentanil, remifentanil has been examined for its ability to induce unconsciousness. McDonnell *et al* demonstrated anaesthetic ED_{50} and ED_{90} doses of alfentanil of 111 and 165 μg/kg respectively in unpremedicated patients [67]. These data are in good accord with the findings of Nauta *et al*, and de Lange *et al* who reported values of 119 μg/kg in unpremedicated general surgical patients, and 41-50 μg/kg in lorazepam premedicated patients for cardiac surgery [68, 69]. Because of the rapidity of onset of action of alfentanil ($t_{1/2}k_{eO}$: 1.6 minutes), these authors pretreated patients with a small dose of muscle relaxant prior to the opioid to avoid muscle rigidity, and an anticholinergic agent to minimise episodes of bradycardia.

Similar studies have been conducted evaluating remifentanil as an induction agent. Joshi and colleagues administered doses of remifentanil or alfentanil over 2 minutes, and assessed after a further one minute [70]. For remifentanil, doses ranged from 2 to 10 μg/kg. The drug was only effective in causing loss of consciousness at doses of 6 μg/kg or greater; and the ED_{50} dose was 9.5 μg/kg/ 2 minutes infusion period (95% CI: 7.5-52.8). the comparable figure for alfentanil was 154 μg/kg/2 minute infusion (CI: 112-469). However there was a high incidence of muscle rigidity and purposeless movement with both drugs, rendering them inappropriate for induction of anaesthesia.

Summary

The clinical anaesthetist is therefore left with a number of i.v. drugs for both induction and maintenance of anaesthesia. None yet appears to achieve the ideal pharmacological profile described by the late John Dundee (Table 1), nor the general ease of administration of the volatile agents – despite the recent introduction of 'target-controlled infusion' systems [71]. Hence for the foreseeable future, I believe that we shall induce and maintain anaesthesia by the combination of both i.v. and gaseous agents. Perhaps future developments will prove me wrong!

Table 1 *Properties of the ideal intravenous anaesthetic agents*

Physical properties	*Pharmacological properties*
water soluble	minimal cardiorespiratory depression
stable in solution	does not cause histamine release; or predispose to hypersenstivitiy reactions
long shelf life	induction in one ar-brain cirulation time
no pain on intravenous injection	metabolism to pharmacologically inactive metabolites
non-irritant on subcutaneous injection	no myoneural blockade
pain on arterial injection	
no sequelae from arterial injection of small doses	
low incidence of venous thrombosis	
small volume of an isotonic solution required for induction	

References

1. Westrin P, Jonmarker C, Werner O. Dissolving methohexital in a lipid emulsion reduces pain associated with intravenous injection. Anesthesiology 1992; 76: 930-934
2. Vanacker B, Wiebalck A, Van Aken H, Sermeus L, Bouillon R, Amery A.

Induktionsqualitat und nebennierenrindenfunktion: Ein klinischer vergleich von Etomidat-lipuro und hypnomidate. Der Anaesthesist 1993; 42: 81-89

3. Kulka PJ, Bremer F, Schuttler J. Narkoseeinleitung mit etomidat in lipidemulsion. Der Anaesthesist 1993; 42, 205-209
4. Nebauer AE, Doenicke A, Hoernecke R, Angster R, Maker M. Does etomidate cause haemolysis? British Journal of Anaesthesia 1992; 69: 58-60.
5. Doenicke A, Roizen MF, Nebauer AE, Kugle A, Hoernecke R, Beger-Hintzen H. Comparison of two formulations of etomidate, 2-hydroxypropyl-β-cyclodextrin (HPCD) and propylene glycol. Anesthesia and Analgesia 1994; 79, 933-939
6. Habazettl H, Vollmar B, Rohrich F, Conzen P, Doenicke A, Baethmann A. Anasthesiologische wirksamkeit von propanidid als liposomendispersion. Der Anaesthesist 1992; 41: 448-456
7. Klockgether-Radke A,. Kersten J, Schroder T, Stafforst D, Kettler D, Hellige G. Anasthesie mit propanidid in liposomaler zubereitung. Der Anaesthesist 1995; 44: 573-580
8. Sear JW. Recovery from anaesthesia: which is the best descriptor of a drug's recovery profile? Anaesthesia 1996; 51: 997-999
9. Doenicke AW, Roizen MF, Rau J, Kellermann W, Babl J. Reducing pain during propofol injection: the role of the solvent. Anesthesia and Analgesia 1996; 82: 472-476
10. Doenicke A, Roizen MF, Rau J, Kugler J. Two different solvents for propofol: is hypnotic effect different? Anesthesia and Analgesia 1996; 82: s94 (IARS abstract)
11. Bielen SJ, Lysko GS, Gough WB. The effect of a cyclodextrin vehicle on the cardiovascular profile of propofol in rats. Anesthesia and Analgesia 1996; 82: 920-924
12. Schuttler J, Schuttler M, Kloos S, Nadstawek J, Schwilden H. Optimierte dosierungsstrategien fur die totale intravenose anaesthesie mit propofol und ketamin. Der Anaesthesist 1992; 40: 199-204
13. Adams HA, Thiel A, Jung A, Fengler G, Hempelmann G. Untersuchumgen mit S(+) ketamin an probanden. Der Anaesthesist 1992; 41: 588-596
14. White PF, Schuttler J, Shafer A, Stanski DR, Horai Y, Trevor AJ. Comparative pharmacology of the ketamine isomers. British Journal of Anaesthesia 1985; 57: 197-203
15. Hering W, Geisslinger G, Kamp DH, Dinkel M, Tschaikowsky K, Rugheimer E, Brune K. Changes in the EEG power spectrum after midazolam anaesthesia combined with racemic or S (+) ketamine. Acta Anaesthesiologica Scandinavica 1994; 38: 719-723
16. Schuttler J, Stanski DR, White PF, Trevor AJ, Horai Y, Verotta D, Sheiner LB. Pharmacodynamic modeling of the EEG effects of ketamine and its enantiomers in man. Journal of Pharmacokinetics and Biopharmaceutics 1987; 15: 241-253
17. Geisslinger G, Hering W, Thomann P, Knoll R, Kamp H-D, Brune K. Pharmacokinetics and dynamics of ketamine enantiomers in surgical patients using a stereospecific analytical method. British Journal of Anaesthesia 1993; 70, 666-671
18. Kharasch ED, Labroo R. Metabolism of ketamine stereoisomers by human liver microsomes. Anesthesiology 1992; 77: 1201-1207
19. White PF, Ham J, Way WL, Trevor AJ. Pharmacology of ketamine isomers in surgical patients. Anesthesiology 1980; 52: 231-239
20. Doenicke A, Angster R, Maker M, Adams HA, Grillenberger G, Nebauer AE. Die wirkung von S(+) ketamin auf katecholamine und cortisol im serum. Der Anaesthesist 1992; 41: 597-603

21. Adams HA, Bauer R, Gebhardt B, Menke W, Baltes-Gotz B. TIVA mit S(+) ketamin in der orthopadischen alterschirugie. Der Anaesthesist 1994; 43: 92-100
22. Crozier TA, Sumpf E. Der Einfluα einer totalen intravenosen anasthesie mit S(+) ketamin/propofol auf hamodynamische, endokrine und metabolische stressreaktion im vergleich zu alfentanil/propofol bei laparotomien. Der Anaesthesist 1996; 45: 1015-1023
23. Cashin MF, Moravek V. The physiological action of cholesterol. American Journal of Physiology 1927; 82: 294-298
24. Selye H. The anesthetic effect of steroid hormones. Proceedings of the Society of Experimental Biology and Medicine 1941; 46: 116-121
25. Laubach GD, P'An SY, Rudel HW. Steroid anesthetic agent. Science 1955; 122: 78
26. Taylor N, Shearer WM. The anaesthetic properties of 21-hydroxypregnanedione sodium hemisuccinate (hydroxydione), a pharmacological and clinical study of 130 cases. British Journal of Anaesthesia 1956; 28: 67-00
27. Galley AH, Rooms M. An intravenous steroid anaesthetic. Experiences with Viadril. Lancet 1956; i: 990-994
28. Montmorency FA, Chen A, Rudel H, Glas WW, Lee LE. Evaluation of cardiovascular and general pharmacologic properties of hydroxydione. Anesthesiology 1958; 19: 450-456
29. Merryman W, Bioman R, Barnes L, Rothchild I. Progesterone 'anaesthesia' in human subjects. Journal of Clinical Endocrinology 1954; 14: 1567-1569
30. Figdor SK, Kodet MJ, Bloom BM, Agnello EJ, P'An SY, Laubach GD. Central activity and structure in a series of water-soluble steroids. Journal of Pharmacology and Experimental Therapeutics 1957; 119: 299-309
31. Gyermek L. Pregnanolone: A highly potent, naturally occurring hypnotic-anesthetic agent. Proceedings of the Society of Experimental Biology and Medicine 1967; 125: 1058-1062
32. Mok WM, Herschkowitz S, Krieger NR. In vivo studies identify 5α-pregnan-3α-ol-20-one as an active anesthetic agent. Journal of Neurochemistry 1991; 57: 1296-1301
33. Mok WM, Herschkowitz S, Krieger NR. Evidence that 3α-hydroxy-5α-pregnan-20-one is the metabolite responsible for anesthesia induced by 5α-pregnanedione in the mouse. Neuroscience Letters 1992; 135: 145-148
34. Krieger NR, Mok WM. Steroid brain levels at specified behavioral endpoints for general anesthesia. Annals of the New York Academy of Science 1991; 625: 556-557
35. Norberg L, Wahlstrom G, Backstrom T. The anaesthetic potency of 3α-hydroxy-5α-pregnan-20-one and 3α-hydroxy-5β-pregnan-20-one determined with an intravenous EEG-threshold method in male rats. Pharmacology and Toxicology 1987; 61: 42-47
36. Carl P, Hogsklide S, Nielsen JW, Sorensen MB, Lindholm M, Karlen B, Backstrom T. Pregnanolone emulsion: A preliminary pharmacokinetic and pharmacodynamic study of a new intravenous anaesthetic agent. Anaesthesia 1990; 45: 189-197
37. Larsson-Backstrom C, Lutteman Lustig L, Eklund A, Thorstensson M. Anaesthetic properties of pregnanolone in mice in an emulsion preparation for intravenous administration: a comparison with thiopentone. Pharmacology and Toxicology 1988; 63: 143-149
38. Hogskilde S, Wagner J, Carl P, Sorensen MB. Anaesthetic properties of pregnano-

lone emulsion. A comparison with alphaxalone/alphadolone, propofol, thiopentone and midazolam in a rat model. Anaesthesia 1987; 42: 1045-1050

39. Hogsklide S, Wagner J, Strom J, Sjontoft E, Olesen HP, Sorensen MB. Cardiovascular effects of pregnanolone emulsion: an experimental study in artificially ventilated dogs. Acta Anaesthesiologica Scandinavica 1991; 35: 669-675
40. Wouters PF, Van de Velde MA, Marcus MAE, Deruyter HA, Van Aken H. Hemodynamic changes during induction of anesthesia with eltanolone and propofol in dogs. Anesthesia and Analgesia 1995; 81: 125-131
41. Hering WJ, Ihmsen H, Langer H, Uhrlau C, Dinkel M, Geisslinger G, Schuttler J. Pharmacokinetic-pharmacodynamic modeling of the new steroid hypnotic eltanolone in healthy volunteers. Anesthesiology 1996; 85: 1290-1299
42. Beskow A, Gralls M, Werner O, Westrin P. Pharmacokinetics of eltanolone after i.v. bolus injection in children aged 6-10 years. British Journal of Anaesthesia 1996; suppl 2: A311 (ESA abstract)
43. Gralls M, Nimmo W, Watson N. Pharmacokinetics of eltanolone after i.v. bolus injection in elderly patients. British Journal of Anaesthesia 1996; suppl 2: A286 (ESA abstract)
44. Wolff J, Carl P, Bo Hansen P, Hogskilde S, Christensen MS, Sorensen MB. Effects of eltanolone on cerebral blood flow and metabolism in healthy volunteers. Anesthesiology 1994; 81: 623-627
45. Powell H, Morgan M, Sear JW. Pregnanolone: a new steroid intravenous anaesthetic. Dose-finding study. Anaesthesia 1992; 47: 287-290
46. Hering W, Biburger G, Rugheimer E. Induction of anaesthesia with the new steroid intravenous anaesthetic eltanolone (pregnanolone). Dose finding and pharmacodynamics. Der Anaesthetist 1993; 42: 74-80
47. van Hemelrijck J, Muller P, Van Aken H, White PF. Relative potency of eltanolone, propofol, and thiopental for induction of anesthesia. Anesthesiology 1994; 80: 36-41
48. Beskow A, Werner O, Westrin P. Intravenous induction of general anaesthesia with eltanolone in children 6-15 years of age. Acta Anaesthesiologica Scandinavica 1997; 41: 242-247
49. Sear JW. Eltanolone for induction of anaesthesia in the surgical patient: A comparison of dose requirements in young and elderly patients. Acta Anaesthesiologica Scandinavica 1997 – in press.
50. Sear JW, Prys-Roberts C. Dose-related haemodynamic effects of continuous infusions of Althesin in man. British Journal of Anaesthesia 1979; 51: 867873
51. Sear JW, Prys-Roberts C, Gray AJG, Walsh EM, Curnow JSH, Dye J. Infusions of minaxolone to supplement nitrous oxideoxygen anaesthesia. A comparison with Althesin. British Journal of Anaesthesia 1981; 53: 339350
52. Sear JW, Jewkes C, Wanigasekera V. Hemodynamic effects during induction, laryngoscopy and intubation with eltanolone (5β-pregnanolone) or propofol. A study in ASA I and II patients. Journal of Clinical Anesthesia 1995; 7: 126-131
53. Tassani P, Janicke U, Ott E, Groh J, Conzen P. Hemodynamic effects of anesthetic induction with eltanolone-fentanyl versus thiopental-fentanyl in coronary artery bypass patients. Anesthesia and Analgesia 1995; 81: 469-473
54. Riou B, Ruel P, Hanouz J-L, Langeron O, Lecarpentier Y, Viars P. In vitro effects of eltanolone on rat myocardium. Anesthesiology 1995; 83: 792-798
55. Wouters PF, Van de Velde M, Marcus M, Van Aken H. Negative inotropic properties of pregnanolone in dogs due to the lipid emulsion. Anesthesia and Analgesia 1995; 80: S559 (IARS abstract)

56. Mouren S, Abdenour L, Souktani R, Coriat P. Effects of eltanolone on myocardial performance and coronary flow in intact and catecholamine-depleted isolated rabbit hearts. Anesthesiology 1996; 85: 1378-1385
57. Spens HJ, Drummond GB, Wraith PK. Changes in chest wall compartment volumes on induction of anaesthesia with eltanolone, propofol and thiopentone. British Journal of Anaesthesia 1996; 76: 369-373
58. Rajah A, Powell H, Morgan M. Eltanolone for induction of anaesthesia and to supplement nitrous oxide for minor gynaecological surgery. Anesthesia 1993; 48: 951-954
59. Kallela H, Haasio J, Korttila K. Comparison of eltanolone and propofol in anesthesia for termination of pregnancy. Anesthesia and Analgesia 1994: 79: 512-516
60. Eriksson H, Haasio J, Korttila K. Comparison of eltanolone and thiopental in anaesthesia for termination of pregnancy. Acta Anaesthesiologica Scandinavica 1995; 39: 479-484
61. Figdor SK, Kodet MJ, Bloom BM, Agnello EJ, P'An SY, Laubach GD. Central activity and structure in a series of water-soluble steroids. Journal of Pharmacology and Experimental Therapeutics 1957; 119: 299-309
62. Gemmell DK, Campbell AC, Anderson A, Byford A, Hill-Venning C, Marshall RJ. ORG 20599: a new water soluble aminosteroid intravenous anaesthetic. British Journal of Pharmacology 1994: 111: suppl., 189P (BPS abstract)
63. Gemmell DK, Byford A, Anderson A, Marshall RJ, Hill DR, Campbell AC, Hamilton N, Hill-Venning C, Lambert JJ, Peters JA. The anaesthetic and GABA modulatory actions of ORG 21465, a novel water soluble steroidal intravenous anaesthetic agent. British Journal of Pharmacology 1995; 116: suppl 443P (BPS abstract)
64. Ihmsen H, Hering W, Albrecht S, Dingemanse J, Zell M, Schwilden H, Schuttler J Pharmacokinetics of the new benzodiazepines RO 48-6791 and RO 48-8684 in comparison with midazolam in young and elderly volunteers. Anesthesiology 1996; 85: A 317 (ASA abstract)
65. Hering W, Ihmsen H, Albrecht S, Schwilden H, Schuttler J. Ro 48-6791 – ein kurzwirksames benzodiazepin. Untersuchungen zur pharmakokinetik und pharmakodynamik bei jungen und alteren probanden im vergleich mit midazolam. Der Anaesthesist 1996; 45: 1211-1214
66. Egan TD. Remifentanil pharmacokinetics and pharmacodynamics. A preliminary appraisal. Clinical Pharmacokinetics 1995; 29: 80-94
67. McDonnell TE, Bartkowski RR, Williams JJ. ED_{50} of alfentanil for induction of anesthesia in unpremedicated young adults. Anesthesiology 1984; 60: 136-140
68. Nauta J, de Lange J, Koopman D, Spierdijk J, van Kleef J, Stanley TH. Anesthetic induction with alfentanil; a new short acting narcotic analgesic. Anesthesia and Analgesia 1982; 61 267-272
69. de Lange S, Stanley TH, Boscoe MJ. Alfenatil-oxygen anaesthesia for coronary artery surgery. British Journal of Anaesthesia 1981; 53: 1291-1296
70. Joshi P, Jhaveri R, Bauman V, McNeal S, Batenhorst R, Glass PSA. Comparative trial of remifentanil and alfentanil for anesthesia induction. Anesthesiology 1993; 79: A 379 (ASA abstract)
71. White M, Kenny GNC. Intravenous propofol anaesthesia using a computerised infusion system. Anaesthesia 1990; 45: 204-209

Entonox and its Development

John A S Ross, Ian L Marr* and Michael E Tunstall

DEPARTMENTS OF ENVIRONMENTAL AND OCCUPATIONAL MEDICINE AND CHEMISTRY*, UNIVERSITY OF ABERDEEN, UK

History of Nitrous Oxide and Oxygen Mixtures

Nitrous oxide, first synthesised by Joseph Priestley in 1776 and suggested to be an analgesic by Humphrey Davy in 1800, made its way into established clinical practice in the 1860s as an anaesthetic agent, initially for dentistry and then for general surgery. However, because nitrous oxide is not potent enough reliably to induce anaesthesia at atmospheric pressure, its use as the sole anaesthetising agent was accompanied by hypoxia and the coma associated with anoxia was used to assist its action. From the first the danger of hypoxia was identified and in 1868 Edmund Andrews, a Chicago professor of surgery, used mixtures of one part oxygen in two parts nitrous oxide premixed and administered either from a gasometer in hospital use or from a large rubber bag elsewhere. In city practice he found the large bulk of the rubber bag did not cause problems as it could be transported in a carriage without attracting attention (1). However, by 1872 Andrews was referring to 'the blue asphyxiated look characteristic of full anaesthesia by nitrous oxide' and was advocating the administration of 100% nitrous oxide dispensed from a liquid filled cylinder and breathed by the patient from a bag (2). In 1878 Paul Bert stated that pure nitrous oxide could not be given for more than 2 minutes if asphyxia was to be avoided and in 1879 Fontaine modified Bert's concept of a hyperbaric chamber to allow anaesthetising doses of nitrous oxide to be given in air without concomitant hypoxia (3).

Stanislaus Klikowitsch is accepted as the pioneer of nitrous oxide analgesia in obstetrics. In 1881 he described his experience of the intermittent administration of a mixture of 80% nitrous oxide and 20% oxygen in the labours of 25 women. To avoid expense, he manufactured his own nitrous oxide and it was administered after premixing with oxygen from a container at atmospheric pressure to avoid hypoxia. It is to Klikowitsch that we owe the first description of effective self-administration of nitrous oxide mixtures for the relief of labour pains. Inhalation should start 30-60 seconds before the expected pain and between 2-6 inhalations should give the expected effect. The important point was made that, encouraged by success, the mother would quickly learn to take deep breaths, to hold the mouth-piece herself and to breath from it at the beginning of subsequent pains. This is the first description of patient controlled analgesia. Klikowitsch also demonstrated, using intra-uterine pressure manometry, that 80% nitrous oxide did not impair the frequency, duration or strength of uterine contractions (4).

In spite of this excellent work on the sedative and pain relieving properties of nitrous oxide in an awake population, the drive underlying the development of the use of nitrous oxide was the provision of general anaesthesia in an age when the alternatives were perceived as either difficult to use in the case of ether, or, in the case of chloroform, potentially and immediately lethal. The inadequate potency of nitrous oxide was systematically combined with the usually reversible cerebral failure induced by transient anoxia in order to produce brief periods of oblivion during which dentistry and minor surgery could be performed. In 1945, Barach and Rovenstine concluded that the use of anoxia as an aid to nitrous oxide anaesthesia had 'been the cause of (A) death from asphyxia, (B) psychoses from permanent brain damage, (C) personality defects which may or may not be recognised, (D) impairment in circulatory and respiratory function which may contribute to pulmonary atelectasis, pulmonary oedema, or cardiac failure'. They described and advocated the use of 80% nitrous oxide in oxygen premixed in cylinders at 700 pounds pressure stating that this was the only way technically to provide a reliable non-hypoxic gas mixture at that time (5). The disadvantages of the partial asphyxia technique in dental anaesthesia was nicely demonstrated and summarised by Bourne in 1960 who also identified the interaction of anoxia with fainting in the dental chair (6).

The turn of the century saw the development of machines which allowed the self-administration of nitrous oxide in air, hypoxic gas mixtures, initially by Guedel in 1912 and then by Minitt in 1933. The Minnitt Gas-Air Analgesia Apparatus was approved by the Central Midwives Board in 1936 for use in obstetrics by midwives working alone for mothers who had been passed by a doctor as fit for the technique and allowed the self administration of 50% nitrous oxide in air; in other words 10.5% oxygen. A number of other devices providing nitrous oxide in air were approved but the concern on the exposure of mother and foetus to hypoxia, initially expressed by Eastman in 1936 (7), was finally brought to a head by Cole and Nainby-Luxmore's 1962 paper on the hazards of gas and air in obstetrics identifying the poor performance of the equipment used which commonly provided less than 5% oxygen (8). Approval for the use of gas and air mixtures in obstetrics was withdrawn in 1970.

Meanwhile the firm of Aga had produced a machine, the Calmator, designed by Andreas Warming of Copenhagen, which gave nitrous oxide with oxygen and the use of this equipment was described by Seward in 1949 (9). 75% nitrous oxide in oxygen was better than 85% which caused loss of control and partial anaesthesia. The 75% mixture was also better than 50% gas and air and reversed foetal distress associated with the hypoxia of this latter technique. Although never approved by the Central Midwife's Board, this equipment was used without additional supervision by trained midwives. 1958 saw the introduction of the Lucy Baldwin apparatus, a modified Walton dental anaesthetic apparatus, which delivered 50% nitrous oxide in oxygen for normal labours but could be set to give 70% for the more painful. 80% could be given by undoing a lock but this was never recommended for use by the midwife since anaesthesia could be produced.

History of Entonox

It was against this background of dissatisfaction with existing, widely accepted and used equipment, the final crystallisation of the dangers of hypoxia and the realisation that high concentrations of nitrous oxide were undesirable in obstetrics that Tunstall wrote to the Lancet in 1961 stating that it was possible to mix nitrous oxide in oxygen in cylinders at a storage pressure of 2000 pounds in the gaseous phase. Indeed, it was at his prompting that the British Oxygen Company had performed the experimental work required since theory seemed to be against the possibility. Subsequently, Dr J.W. Haworth had demonstrated that, at a pressure of 2000 pounds oxygen had a considerable solvent effect on nitrous oxide and this was attributed to the Poynting effect (10). Tunstall had used premixed 50% nitrous oxide in oxygen for the relief of pain in childbirth at St Mary's Hospital Portsmouth and in the following year used the same mixture in dental anaesthetic practice. The advent of halothane in 1956 had provided a non-irritant but potent volatile agent which made the nitrous oxide and partial asphyxia technique unnecessary and various techniques were described for the use of Entonox in dental practice with halothane (11, 12).

In Entonox the vapour pressure of nitrous oxide exceeded that which was thought at the time to be possible. Since saturated vapour pressure is temperature dependent, concern was soon expressed that Entonox might not prove stable under conditions of storage at low temperature, in the boot of a car or on the back of a lorry for example (13). It was soon found that, in a cylinder filled to 13.7 MPa, nitrous oxide condensed out at a temperature of −7 to −8 °C (13, 14). As a result, simple handling measures were instituted to ensure good mixing within the cylinder if it had been allowed to cool (14) and in 1965 the Central Midwifery Board approved Entonox for use by unsupervised midwives.

Subsequent to this work, a trial was conducted under the auspices of the Medical Research Council in order to identify the relative benefits of 70% as opposed to 50% nitrous oxide in oxygen for pain relief in labour since concern had been expressed that anaesthesia might be inadvertently induced in mothers using the mixtures. It was found that while there was little difference in pain relief for the two mixtures, significantly more mothers lost consciousness with the higher concentration and it was concluded that 50% nitrous oxide in oxygen was safe for unsupervised midwives to use . The comment was made, however, that 70% nitrous oxide might be of benefit in abnormal labour. Indeed. this had been the experience of several previous studies (16).

Since 1970 Entonox has become routinely used as an analgesic in the emergency services (17, 18) and has achieved a role in analgesia for painful procedures within the hospital (19, 20, 21) and in paediatrics (22).

Physical Principles behind Entonox

The vapour pressure of nitrous oxide at 22 °C is 5.508 MPa and so by Dalton's law of partial pressures the maximum pressure of a cylinder containing 50% nitrous oxide at this temperature should be 11.016 MPa. In fact at 13.705 MPa both oxygen and nitrous oxide exist in an homogenous mixture at this temperature with the partial pressure of nitrous oxide at 6.853 MPa, 24.4% higher than expected from consideration of an ideal gas, and this admixture is stable down to around −8 °C. The behaviour of nitrous oxide in Entonox has been attributed to the increase in saturated vapour pressure of a liquid caused by the presence of an indifferent gas. This effect is named after John Henry Poynting who first described the interaction (23). The Poynting correction, however, assumes ideal behaviour and is unlikely to account for the entire effect since at high pressure neither oxygen nor nitrous oxide behave ideally. It is more realistic to consider that the two gases develop a solute:solvent relationship and the solvent properties of supercritical fluids are well known (24, 25, 26)

The critical pressure of oxygen is 5.043 MPa and the critical temperature is −118.4 °C. The contents of an Entonox cylinder therefore contain oxygen as a supercritical fluid but nitrous oxide below its critical point. In a supercritical fluid the density of the vapour is similar to that of a liquid and as pressure increases so does the density. The solvating power of a supercritical fluid is close to that of a liquid and is density dependent. So as density increases with compression, oxygen is capable of dissolving increasing amounts of nitrous oxide as pressure exceeds its critical point (Figure 1). This concept predicts that as cylinder pressure falls the temperature at which nitrous oxide condenses might rise and this is indeed the case. In a piece of definitive work, Bracken and his colleagues defined the physical characteristics of nitrous oxide and oxygen mixtures (27). While a cylinder filled to 13.7 MPa was stable at temperatures down to −7 °C, one filled to 11.7 MPa became unstable at a temperature of −5.5 °C (Figure 1). At lower filling pressures the mixture became more stable with regard to temperature. It seems that as the pressure of Entonox increases the dew point moves initially to the right as the partial pressure of nitrous oxide rises and approaches saturation. As pressure increases, however, so does the solvating power of oxygen and increasing partial pressures of nitrous oxide become stable in oxygen. This causes the dew point curve to become steeper and depart from that expected from ideal behaviour. This effect is seen in all oxygen, nitrous oxide mixtures.

Inhalational Analgesia

The use of volatile and gaseous agents in the relief of pain is termed inhalational analgesia. Nitrous oxide, diethyl ether, methoxyflurane and

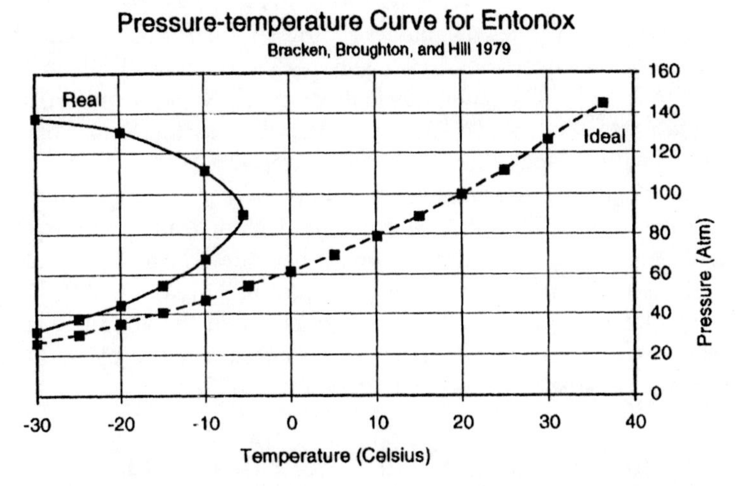

Figure 1 *This shows the actual phase envelope (solid line) of Entonox in comparison with what might be theoretically envisaged if there was no interaction between oxygen and nitrous oxide (broken line). Within the solid line equilibrium exists between the gas and liquid phases. Below and to the right the agent is in the gas phase. Above and to the left of the line it is in the liquid phase.*

trichlorethylene have demonstrable analgesic properties at subanaesthetic doses in experimental models of pain in man (28, 29, 30). The effect of halothane on pain seems controversial in the same and similar studies but at 0.2 and 0.33 MAC halothane did not raise pain thresholds (30, 31) while at 0.5 and 0.66 MAC there was a significant analgesic effect in two studies using heat algesimetry and exercise in an ischaemic arm as stimuli (29, 31) while studies using pretibial pressure pain thresholds produced conflicting results (28, 29). At 0.2 MAC the newer agents, enflurane, isoflurane and sevoflurane have been demonstrated not to effect pain threshold measured by radiant heat algometry (30). A wider battery of tests has confirmed a lack of effect for isoflurane at 0.2 MAC (0.24%) and lower concentrations (32).

In distinction to the effects identified in the laboratory, many of these agents have been used in the practical relief or prevention of pain. Perhaps the most impressive demonstration of inhalational analgesia was accomplished by Artusio using diethyl ether (33). Artusio divided stage I ether anaesthesia into three planes. During all three planes the patient was awake and communicative. In plane 1 there was neither amnesia nor analgesia. In plane 2 there was complete amnesia, as assessed by questioning after surgery, and partial analgesia. In plane 3 analgesia was complete and the patient was again amnesic. Artusio reported 132 operations in which the patient was kept in plane 3 and these included 25 intra-abdominal procedures and 110 mitral valvotomies. Opiate analgesia was not used before or during the procedure. Reier described a very similar use of methoxyflurane in the production of stage I anaesthesia for a number of orthopaedic and gynaecological procedures and cardioversion (34). Nitrous oxide nitrous usage has been outlined in very similar terms by Parbrook who, while admitting a degree of inter-individual variability, described 4 zones of analgesia. Unlike ether and methoxyflurane, nitrous oxide was analgesic at low dose; 6-25% or zone 1. Amnesia was encountered in zone 2 (26-45%) at concentrations above 30% and was marked in zone 3 (46-65%). Zone 4, 66-85%, was described as light anaesthesia and was associated with complete analgesia and amnesia.

Inhalational analgesia has been widely used to relieve the pain of labour. Since the description of Entonox, chloroform, cyclopropane, methoxyflurane, enflurane, desflurane and isoflurane have all been described as helpful agents in childbirth used at concentrations higher than 0.2 MAC (36-41). Trichlorethylene was already in use (42). The equipment used for volatile agents has always been some variety of vaporiser, either preset for use by midwives or operated by an anaesthetist or medical attendant.

Combination Inhalational Analgesia in Labour

Being struck by the efficacy of isoflurane in particular among these agents and the stated occasional need for 70% nitrous oxide as an analgesic an labour, we studied the combination of 0.25% isoflurane in Entonox self administered during labour using an Oxford Miniature Vaporiser (43). 0.25% isoflurane is equivalent to 0.2 MAC and so its combination with Entonox gives a mixture with a nominal anaesthetic potency of 0.7 MAC, the same as for 70% nitrous oxide in oxygen. Each agent was inhaled for 5 contractions after which the degree of pain relief obtained was assessed by linear analogue scoring. In one group of mothers Entonox was used first (n=20) and in another isoflurane in Entonox was the first gas (n=19). The degree of sedation produced was assessed by the attendant midwife before and after the trial of both gases. Pain relief was superior when isoflurane was added (Figure 2) and the level of sedation was acceptable with no evidence of an undue isoflurane effect. Fourteen subjects chose to use the isoflurane mixture through to delivery without problem. These data were later supported by the work of Wee who used 0.2% isoflurane in Entonox and found that linear analogue pain score were significantly lower for mothers using isoflurane (44).

The use of draw-over vaporisers in the clinical context of a labour ward has some problems. Agitation (or dropping) of the equipment may alter the concentration delivered, the option of varying the concentration requires an anaesthetist to be present and it can be guaranteed to run out at an awkward time. The premixture of nitrous oxide and oxygen to form Entonox enabled nitrous oxide analgesia in labour which was simple to administer, was portable and was safe since the concentrations of oxygen and nitrous oxide could not vary. We have similarly premixed 0.25% isoflurane in Entonox at cylinder pressures of 13.7 MPa. Our house name for this mixture is Isoxane. Similar considerations apply to this mixture as apply to Entonox. The saturated vapour pressure of isoflurane at 20 °C is 31.7 kPa. The highest pressure achievable for a mixture of 0.25% isoflurane at this temperature therefore should be 12.7 MPa before the dew point for the agent is exceeded and liquid isoflurane becomes precipitated. In fact, at room temperature, it is possible to

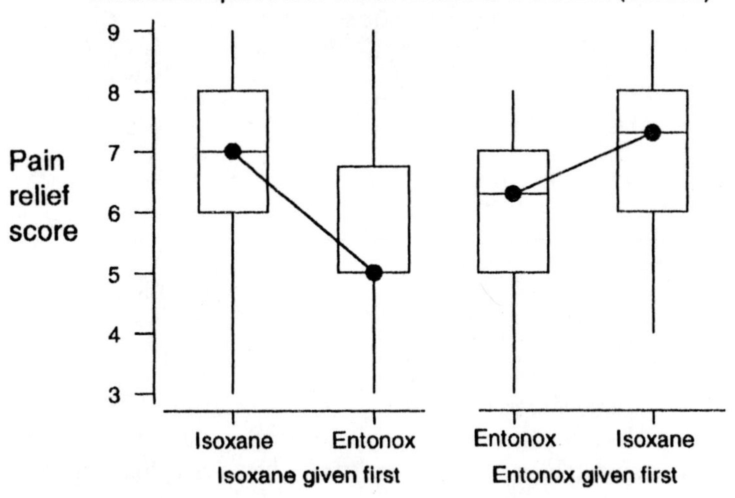

Figure 2 *Pain relief was assessed in labour using a visual analogue score. Scores were assessed for each agent after five contractions and two groups of women were studied. On group received Entonox first and the other Isoxane first. Median scores are presented with the interquartile range. Whiskers run from the interquartile values to the next adjacent point. Isoxane offers significantly greater pain relief.*

have about 0.4% isoflurane in Entonox at 13.7 MPa with no evidence of a liquid phase. We have also found that 0.25% isoflurane in Entonox at 13.7 MPa (a theoretical vapour pressure of 34.25 kPa) is stable at temperatures down to $-3\,^{\circ}C$, a temperature at which the saturated vapour pressure of isoflurane is less than 10 kPa (45). Similar observations have been made for halothane and enflurane by Uyanik (46) who also clearly demonstrated that the higher than expected gaseous partial pressure of the agents in Entonox were not explained from existing hypotheses, including Poynting's, which assume ideal behaviour. This observation of Uyanik's has encouraged us to look at some of the more recent volatile agents as an alternative to isoflurane and we have measured the separation temperature in Entonox of 0.2 MAC equivalent of enflurane, sevoflurane and desflurane in addition to isoflurane (Table 1). Although higher than ideal concentrations are possible for all four agents, isoflurane has the best storage characteristics in terms of temperature.

Having produced a stable gaseous mixture of isoflurane in Entonox, the next stage is to administer it and we have used the standard Entonox demand valve for this. In doing so we have noted an effect due to the localised cooling of the gas mixture as it expands through the valve. This causes condensation or sputter of small amounts of isoflurane immediately downstream but this condensate is revaporized as it is pushed forward by continuing gas flow and the concentration actually inhaled is as expected. The sputter from a 0.25%

Table 1

	0.2 MAC (kPa)	*Maximum cylinder concentration at 13.7 MPa and 20 °C with 'ideal' gases*	*Separation temp. of 0.2 MAC in Entonox*
Isoflurane	0.25	0.23	−3.0 °C
Englurane	0.37	0.17	0 °C
Sevoflurane	0.5	0.16	4 °C
Desflurane	1.45	0.68	6–7 °C

The table shows the approximate concentration of 0.2 MAC for each agent, the theoretical maximum cylinder concentration possible considering the system to be behaving according to the ideal gas laws. In fact it is possible to mix 0.2 MAC of these agents in high pressures of Entonox and the separation temperatures of the mixtures are shown.

mixture of isoflurane causes rapid oscillations of concentration of not more than 0.1% from the mean at maximum cylinder pressure and the effect reduces as cylinder pressure falls to become undetectable at a cylinder pressure of about 5.5 MPa.

The Use of Isoxane in Labour

We have conducted a preliminary trial of this mixture in order to determine what types of labour would be thought suitable for the agent by midwives and to identify any unexpected safety problems. With the approval of the Grampian Area Joint Ethics Committee and after the approval of the Obstetric Senior Staff Committee and Nursing Unit Management had been given, Isoxane was made freely available to midwives at Aberdeen Maternity Hospital for a period of time. Midwives were instructed, on the basis of previously published work, that the addition of 0.25% isoflurane to Entonox was a more effective analgesic in labour than Entonox alone without being unduly sedative and that Isoxane was a high pressure gas mixture of this composition which could be used in the same manner as Entonox. Isoxane, therefore, was used as thought necessary by the midwife and as agreed by the mother.

After the study period, case controls were identified from the Aberdeen Maternity and Neonatal Databank and matched for age, height, parity and geographical location (in Aberdeen City or from the country areas). Controls were taken from periods of time during which Isoxane was not available. For each mother details of the course of labour, analgesia during labour, method of delivery, blood loss and Apgar score of the baby were taken from the database. No details on the number of doses of opiate given during labour were available and this information was culled retrospectively from the maternity notes of mothers receiving Isoxane.

More mothers in the Isoxane group had labour induced with artificial rupture of membranes (62) as compared to control (44) and the time from rupture of membranes to delivery was significantly longer for mothers in the Isoxane group (median 7.9 hrs) compared to control (median 5.7 hrs). More mothers received opiate in the Isoxane group (183) than control (139) and there was also a higher use of transepidermal neural stimulation (TENS) indicating that these mothers experienced a more painful labour than control. These results indicated that Isoxane was used for the more complex and trying labours.

Isoxane was used for 0.1-14.5 hours (median 2.4, 95% c.i. 1.9-2.8). No mother lost consciousness during labour and no relationship was found between duration of use of Isoxane and sedation. Forty-six mothers stopped using Isoxane before the onset of the 2nd stage of labour. The reasons for stopping, followed by the numbers in parentheses, are as follows:- epidural established (24), smell (5), dislike (1), nausea (2), dizziness (2), apparatus problems (2), drowsiness (2), preferred Entonox (1), no reason recorded (4), and supply running out (3).

The study was designed to determine what kind of labour would attract the use of Isoxane and not to assess primarily measures of outcome. However, these are of interest with regard to the overall safety of the technique. There were more deliveries by caesarian section in the control group but operative delivery was as a planned event in most cases. In contrast, all caesarian sections in the Isoxane group had uterine dysfunction as an indication ($p<0.001$). This indication was present in only 8 of the control group. Although there was no real difference in the frequency of forceps delivery foetal distress was more frequent as an indication in the control group (43%) in contrast to the Isoxane group (25%) ($p=0.03$). Again uterine dysfunction was more common in the Isoxane group but the difference was not significant at the 10% level. This incidence of dysfunctional labour further supports the conclusion that Isoxane was used for the more trying labours.

There was no difference overall in blood loss between the two groups. Analysis of blood loss by method of delivery revealed a slightly higher blood loss in the Isoxane group for spontaneous deliveries (95% c.i. for difference 0-50 ml) and for caesarian section (95% c.i. of difference 100-300 ml). The difference in blood loss for spontaneous delivery is not clinically significant and that for caesarian section is probably related to the differing indications for operative delivery in the two groups.

There was no difference in five minute Apgar score between the two groups. The Apgar score at one minute was higher for the control group ($p=0.012$) although the median values were the same. This difference was only present for mothers who had received opiate. Resuscitation was required in 23 babies in the control group, and in 33 babies in the Isoxane group, ($p<0.044$). Although more resuscitation with the bag and mask in association with naloxone was required for spontaneous vertex deliveries in the Isoxane group, this caused no undue problems. The incidence of neonatal resuscitation in the 23 labours where Isoxane had been discontinued for more than 60 minutes was considered separately. Sixteen required no resuscitation, one required naloxone and six needed naloxone and mask ventilation. The incidence of resuscitation in this sub-group was the same as in the rest of the Isoxane group and the one minute Apgar scores were also the same.

Isoxane was administered as indicated for pain relief in labour in 235 cases. There was a tendency for mothers having more prolonged and painful labours than control to be given the gas mixture. This is not surprising and reflects the indications for the use of Isoxane. No obvious disadvantages to the mother and baby were found and there seemed to be no untoward side-effects. Minor differences in blood loss from control for mothers receiving Isoxane were more likely to be due to the nature of their labour than to the inhalation of isoflurane. Lower Apgar scores at one minute and a higher incidence of resuscitation for the babies of mothers receiving Isoxane and opiates for pain relief were probably due to the nature of the labours that attracted these agents rather than the agents themselves. Indeed, the Apgar score at one minute in deliveries where the mother had stopped breathing Isoxane some time before birth, and in whom the agent would have been excreted, were as low as in the group breathing Isoxane right up to delivery where opiate had been given within 5 hours. Apgar scores at five minutes did not differ from control and there were no undue problems in the newly born. From this study Isoxane would seem to be an acceptable method of pain relief in labour

Inhalation Analgesia for Repeated Painful Procedure

While conducting the study on the use of Isoxane in labour we were approached by the nurses of the acute pain team at Aberdeen Royal Infirmary to supervise the administration of Entonox to patients requiring repeated painful procedures such as physiotherapy to free the movement of joints after prolonged immobilisation and wound dressings of various kinds. The toxicity following continuous exposure to nitrous oxide caused by oxidation of the methionione synthase and vitamin B_{12} complex is well established. The toxicity of repeated short term administration is also documented in a man who was given Entonox for 15-20 minutes three time a day to facilitate physiotherapy (47). Megaloblastic change in the bone marrow was noted after 24 days and, after a recovery period of two weeks, again after two weeks of giving Entonox twice a day for 20 minutes. More recent work has indicated that the inhibition of methionine synthase as indicated by raised blood homocysteine levels was still present one week after a single exposure to nitrous oxide (48) and it is not surprising that the toxic effects are cumulative.

There is no evidence for such cumulative toxicity for the volatile anaesthetics and we considered that low concentrations of such agents might be effective for this application. The blood:gas partition coefficient of nitrous oxide is low and this confers a rapid onset of action. Any replacement would ideally require a similar property and we chose desflurane for this reason which has a partition coefficient of 0.42 (49). Unfortunately at 0.5 MAC, the anaesthetic equivalent of Entonox, mixtures of desflurane separate out very easily at high pressure. Accordingly we mixed desflurane 1.3% with 0.3% isoflurane to achieve this MAC level but found that liquid precipitation occurred at a cylinder temperature of 4 °C. By dropping the agent concentrations to 1% desflurane and 0.25% isoflurane, however, a mixture is produced which is

Table 2

Age	*Indication*	*Condition*	*Number of administrations*	*Mean latency onset (minutes)*	*Mean duration of use (minutes)*
31	Physiotherapy to elbow	Fit	19	1.9	17.3
75	Leg ulcers	Right Heart Failure	7	1.7	13
56	Leg ulcers	Peripheral vascular disease	4	4.0	13.8
66	Leg ulcers	Rheumatoid arthritis	15	3.1	21.8
33	Perineal wound post-partum	Fit	4	2.0	12.25
24	Pyoderma gangrenosum	Renal failure on dialysis	14	1.6	30.3
24	Pyoderma gangrenosum	Renal failure on dialysis	6	2.3	17.0
29	Physiotherapy to elbow	Fit	9	4.22	13
76	Leg ulcers	Peripheral vascular disease	16	4.12	21.4
41	Physiotherapy to wrist	Leukaemic, chemotherapy burn	5	2.2	14.4
52	Removal of K nails	Infected compound fracture femur	1	5	23

Patient details from preliminary trials of 1% desflurane with 0.25% isoflurane. The patient's age and underlying condition are detailed together with how long the mixture was breathed before it was possible to start the procedure (latency). The mean duration of administration and the number of times the mixture was given are also detailed.

stable down to −3 °C. This combination has been mixed to 137 atmospheres in cylinders with 60% oxygen with the balance gas being nitrogen. After appropriate local ethics committee approval, the mixture has been self administered without problem by 39 patients using the standard Entonox demand valve and we present some preliminary data from the first 11 of these cases.

A wide variety of cases used the mixture (Table 2) and the general approach to each case was to relieve rest or night pain with an appropriate analgesic regime and to use the gas mixture for the added pain of a wound dressing or physiotherapy. If the procedure caused pain which lasted some time after then an analgesic premedication was given which conferred an appropriate degree and duration of analgesia. The gas mixture was breathed until the patient felt that the procedure could start and this took 1.5-5 minutes. The gas was given on up to 19 separate occasions for one patient

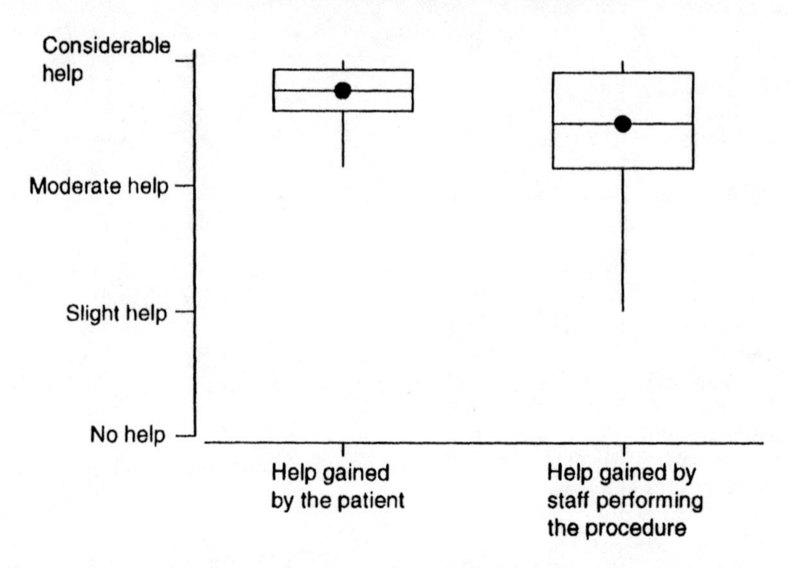

Figure 3 *The degree of help was scored on a simple 4 point system. The data displayed are the median and interquartile range from a group of 11 patients who received a mixture of 1% desflurane and 0.25% isoflurane.*

receiving physiotherapy and the mean duration of use varied from 13 minutes to half an hour. As and if the patient became drowsy the seal of the mask on the face was lost and wakefulness returned but in no case was verbal contact lost with the patient. In most of these cases a degree of co-operation was required between the staff and patient and this was maintained while the gas was being given.

We quantified the effect of the gas in terms of the amount of help gained both by the patient and by the staff performing the procedures using a simple four point scale: no help; slight help; a moderate amount; considerable help (Figure 3). On all occasions the patient found the gas of benefit. In only one patient did the physiotherapist find the gas of only slight help. This patient had an immobile elbow which was later found to be completely fixed when examined under general anaesthesia. Of these patients seven had previously received Entonox. Four unequivocally preferred the trial gas while the others oscillated between finding no difference between the two agents or preferring the trial gas.

The limited amount of work done does not permit firm conclusions to be made regarding the use of this mixture of isoflurane and desflurane. However, the mixture has proved useful and as acceptable to patients as Entonox if not more so. There is not the same risk of cumulative toxicity and, given the mixture is less potent than Entonox, it may be suitable for use by nurses without direct medical supervision.

Conclusion

Entonox has now been available for about 35 years. The effect that makes this mixture of nitrous oxide and oxygen possible, the solvent effect of a supercritical gas phase, also allows other clinically useful gas mixtures to be dispensed in cylinders at high pressure and the use of these mixtures may go some way to offset the deficiencies of Entonox in terms of its limited potency and potential for toxicity on repeated administration. Entonox has proved to be a safe agent in nursing and 0.25% isoflurane in Entonox, Isoxane, also seems to be safe in midwifery practice. A lower potency mixture of desflurane and isoflurane promises also to be acceptable and to offer some solution to the difficult problem of the discomfort of repeated painful procedures which are not thought to need a general anaesthetic when the toxic effects of nitrous oxide need to be avoided.

References

1. Andrews, E. The oxygen mixture, a new anaesthetic combination. Chicago Medical Examiner 9: 656-661. 1868.
2. Andrews, E. Liquid nitrous oxide as an anaesthetic. Chicago Medical Examiner 13: 34-36. 1872.
3. Frost, E. A. M. A history of nitrous oxide. In: Nitrous Oxide/ N_2O. Eger, E.I. ed. Edward Arnold. London. 1985. 1-22.
4. Richards, W., Parbrook, G.D., Wilson, J. Stanislav Klikovich (1853-1910). Anaesthesia 31: 933-940. 1976.
5. Barach, A.L., Rowenstein, E.A. The hazard of anoxia during nitrous oxide anaesthesia. Anesthesiology 6: 449-461. 1945.
6. Bourne, J. G. Nitrous oxide in dentistry. Lloyd-Luke. London. 1960.
7. Eastman, N.J. Fetal Blood Studies. American Journal of Obstetrics and Gynecology 31: 563-572 .1936.
8. Cole, P.V., Nainby-Luxmore, R.C. The hazards of gas and air in obstetrics. Anaesthesia 17: 505-518. 1962.
9. Seward, E.H. Obstetric analgesia: a new machine for the self-administration of nitrous oxide-oxygen. Proceedings of the Royal Society of Medicine 42: 745-746. 1949.
10. Tunstall, M.E. Obstetric analgesia: the use of a fixed nitrous oxide and oxygen mixture from one cylinder. Lancet ii: 964. 1961.
11. Latham, J., Parbrook, G.D. the use of pre-mixed nitrous oxide and oxygen in dental anaesthesia. Anaesthesia 21: 473-479. 1966.
12. Bracken, R.C., Brookes, R.C., Goldman, V. New equipment for dental anaesthesia using premixed gases and oxygen. British Journal of Anaesthesia 40: 903-906. 1968.
13. Cole, P.V. Nitrous oxide and oxygen from a single cylinder. Anaesthesia 19: 3-11. 1964.
14. Gale, C.W., Tunstall, M.E., Wilton-Davies, C.C. Premixed gas and oxygen for midwives. British Medical Journal 1: 732-736. 1964.
15. Cole, P.V., Crawford, J.S., Doughty, A.G., Epstein, H.G., Hill, I.D., Rollason, W.N., Tunstall, M.E. Specifications and recommendations for nitrous oxide/ oxygen apparatus to be used in obstetric analgesia. Anaesthesia 25: 317-327. 1970.

16. Baird, D. et al. Clinical trials of different concentrations of oxygen and nitrous oxide for obstetric analgesia. Report to the Medical Research Council of the committee on nitrous oxide and oxygen analgesia in midwifery. British Medical Journal 1: 709-713. 1970.
17. Snook, R. Resuscitation at road accidents. British Medical Journal 4: 348-350. 1969
18. Baskett, P.J., Withnell, A. Use of Entonox in the ambulance service. British Medical Journal 2: 41-43. 1970.
19. Laing, M.R., Clark, L.J. Analgesia and removal of nasal packing. Clinical Otolaryngology 15: 339-342.
20. Marsden A.K. Entonox in the emergency department. Injury 10: 311-312.
21. Bennett, J.A., Salmon, P.R., Branch, R.A., Baskett, P.J.F., Read, A.E.A. The use of inalational analgesia during fibre-optic colonoscopy. Anaesthesia 26: 294-297.
22. Wattenmaker, I., Kasser, J.R., McGravey, A. Self-administered nitrous oxide for fracture reduction in children in an emergency room setting. Journal of Orthopaedic Trauma 4: 35-38. 1990.
23. Poynting, J.H. Change of state: solid:liquid. Philosophical Magazine 12: 32-48. 1881.
24. Hannay, J.B., Hogarth, J. On the solubility of solids in gases. Proceedings of the Royal Society 30: 178-188. 1880.
25. Larson, A.T., Black, C.A. The concentration of ammonia in a compressed mixture of hydrogen and nitrogen over liquid ammonia. Journal of American Chemical Society 47: 1015-1020. 1925.
26. Booth, H.S., Bidwell, R.M. Solubility measurements in the critical region. Chemical Reviews 44:477-513. 1949.
27. Bracken, A.B., Broughton, G.B., Hill, D.W. Equilibria for mixtures of oxygen with nitrous oxide and carbon dioxide and their relevance to the storage of N_2O/O_2 cylinders for use in analgesia. Journal of Physics D: Applied Physics 3:1747-1758. 1970.
28. Dundee, J.W., Moore, J. Alterations in response to somatic pain associaited with anaesthesia. IV: the effect of sub-anaesthetic concentrations of inhalation agents. British Journal of Anaesthesia , 32: 453-459. 1960.
29. Robson, J.G., Davenport, H.T., Sugiyama, R. Differentiation of two types of pain by anaesthetics. Anesthesiology 26: 31-36. 1965.
30. Tomi, K., Mashimo, T., Tashiro, C., Pak, M., Nishimura, S., Nishimura, M., Yoshiya, I. Alterations in pain threshold and psychomotor response associated with subanaesthetic concentrations of inhalation anaesthetics in humans. British Journal of Anaesthesia 70: 684-686. 1993.
31. Houghton, I.T., Redfern, P.A., Utting, J.E. The analgesic effect of halothane. British Journal of Anaesthesia 45: 1105-1110. 1973.
32. Petersen-Felix, S., Arendt-Nielsen, L., Pak, P., Fischer, M., Bjerring, P,. Zbinden, A.M. Analgesic effect in humans of subanaesthetic isoflurane concentrations evaluated by experimentally induced pain. British Journal of Anaesthesia 75: 55-60. 1995.
33. Artusio, J.F. Di-ethyl ether analgesia: detailed description of first stage of ether anaesthesia in men. Journal of Pharmacology and Experimental Therapeutics 111: 343-348.
34. Reier, C.E. Methoxyflurane analgesia: a clinical appraisal and detailed description of stage I in man. Anesthesia and Analgesia 49:318-322. 1970.

35. Parbrook, G.D. The levels of nitrous oxide analgesia. British Journal of Anaesthesia 39: 974-982.
36. Moya, F. Use of a chloroform inhaler in obstetrics. New York State Journal of Medicine 61: 421-429. 1961
37. Shnider, S.M., Moya, F., Thorndike, V., Bossers, A., Morishima, H., James, L.S. Clinical and Biochemical Studies of Cyclopropane analgesia in obstetrics. Anesthesiology 24: 11-17. 1963.
38. Major, V., Rosen, M,. Mushin, W.M. Methoxyflurane as an obstetric analgesic: a comparision with trichlorethylene. British Medical Journal 2: 1554-1561. 1966.
39. Abboud, T.K., Shnider, S.M., Wright, R.G., Stephen, H.R., Craft, J.B., Henriksen, E.H., Johnson, J., Jones, M.J., Hughes, S.C., Levinson, G. Enflurane analgesia in obstetrics. Anesthesia and Analgesia 60:133-137. 1981.
40. McLeod, D.D., Ramayya, G.P., Tunstall, M.E. Self-administered isoflurane in labour. A comparative study with Entonox. Anaesthesia 40: 424-426. 1985.
41. Abboud, T.K., Swart, F., Donovan, M.M., Peres Da Silva, E., Yakal, K. Desflurane analgesia for vaginal delivery. Acta Anaesthesiologica Scandinavica 39: 256-261. 1995.
42. Gordon, R.A., Morton, V., Trichlorethylene in obstetrical analgesia and anaesthesia. Anesthesiology 12:680-687. 1951.
43. Arora, S., Tunstall, M., Ross, J. Self-administered mixture of entonox and isoflurane in labour. International Journal of Obstetric Anesthesia 4: 188-202.1992.
44. Wee, M.Y.K., Hasan, M.A. Thomas, T.A. Isoflurane in labour. Anaesthesia 48:369-372. 1993.
45. Rodgers, R.C., Hill, G.E. Equations for vapour pressure versus temperature: derivation and use of the Antoine equation for use on a hand-held programmable calculator. British Journal of Anaesthesia 50: 415-424. 1978.
46. Uyanik, A., Marr, I.L., Ross, J.A.S., Tunstall, M.E., Preparation and high pressure behaviour of pre-mixed volatile liquid anaesthetics in Entonox. British Journal of Anaesthesia 73: 712P. 1994.
47. Nunn, J.F., Sharer, N.M., Gorchein, A., Jones, J.A., Wickramasinghe, S.N. Megaloblastic haemopoeisis after multiple short-term exposures to nitrous oxide. Lancet 1: 1379-1381. 1982.
48. Ermens, A.A.M, Refsum, H., Rupreht, J., Spijkers, L.J.M., Guttormsen, A.B., Lindemans, J., Ueland, P.M., Abels, J. Monitoring cobalamin inactivation during nitrous oxide anesthesia by determination on homocysteine and folate in plasma and urine. Clinical Pharmacology and Therapeutics 49:385-393. 1991.
49. Eger, E.I. Desflurane animal and human pharmacology: aspects of kinetics, safety, and MAC. Anesthesia and Analgesia 75:S3-9. 1992.

PET Scanning – What Can it Tell us about Anaesthesia?

Per Hartvig[1], Jesper Andersson[1], Arne Wessén[2], Mats Enlund[2], Lars Wiklund[2], Sven Valind[1] and Bengt Långström[1]

UPPSALA UNIVERSITY PET CENTRE[1] AND DEPARTMENT OF ANAESTHESIOLOGY AND INTENSIVE CARE[2], UPPSALA UNIVERSITY HOSPITAL, S-751 85 UPPSALA, SWEDEN

Introduction

'Sopiens suscitans' is the motto of the Swedish Society of Anaesthesia and Intensive Care. The translation 'putting to sleep, waking up' indicates in a wide sense that a safe sleep is needed for awakening. This principle of a safe sleep is of major concern in all aspects of anaesthesia. Knowledge on the influence of anaesthesia on cerebral blood flow and metabolism is of fundamental importance in the assessment of safety. The basic requirement is that the decrease of cerebral metabolism should exceed that of the cerebral blood flow thereby not forcing brain tissue into a hypoxic state. The effects of different doses of new anaesthethic agents on cerebral blood flow and metabolism should therefore be studied at an early stage in drug development.

The effects on cerebral blood flow of most anaesthic drugs have been assesssed whereas information on cerebral metabolism is more scarce. Earlier studies on cerebral metabolism are indirect and in most instances estimates are only involving the global hemisphere. Obviously more precise methods are needed which non-invasively measure regional oxygen consumption and oxygen extraction in the brain. This need is furnished by positron emission tomography. However, there are to date only a few studies reporting on effects of anaesthesia on cerebral blood flow and oxygen metabolism using this technique.

Studies on Cerebral Hemodynamics

The effect of anaesthetics on cerebral hemodynamics in man has previously been studied with different tracer gas washout techniques first described by Kety and Schmidt (1). Nitrous oxide, ^{133}Xe-xenon and argon have been used as tracer gases. The cerebral metabolic rate of oxygen was calculated from the product of arterio-venous oxygen difference and the cerebral blood flow.

However, these earlier techniques can only be used for measurements of global hemodynamics. Most intravenous anaesthetic agents cause a reduction in both cerebral blood flow and oxygen metabolism as was found for the steroid anaesthetic drug eltanolone (2), while effects of inhalational agents vary. The global information obtained may, however, overlook significant regional changes.

Positron Emission Tomography

Positron emission tomography, PET, has enabled more extensive studies on human cerebral physiology and is at present the technique of choice for imaging brain function and metabolism (3). It provides excellent opportunities to assess not only global but also regional cerebral effects of anaesthesia in humans.

PET is a noninvasive tracer technique which quantitates the concentration over time of a radiolabelled tracer in physiological or biochemical processes in the tissue of living animal or man. The method has also been characterized as an *in vivo* autoradiographic technique for assessment of radiotracer kinetics. By the proper choice of radiotracer almost any process within living tissue can be quantitated and a physiologic measure is obtained in absolute terms. Some examples of physiological processes studied with PET are given in Table 1.

Table 1 *Some physiological processes studied with PET*

Physiological process	*Radiolabelled tracer*
Blood volume	Carbon monoxide
Blood flow	Water, butanol
Tissue pH	Sodium bicarbonate
Energy substrate utililization	Oxygen, glucose, fatty acids
Amino acid transport /utilization	Amino acids
Enzyme concentration	Enzyme inhibitor
Enzyme activity	Substrate for the enzyme
Neurotransmitter synthesis	Precursor in transmitter synthesis
Receptor localization /kinetics	Receptor ligand

The radionuclides most frequently used are ^{15}O; ^{13}N; ^{11}C and ^{18}F with physical half-lives of 2; 10; 20 and 110 minutes, respectively. Radionuclides with longer physical half-lives may be used for studies on slower processes within the tissue. The radionuclide has to be incorporated in the tracer molecule by fast and reproducible radioorganic synthesis. Due to the physical decay, the time for production of most radiotracers is limited to about three half-lives. This includes time for pharmaceutical preparation, control of identity as well as chemical and radiochemical purity. The radiolabelled tracer is injected intravenously or inhaled in gaseous form. The radioactivity signals are collected by external detectors for predetermined intervals in tomographic slices interspaced with approximately 6 mm. After reconstruction of images the concentration and kinetics of the tracer molecule within the tissue are calculated and used in further analysis using mathematical models in order to quantify in absolute terms a parameter of the studied physiologic process.

The obvious advantage of PET is the possibility of *in vivo* quantitation of physiologic processes in the healthy and diseased states. Furthermore, trace amounts of radiotracers administered do not affect parameters under study. The rapid disappearance of radioactivity makes repeated studies possible which further strengthen the interpretation and validity of the results. On the other hand, the limited spatial resolution of PET might hamper accurate quantitation of radiotracer concentration in discrete areas of the tissue. Another disadvantage of the PET is the co-determination of metabolites with withheld radiolabel. This prevents the analysis of radiotracers with an extensive metabolism as e.g. ^{11}C-glucose.

Measurement of Cerebral Blood Flow and Metabolism

At the Uppsala University PET Centre, the regional cerebral blood volume (rCBV), blood flow (rCBF) and the fraction of oxygen in arterial blood extracted (rOER) by the brain along with oxygen utilization ($rCMRO_2$) have been measured by carbon monoxide radiolabelled with ^{15}O- CO, $^{15}O\text{-}H_2O$ and $^{15}O\text{-}O_2$, respectively. Blood volume may change somewhat during anaesthesia, and this information is of importance for further calculation of e.g. oxygen metabolism. The short half-life of ^{15}O ($T_{1/2}$ =2.05 min) permits measurements to be made in sequence with some 10 mins. between successive tracers.

Following inhalation of ^{15}O-CO, the tracer is equlibrated between arterial and venous blood and binds to hemoglobin in the erythrocytes. Hence the radiotracer will be confined to the blood. The amount of carbon monoxide inhaled is far less than that following smoking of a cigarette. The regional blood volume in the brain (rCBV) is calculated by relating the radioactivity concentration in the brain tissue to that in blood samples.

Regional cerebral blood flow, rCBF, may be measured following an intravenous bolus injection of $^{15}O\text{-}H_2O$ and using an autoradiographic technique (4). $^{15}O\text{-}H_2O$ is considered as freely diffusible and will be transported via the blood into the brain, where it is equilibrated with the tissue. The combination of measurements in the brain and in the arterial blood allows for calculation of absolute flow values, in this case, rCBF. Radioactivity data with PET are collected in five second intervals, corrected for radioactivity decay and added together to yield an image containing information from time of arrival of the bolus in the brain to 45 seconds past arrival. Arterial blood is continuously withdrawn by an automatic device that is previously calibrated to yield radioactivity concentrations. Blood radioactivities are corrected for delay and dispersion (5).

The use of $^{15}O\text{-}O_2$ for measurement of brain oxygen metabolism is slightly more complicated, since it is rapidly metabolized yielding ^{15}O-water. Therefore the measures obtained following inhalation of $^{15}O\text{-}O_2$ need correction for metabolism using data from the previously performed $^{15}O\text{-}H_2O$ study. Measurements may be done by continuous inhalation of $^{15}O\text{-}O_2$ through a mouth piece or an oral airway via a ventilator. A constant radioactivity concentration is administered during seven to ten minutes for equilibration followed by scanning with PET for five minutes. Arterial blood is collected at intervals and the radioactivity measured. The blood flows estimated are used in the steady state operational equation for evaluation of oxygen extraction and metabolism (6). Data are corrected for $^{15}O_2$ in the vascular compartment using blood volume figures from baseline conditions (7). After this, estimates of oxygen metabolism ($rCMO_2$) but also oxygen extraction (rOER) may be calculated. Oxygen extraction is a dimensionless entity, usually expressed as percentage, describing the proportion of locally available oxygen that is consumed in the tissue. The normal extraction is in the range 30-50%. An elevated oxygen extraction usually indicates hypoxia or ischemia of the tissue.

Complete definition of the hemodynamic state requires serial PET examinations with ^{15}O-water and ^{15}O-oxygen and possibly with ^{15}O-carbon monoxide. Each of these studies lasts a few minutes with a pause of approximately ten minutes in between to allow for decay of radioactivity of the radiotracer. The total time for the entire series of measurements is in the range 30 to 40 minutes. During this time, the subject has to be in steady state with respect to variables that might influence hemodynamics. In particular, the anaesthethic depth which influences regional oxygen metabolism and P_aCO_2 influencing blood flow and oxygen extraction must be constant.

Applications of Positron Emission Tomography in Studies on Anaesthethic Agents

PET has previously been used in anaesthesia research, e.g. in studies on the effect of isoflurane in the hypocapnic baboon (8). In humans, PET was employed in studies on cerebral glucose metabolism during propofol anesthesia (9). Light propofol anaesthesia produced a marked global depression of glucose metabolism with marked differences within cortical regions. PET has also been used in patients to study cerebral glucose metabolism in normal brain versus glioma during barbiturate anaesthesia (10).

Cerebral blood flow and oxygen metabolism during eltanolone anaesthesia

Eltanolone is a novel intravenous steroid-based anaesthetic agent. The drug has been tested in large clinical trials but the development was recently withdrawn. The safety of eltanolone anaesthesia measured as effects on cerebral blood flow, cerebral metabolic rate of oxygen and oxygen extraction were evaluted in man using PET (11) (Figure 1). Series of measurements with $^{15}O\text{-}O_2$ and $^{15}O\text{-}H_2O$ were carried out in the awake state, during anaesthesia with eltanolone infusion and during recovery. Cerebral blood flow was

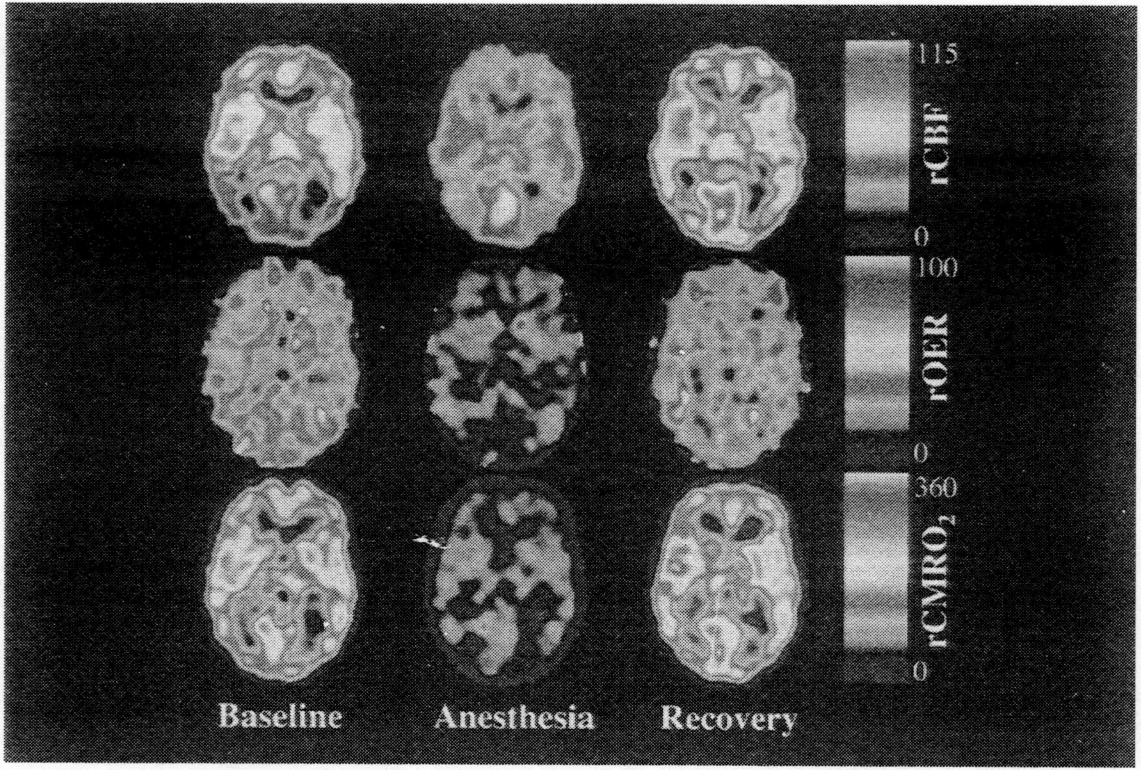

Figure 1 *Regional cerebral blood flow, rCBF; oxygen extraction ratio, rOER and rate of oxygen metabolism, $rCMRO_2$ in healthy volunteers before, during and in the recovery of eltanolone anaesthesia.*

reduced by 31 ± 15 % during anaesthesia in almost all cortical regions. Oxygen metabolism was reduced by 52 ± 8 % but increased to near baseline values during recovery. Oxygen extraction remained homogenous and was lowered by 32 ± 22 % during anaethesia and returned to 82% of baseline during recovery. Similar to other intravenous anaesthetic agents, eltanolone decreased both blood flow and oxygen metabolism. The more pronounced depression of oxygen metabolism as compared with blood flow indicated some protective effect against hypoxia. Hence, there was no sign of ischemia in any brain region. Therefore, eltanolone was considered safe and particularly useful in neuroanaesthesia and intensive care (11).

Cerebral blood flow and oxygen metabolism in hypotensive anaesthesia

Hypotensive anaesthesia has since long time been controversial due to the risk of hypoxic brain damage. Several studies have been unable to demonstrate signs of cerebral hypoxia, whereas increase in biochemical markers for brain injury have been shown in some patients undergoing hypotensive anaesthesia (13). To further illuminate risks to cerebral oxygenation and hemodynamics, cerebral oxygen metabolism rate, cerebral blood flow and oxygen extraction ratio were measured in Rhesus monkeys during propofol and isoflurane anaesthesia in the normo- and hypotensive states (12). Cerebral blood flow varied widely between brain regions in isoflurane anaesthesia, particularly in comparison with propofol anaesthesia. On the other hand, oxygen metabolism decreased globally and dose-dependently during both isoflurane and propofol anaesthesia. The metabolism flow coupling was intact with propofol but not during isoflurane anaesthesia. Hypotension reduced regional blood flow but not oxygen metabolism and extraction. The latter parameter increased for both study drugs when changing from normo- to hypotension. However, it was suggested that hypocapnia rather than hypotension was responsible for the increased oxygen extraction ratio. Hypoxia was also observed in two animals possibly as a result of hyperventilation. Cerebral oxygenation was thus adequate during propofol and isoflurane anaesthesia. Hypotension reduced blood flow but did not result in hypoxia. The sensitivity and specificity of PET also enabled the detection of a cerebellar hypoxia with an oxygen extraction over 70% in cortical regions following ketamine induction in some of the monkeys. Ketamine is a metabolic stimulant and is also reported to increase cerebellar blood flow. The demand-supply relationship might have been unfavourable resulting in an apparent hypoxic state.

Ketamine mechanism of action studied with positron emission tomography

The *N*-methyl-D-aspartate (NMDA)-sensitive glutamate receptor has a prominent role in basic brain function and in synaptic plasticity. The NMDA receptor activation is suggested to be operative in the consolidation of learning and memory (14). Activation of NMDA receptors appears also to be an important mechanism in the pathogenesis of ischemia and degenerative neuronal damage and in the generation of epileptic pathology. Furthermore, anaesthethic drugs irrespective of chemical class have been suggested to directly or indirectly act on nitric oxide and the NMDA-receptor controlled ion calcium channel (15). Ketamine is a phencyclidine-like dissociative anaesthetic drug which binds specifically in the NMDA-receptor gated ion channel and blocks the receptor in a non-competitive manner. The occupancy on the phencyclidine site of *(S)*- and *(R)*-enantiomers of ketamine correlates positively with their dose-response curves for analgesia and dysperception in humans (16). Therefore ketamine may reflect NMDA receptor inhibition and prompted the use of ^{11}C-labelled ketamine as a probe for NMDA receptor function in studies with PET.

By the use of the PET technique in the Rhesus monkey, we have *in vivo* confirmed a dose-dependent inhibition of binding in the brain of both the *(S)*- and *(R)*-enantiomer of ketamine radiolabelled with ^{11}C in the methyl group (17). The receptor affinity of the *(S)*-enantiomer was in the µmol/L range and was two to three times higher than that of the *(R)*-enantiomer. The *(S)*-enantiomer also competed for binding sites with the *(R)*-enantiomer and with dizolcipine, a model drug for receptor blockade in the NMDA-receptor controlled ion channel.

This validation in the monkey made possible measurements of clinical effects of low subdissociative doses of *(S)*-ketamine in relation to receptor occupancy within the NMDA-receptor complex (20). In a randomized, double blind cross-over study in healthy volunteers, plasma concentration, maximal brain concentrations as well as specific regional binding in the brain after administration of 0, 0.1 and 0.2 mg/kg doses of *(S)*-ketamine were related to induced effects such as analgesia, amnesia and mood changes. Brain concentrations and the specific binding in the brain were assessed by simultaneous administration of *(S)*-[N-methyl-^{11}C]ketamine quantified by positron emission tomography. The specific binding decreased marginally, but significantly, in different brain regions after administration of subanaesthetic doses of *(S)*-ketamine. However, all pharmacological effects were pronounced and dose-dependent. This points to a very steep dose-response relation and *in vitro* studies have shown that a 50% receptor occupancy correspond to deep anaesthesia whereas 20% receptor occupancy may give a prominent pain relief (16). A similar decrease of specific binding was seen following the 0.2 mg/kg.

Further experiments with PET have also indicated that ^{11}C-labelled *(S)*-ketamine concentrations varied similarly with regional blood flow and was not able to discriminate small changes in NMDA-receptor channel activation induced by ischemic pain in man. Epilepsia may activate NMDA receptors. However, in human epilpetic brain tissue *(S)*-ketamine binding decreased possibly as a result of lower blood flow and neurodegeneration. Therefore, ^{11}C-labelled *(S)*-ketamine does not seem to be a useful probe for functional studies with PET on neuronal activation in different brain areas (18).

Conclusions

PET is an autoradiographic technique quantitating the distribution and binding of a radiolabelled tracer molecule. It does not tell us anything about the function of the tracer molecule. Proper design of studies may, however, reveal the physiological interference on tissue biochemistry of the compound by measurements of effects on e.g. blood flow, oxygen metabolism and extraction. PET may also deepen the understanding of the pharmacologic mechanism of action in protocols both assessing the pharmacodynamics and pharmacokinetics of the drug. Therefore, PET is a powerful method in the assessment of safety of brain hemodynamics and oxygenation during anaesthesia. PET is a basic requirement for 'Sopiens suscitans'.

Acknowledgements

The studies were financially supported by Zeneca Ltd.,United Kingdom; Pharmacia Human Pharmacology Laboratory, Uppsala, Sweden; Warner Lambert, Solna, Sweden and by the Swedish Medical Research Counsil, project nos 06579 and 08645.

References

1. Kety SS, Schmidt CF. The determination of cerebral blood flow in man by use of nitrous oxide in low concentrations. Am J Physiol 143:53-66, 1945
2. Wolff J, Carl P, Hansen PB, Hogskilde S, Christensen MS, Sörensen MB. Effects of eltanolone on cerebral blood flow and metabolism in healthy volunteers. Anesthesiology 81: 623- 627, 1994
3. Wegener WA, Alavi A. Positron emission tomography in the investigation of neuro-psychiatric disorders: update and comparison with magnetic resonance imaging and computerized tomography. Int J Rad Appl Instrument B. 18: 569-582, 1991
4. Herscovitch P, Markham J, Raichle ME. Brain flow measured with intravenous $H_2{}^{15}O$ Theory and error analysis. J Nucl Med 24: 782-789, 1983
5. Meyer E. Simultaneous correction for tracer arrival delay and dispersion in CBF measurements by the $H_2{}^{15}O$ autoradiographic method and dynamic PET. J Nucl Med 30: 1069-1078, 1989.
6. Frackowiak RS, Lenzi GL, Jones T, Heather JD. Quantitative measurement of regional cerebral blood flow and oxygen metabolism in man using ^{15}O and positron emission tomography: theory, procedure and normal values. J Comput Assist Tomogr. 4: 727-736, 1980
7. Lammertsma AA, Wise RJ, Heather JD, Gibbs JM, Leenders KL, Frackowiak RS, Rhodes CG, Jones T. Correction for the presence of intravascular oxygen-15 in the steady state technique for measuring regional oxygen extraction ratio in the brain: 2. Results in normal subjects and brain tumor and stroke patients. J Cereb Blood Flow Metab 3:425-431, 1983
8. Archer DP, Labrecque P, Tyler JL, Meyer E, Evans AC, Villemure JG, Casey WF,

Diksic M, Hakim AM, Trop D. Measurement of cerebral blood flow and volume with positron emission tomography during isoflurane administration in the hypocapnic baboon. Anesthesiology 72: 1031-1037, 1990

9. Alkire MT, Haier RJ, Barker SJ, Shah NK, Wu JC, Kao YJ. Cerebral metabolism during propofol anesthesia in humans studied with positron emission tomography. Anesthesiology 82: 393-403, 1995.
10. Blacklock JB, Oldfield EH, Di CG, Tran D, Theodore W, Wright DC, Larson SM. Effect of barbiturate coma on glucose utilization in normal brain versus gliomas. Positron emission tomography studies. J Neurosurg 67: 71-75, 1987
11. Wessén A, Widman M, Andersson J, Hartvig P, Valind S, Hetta J, Långström B. A positron emission tomography study of cerebral blood flow and oxygen metabolism in healthy male volunteers anaesthetized with eltanolone. Acta Anaesthesiol. Scand In press.
12. Enlund M, Andersson J, Valtysson J, Hartvig P, Wiklund L. Cerebral normoxia in Rhesus monkey during isoflurane- or propofol-induced hypotension and hypnocapnea, despite disparate blood flow patterns. A positron emission tomography study. Acta Anaesthesiol. Scand. In press.
13. Enlund M, Mentell O, Engström Horneman G, Ronquist G.. Occurence of adenylate kinase in cerebrospinal fluid after isoflurane anaesthesia and orthognatic surgery. Upsala Med Sci. 101: 97-112, 1996
14. Cotman CW, Monaghan DT, Ganong AH. Excitatory amino acid neurotransmitters, NMDA-receptors and Hebb-type synaptic plasticity. A Rev Neurosci 11: 61-80, 1988.
15. Clarke IA, Pockett, Cowden WB. Possible role of nitric oxide in conditions similar to cerebral malaria. The Lancet 340: 894-895, 1992.
16. Øye I, Paulsen O, Maurset A. Effects of ketamine on sensory perception: evidence for a role of *N*-methyl-D-aspartate receptors. J Pharmacol. Exp Therap 260:1209-1213, 1992
17. Hartvig P, Valtysson J, Antoni G, Westerberg G, Långström B, Ratti Moberg E, Øye I. Brain kinetics of *(R)*- and *(S)*- [N-methyl-^{11}C]ketamine in the Rhesus monkey studied by positron emission tomography (PET). Nucl Med Biol 21: 927-934, 1994
18. Hartvig P, Valtysson J, Lindner KJ, Kristensen J, Karlsten R, Gustafsson LL, Persson J, Svensson JO, Øye I, Antoni G, Westerberg G, Långström B. Central nervous system effects of subdissociative doses of (S)-ketamine are related to plasma and brain concentrations measured with positron emission tomography in healthy volunteers. Clin Pharmacol Ther 58:165-173, 1995.

Molecular and Cellular Mechanisms of Anaesthesia

The Neural Processes Involved in Anaesthesia and a Comparison of the Effects of Nitrous Oxide and Halothane on Somatosensory Transmission

A. Angel, R.H. Arnott and Sarah Wolstenholme

CENTRE FOR RESEARCH INTO ANAESTHETIC MECHANISMS, DEPARTMENT OF BIOMEDICAL SCIENCE, THE UNIVERSITY, SHEFFIELD, UK

Inroduction

The state engendered by the administration of anaesthetic agents is highly complex. Cognitive, sensory, motor, autonomic and nociceptive behavioural patterns are either abolished or markedly attenuated. In spite of the rich spectrum of effects shown by anaesthetics the word anaesthesia is still used to describe the desired endpoint of the action of anaesthetics as though this is a single phenomenon. This is presumably as a result of an old idea that anaesthetic agents cause a general depression of central nervous cells and that the state of anaesthesia is merely a global expression of this depression. In addition the word anaesthesia is also loosely, and confusingly, used by some authors when they mean analgesia i.e. a reduction of pain sensation. The major effects of general anaesthetic agents can be listed as follows:

A loss of consciousness

This is possibly a reflection of anaesthetic action on the brainstem reticular formation which has long been implicated in attempts to define consciousness/arousal, waking/sleeping and attentional switching. It occupies the central part of the medulla, pons and midbrain, and projects to the hypothalamus and via specific neurotransmitter containing neuronal subsets, diffusely to the thalamus and cerebral cortex (see Figure 1). The brain stem reticular formation receives an afferent input from all sensory modalities and from the motor cortex. The first indication that this part of the central nervous system could be involved in the generation of general anaesthesia was inferred by Magoun and his collaborators from its role in the integration of sensory and motor processes related to the level of behavioural and electrocortical arousal in animals (see Rhines & Magoun, 1946; Brodal, 1957). They showed that the reticular formation controlled the excitability of spinal motoneurones (Magoun & Rhines, 1946; Rhines & Magoun, 1946). In addition it was shown (Moruzzi & Magoun, 1949) that electrical stimulation within the reticular complex could lead to a generalized electrocortical arousal (high frequency stimulation) or to a generalized electrocortical desynchronisation (low frequency stimulation).

Loss of postural corrections with respect to the gravitational field

Anaesthetised animals lose the ability to maintain a correct posture. This has to be a central phenomenon because the vestibular receptors are, in common

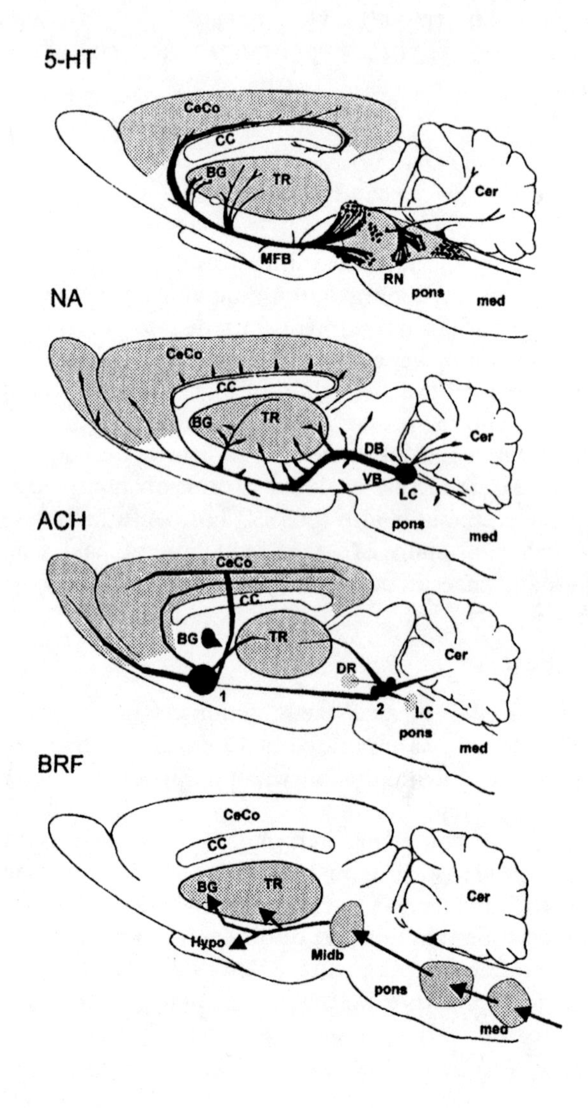

with all other sensory receptors, uninfluenced by anaesthetic agents. It is not known whether the inability of an animal to right itself, when turned over on its back, is due to:

i. a failure of transmission of information to the vestibular nuclei or
ii. a failure of descending information in the vestibulospinal tracts to gain access to spinal α-motoneurones or
iii. a disruption in γ-motoneuronal activity with a consequent loss of gain in the spinal stretch reflex or
iv. a bias of spinal interneurones promoting spinal α-motoneuronal inhibition or indeed
v. a mixture of any or all of these processes.

Loss of volitional movement

All anaesthetics, with the exception of benzodiazepines, at low clinical doses for minor surgery, cause an inability of the anaesthetised animal to make purposive, goal directed movements or, in humans, movements to verbal commands. Again the reasons for this loss are not known. It could be due to a suppression of motor cortical neurons or to spinal mechanisms or to both. Although two common observations are; a) that is it is very difficult to make

---

Figure 1 *Schematic sagittal sections of the 5-hydroxytryptaminergic (5-HT), noradrenergic (NA), cholinergic (ACH) and brainstem reticular formation (BRF) projections in the rat. Note that the raphe nuclei occupy a midline position in the brainstem. The locus coeruleus consists of two small nuclei on the dorsal surface of the anterior pons. The brainstem reticular formation consists of a part in the central area of the posterior medulla, and a part in the central area of the posterior pons. Both these areas project to the midbrain reticular formation. The projection from the midbrain is mainly to the hypothalamus (75%) with the remainder projecting to the thalamic reticular nucleus, which bounds the lateral surface of the thalamic relay nuclei, and to the central intralaminar nuclei. Both of these project to the thalamic relay nuclei and basal ganglia. Note that the specific neurotransmitter reticular systems can influence neurons in the cerebral cortex via two main paths; directly by their diffuse projection to the cerebral cortex and indirectly via their actions on the thalamic relay nuclei and basal ganglia. The main brainstem reticular formation influences the cerebral cortex only indirectly via the thalamic relay nuclei, the intralaminar nuclei and the basal ganglia. These latter project to specific cortical motor and sensory areas and from these cortical areas to 'association' cortex.*

***BG** basal ganglia, **CC** corpus callosum, **Ce Co** cerebral cortex, **Cer** cerebellum, **DB** dorsal noradrenergic bundle, **DR** dorsal raphe nucleus, **Hypo** hypothalamus, **LC** locus coeruleus, **med** medulla, **MFB** medial forebrain bundle, **Midb** midbrain, **pons** pons, **RN** raphe nuclei, **TR** thalamic relay nuclei, **VB** ventral noradrenergic bundle, **1** Basal forebrain cholinergic system (medial septal nucleus, diagonal band nuclei, substantia innominata, magnocellular preoptic field and nucleus basalis) **2** upper brain stem cholinergic system (laterodorsal tegmental nuclei and pedunculopontine nuclei).*

cortical motoneurones respond to electrical stimulation of the cortical surface in anaesthetised animals and b) somatosensory cortical cells in layer V, the major output layer of the cortex, show a marked susceptibility to anaesthetic action (see below).

A reduction in protective reflexes

In general all protective automatic movements to potentially or overtly noxious stimuli are truncated by general anaesthetics. This is generally taken to indicate some degree of analgesia although this interpretation may be mistaken depending upon whether one interprets the automaticity of such reflexes to reflect a motoneuronal pattern generator or the emotive response (i.e. ouch!) to such a stimulus. One only has to ask the simple question " If you pick up a hot pan when do you say ouch, before or after you put it down?" to realize that the speedy motor response takes place before the emotive reaction. Generally anaesthetic agents are weak analgesics, with the exception of the opiate class of anaesthetic agents and nitrous oxide.

Loss of aesthetic sensory experience

Anaesthetised humans cannot hear, see, feel or taste. This is one facet of animal behaviour which can be tested indirectly. All sensory pathways are targetted to specific parts of the cerebral cortex. Stimulation of the afferent limb of the pathways gives rise to an activation of a population of sensory cortical cells and the resultant current flow can be recorded as cerebral evoked responses. The analysis of such responses can give some indication of anaesthetic action (see below).

Loss of memory

Simplistically, if awareness of the environment is severely truncated by anaesthetic agents then less information gains access to the higher structures of the central nervous system. Thus the probability that a reduced information inflow will lay down a memory trace will be lessened. Added to this is the observation that it is extraordinarily difficult to activate cortical association neurons in anaesthetised animals. Since memory loss is total, anaesthetics probably also act upon the cholinergic projection from the brain stem reticular formation since one part of this system, the basal forebrain cholinergic system (see Figure 1), has been implicated in memory mechanisms

An inability to control core temperature

This is probably a central phenomenon since the thermoreceptors are unaffected by anaesthesia. The body thermostat is situated in the hypothalamus with a heat conservation 'area' in the posterior part. The severe skeletomotor upsets seen when anaesthetic agents are administered will by themselves truncate shivering thermogenesis. There is one other aspect of hypothalamic function which is interesting. Most anaesthetics give an increase in antidiuretic hormone levels. Since this hormone is only released as a result of neuronal activity it implies that anaesthetics directly excite some hypothalamic neurons. This increased activity cannot be explained by changes in blood osmolality due to anaesthetic administration, the plasma concentrations employed are usually very small.

Changes in cardiovascular performance

With the exception of ethyl carbamate, anaesthetics usually produce hypotension. This is generally unimportant since cerebral blood flow is autoregulated for mean blood pressures between 70 and 140 mm Hg (9.3 to 18.7 kPa).

Changes in respiratory performance

Anaesthetics usually cause respiratory depression; breathing becomes slowed and unless active steps are taken animals become hypoxic and hypercapnic with a consequent decrease in arterial pH.

Specific anaesthetic related side effects

These can be either central, e.g. enhanced inhibition of spinal motor neurones with barbiturates and increased sympathetic activity with di-ethyl ether, or peripheral actions, e.g. di-ethyl ether causes a depression of vascular smooth muscle and halothane has a depressant action on myocardial cells.

The study of general anaesthesia is also complicated by the absence of a simple structure-activity relationship such as that which holds for most bioactive chemicals. Anaesthetics show a wide range in molecular structure ranging from nitrogen to complex steroids. Although in a superficial sense anaesthetics can be antagonized by convulsant chemicals the only effective 'antagonist' (using the word antagonist to denote reversal of effect) is to increase the ambient pressure to very high levels (around 100 atmospheres, 10.13 MPa). This is not an effect of antagonism by competition with a specific receptor site by a specific anti-anaesthetic chemical but a physico-chemical barrier to the anaesthetic interacting with a putative hydrophobic receptor site. Simplistically the reaction of an anaesthetic with a putative receptor site causes an increase in volume; increasing the ambient pressure reverses this volume increase. Nonetheless the behavioural state of all anaesthetic agents is reversed by high pressures. This gives a tool to tease apart anaesthetic effects per se as opposed to interesting side effects. After all if the process under study is driven in one direction by administration of an anaesthetic agent and is unaffected by high pressure then it is unlikely to be involved in an anaesthetic mechanism.

The remainder of this paper is in two parts. Since so many changes are seen after the administration of anaesthetic agents and the behavioural endpoint is apparently similar for all such agents it is tempting to assume that they all produce this state by a single or similar mechanism of action. However, it is not possible to study all of the above effects in simple experiments. Therefore attention will be focussed on one aspect of anaesthetic action: the reduction of sensory inflow. The first part of this paper will examine the effects of anaesthetic agents on somatosensory transmission and will categorize anaesthetics into four distinct classes. The second part will compare the effects of nitrous oxide (because of its peculiar properties and because this is a Priestley Conference) and halothane on somatosensory transmission.

Methodology

i) Anaesthesia. Female rats (Sheffield strain) in the weight range 190-210 g were anaesthetised with ethyl carbamate (urethane, 25% w/v in 0.9% saline) administered intraperitoneally without premedication. The anaesthetic depth was adjusted to just abolish reflex withdrawal of the hindlimb to a strong pinch (dose 1.25-1.5 g kg^{-1}). At this depth of urethane anaesthesia, blood pressure, alveolar and arterial O_2 and CO_2 tensions are within normal limits (De Wildt et al., 1983). Additional anaesthetic agents were administered either by intraperitoneal injection or by passing the vapour in an oxygen stream past the tracheal cannula via a T-piece.

ii) Preparation. After tracheal cannulation the medulla was exposed by opening the foramen magnum and removing the arch of the first cervical vertebra. The left cortex was then exposed and the dura mater reflected. Any animals that showed signs of cerebral oedema were excluded from the analyses (see Angel et al., 1973). The animals, once prepared, were mounted in a stereotaxic frame and held rigidly by bars inserted into each external auditory meatus to puncture the tympanum and by a bar clamped against the hard palate immediately behind the upper incisors. A leakproof pool was formed over the exposed tissue, by gripping the cut skin edges between an inner perspex ring and an outer circular clip which was filled with liquid paraffin BP which had been saturated with physiological saline at 38 °C. To reduce movements of the exposed nervous tissue to respiratory effort or cardiac pulsation, the animals were suspended, with the body unsupported, between the ear bars and a pin inserted deep to the ligaments covering the dorsal aspect of the pelvic girdle and the trunk was put under slight traction. Rectal temperature was maintained at 37 ± 0.5 °C by circulating water at 40 °C through a hollow copper block placed underneath, but not in contact with, the animal.

iii) Recording. Responses from the cerebral cortex, were recorded via two spring loaded fine silver wires insulated except at their tips which were fused into small balls (diameter 1 mm) placed gently on the pial surface, one (active) placed at the geometric centre of the forepaw projection area and the other (reference) further back on the occipital cortex. The signals from these wires were amplified with resistance-capacity coupled amplifiers with a frequency response flat from 0.1 to 100 Hz. and displayed on a cathode ray-oscilloscope. Analogue data was converted to a digital form (125 samples sec^{-1} for the electrocorticographic and respiratory data and up to 25 samples $msec^{-1}$ for evoked activity) either for on-line processing, such as deriving averaged evoked cortical responses or to use the microprocessor as a storage oscilloscope. In addition the digitized signals could be stored on floppy discs for subsequent statistical evaluation. Glass micropipettes of tip diameter 2-5 μm and with impedances in the range 2-8 MΩ filled with 2% Pontamine Sky Blue in 0.5 M sodium acetate were used to record extracellularly from single neurons. The activity from the electrodes was led via a short piece of silver-iridium wire into a high input impedance, unity gain FET input amplifier. Spike activity was visualized with conventional recording techniques, and monitored via an audio system. Spontaneous and evoked cellular activity was also processed with a microprocessor and data were stored on floppy discs for subsequent statistical manipulations. The microelectrode assembly was mounted on three lathe slides placed mutually at right angles to give positioning in the mediolateral, rostrocaudal and dorsoventral axes. The zero reference point for the stereotaxic co-ordinates in the rostrocaudal and mediolateral planes, set at the beginning of each experiment, was the bregma, the intersection of the sagittal and frontal cranial sutures. The dorsoventral zero was taken as the point at which the micropipette contacted the pial surface. When the surface position had been found, subsequent movement of the microelectrode in the dorsoventral axis was controlled by a stepper-motor drive geared such that four steps of the motor advanced the electrode tip 1μm. The electrode was usually advanced in 2 or 5 μm steps until a cell had been isolated from the background noise and its recorded potential optimized. Respiration was recorded by inserting a fine polythene tube (outer diameter approximately 1mm), connected to a low pressure transducer into the lumen of the tracheal cannula.

iv) Somatic Stimulation. The right forepaw was stimulated electrically via fine entomological pins inserted into the skin on either side of the wrist. Care was taken when inserting the pins not to occlude the blood supply to the hand. Rectangular pulses of 0.1 msec duration and of variable voltage, from 0 to 90 V, isolated from earth (Devices stimulator type 2533) were delivered at a rate of $1\ s^{-1}$. For each experiment the stimulus intensity was set to be supramaximal for the evoked cortical response. At this stimulus level only the A group of nerve fibres would be activated.

v) Anaesthetic administration. Anaesthetics were either administered as a solution by intraperitoneal injection or by passing their vapour, or gas, in an oxygen stream through a T-piece connected to the animal's tracheal cannula.

vi) Statistical analysis. Values for differences of means were evaluated using a two-tailed Student's t-test. Significance was achieved if the p values were < 0.05.

Part 1

Anatomical pathways

A study of the effect of anaesthetic agents on the central nervous system needs to address two key questions. First, if sensory transmission is reduced, is this effect due to a direct action of the anaesthetic agent on all, or part, of the sensory pathway conveying information to the cerebral cortex? Secondly, do anaesthetic agents modify the behaviour of that part of the nervous system which is implicated in the control of consciousness and also in the modulation of information transfer through the specific sensory pathways?

The anatomy of the sensory pathway investigated in this study is very simple. Cutting the dorsal columns in the rat abolishes all short latency cortical electrical activity evoked by tactile or electrical stimulation of the integument. This activity is unaffected by cutting the dorsolateral columns (spinocervical path) or ventrolateral columns (spinothalamic path; Angel, Berridge & Unwin, 1973). Thus the fast, or short latency, pathway activated by peripheral stimulation in the rat is the dorsal column, medial lemniscal, thalamo-cortical system. Afferent information ascends in the dorsal columns to synapse in the dorsal column nuclei. The output of these nuclei crosses over to the other side of the brain and ascends in the medial lemniscus to the ventroposterolateral thalamus (VPL). The output from the thalamus then goes via the thalamocortical radiation fibres to the primary cortical somatosensory receiving area to activate columns of cells located in cortical layers III, IV, V and VI (see Angel, 1993).

The sensory modulatory function of the brain is mainly located in:

- The brain stem reticular formation.
- Its thalamic projection areas, the thalamic reticular nucleus and intralaminar nuclei.
- The specific cortical sensory receiving areas which send topographically organized corticofugal fibres to the sensory relay nuclei, the thalamic reticular and intralaminar nuclei and the medullary and pontine compartments of the brain stem reticular formation.
- The brain stem reticular formation is located in the medial parts of the medulla, pons and midbrain. The cells in the this region are embedded in a network of fibres and are structurally and functionally separate from those aggregates of cells which form the cranial nerve and dorsal column nuclei. This part of the reticular formation can be divided into at least three cellular compartments depending upon the neurotransmitter chemicals they contain. There is a 5-hydroxytryptaminergic projection from the midline raphé nuclei, a noradrenergic projection from the two locus coeruleus nuclei located on the dorsal surface of each side of the anterior pons and a cholinergic projection from the upper brain stem (see Figure 1). These projections are diffuse and project to the cerebral cortex, the thalamus and the basal ganglia. The noradrenergic axon synapses are different from normal terminal synaptic boutons. In the noradrenergic system each nerve fibre may have several thousand expansions, called varicosities, each of which can release the neurotransmitter. Thus activation of noradrenergic neurones will result in the release of a shower of noradrenaline over a wide area in contrast to the focal release of nerve transmitter at 'conventional'

axon terminations. The remaining part of the brain stem reticular formation is organized such that:

- The medullary and pontine components receive sensory information from all the senses.
- They project to the midbrain reticular, mainly to the nucleus cuneiformis.
- The midbrain sends a major projection to the hypothalamus (75% of its axonal output) and to the thalamic reticular and intralaminar nuclei (25%).
- The thalamic reticular nuclei (whose cells are GABAergic) and intralaminar nuclei project to each other and into the main thalamic sensory relay nuclei (see Figure 1).

The thalamic reticular, thalamic relay nuclei and sensory cortical areas are organized into a control loop path in the following manner.

- A specific site in the thalamic reticular nucleus projects into a specific site in a thalamic relay nucleus (reticulo-thalamic projection).
- The specific thalamic relay site projects to a specific cortical sensory area (thalamo-cortical projection).
- This cortical area then projects back to the thalamic relay site from which it receives its projection (cortico-thalamic) and to the thalamic reticular nucleus which projects to the same thalamic site (cortico-reticular).
- There is thus a precisely arranged neural loop of reticulo-thalamic, thalamo-cortical, cortico-thalamic, cortico-reticular and reticulo-thalamic connections. In this loop the thalamic sensory relay neurones are expose to reticular, cortical and cortico-reticular influences all of which can serve a modulatory function on the transfer of sensory information to the cerebral cortex in all the specific sensory paths, i.e. touch, sight, hearing and taste.

Thus the reticular formation is able to exert a powerful influence on the cerebral cortex, directly and indirectly (via thalamic relay nuclei) from the specific chemical projections and indirectly from the remainder of the reticular formation via its thalamic projection. In addition all four systems influence motor function via their actions on the basal ganglia, the transfer of motor information from the cerebellum to the motor cortex via the thalamic motor relay nuclei and the direct cortical action of the specific neurotransmitter systems. This influence will sum with the descending reticular projections to α- and γ-motoneurones in the spinal cord.

Effects of anaesthetic agents on somatosensory transmission

If two fine silver wire electrodes are placed gently in contact with the exposed cortical surface then an electrical record, the resultant of all the cellular currents summed between the two electrodes, will be recorded. This signal is called the electrocorticogram and can be regarded as spontaneous biological "noise". If one of the cortical electrodes is accurately positioned over the geometrical centre for the contralateral forepaw projection to the cerebral cortex and the forepaw is stimulated electrically then a different signal is evoked by each stimulus. In this case it represents the currents generated by all of the cortical cells which are influenced by the forepaw stimulus and is termed the evoked cortical response. These evoked responses are not constant from stimulus to stimulus and are, furthermore, superimposed upon the spontaneous electrocorticogram. To overcome the variation and to minimize the effect of the distortion produced by the biological "noise" it is customary to obtain an average response to a number of consecutive stimuli. For this the evoked responses are digitized and the signal is split into 500 consecutive parts at a sampling rate, in these experiments, of one sample every 40 μsec. The averaged response to a number of stimuli can then be derived with reference to the stimulus which always occurs at time zero. The averaging process sums the stimulus-locked responses in direct proportion to the number averaged whereas the biological "noise", which is a random event with respect to the stimulus, will only sum in proportion to the square root of the number averaged. Thus, in the experiments reported herein, all the averages were to 64 consecutive stimuli and the "noise" component was reduced to one eighth.

Averaged evoked responses consist of a number of deflections but only the first 20 msec of the responses are quantified in this paper. The responses can be quantified by:

- Their latency, i.e. the time from stimulus application to the start of the response. This represents the conduction time along the pathway from the forepaw to the cerebral cortex.
- The first positive wave (Pi) which is the height of the first positive deflection above the baseline.
- The first negative wave (Ni) which is measured from the peak of the Pi to the trough of the ensuing negative wave or inflection (see Figure 2).

All anaesthetics were administered on a baseline of urethane anaesthesia (1.25 g kg^{-1}). Earlier work (Angel, Berridge & Unwin, 1973) showed that the changes in the evoked cortical responses were identical if the animal was anaesthetised with a single anaesthetic throughout the experiment or if the anaesthetics were superimposed upon a urethane baseline. By examining the quantitative changes seen in the evoked cortical responses, three basic groups of anaesthetic action can be described. A fourth group can also be defined and somewhat artificially separated by using anaesthetic agents which possess specific pharmacological actions (see Figure 2).

All the agents which were tested for their anaesthetic action gave a dose dependent increase in the latency of the responses. In the first group (Group 1) the amplitudes of both the first positive and first negative waves showed a dose dependent decrease as anaesthetic depth was increased (Figure 2 (1)). In the second (Group 2) the amplitude of the first positive wave was unaffected by the anaesthetic agents used but the amplitude of the first negative wave was decreased in a dose dependent manner (Figure 2 (2)). In the third group (Group 3) the amplitude of the first positive wave showed a small, dose dependent decrease as anaesthetic dose was increased whereas the amplitude of the first negative wave was either unaltered, or actually increased, as anaesthetic dose was increased (Figure 2 (3)). The fourth group (Group 4) was separated from the first three because neurotransmitter specific agents were used to modify either noradrenergic, cholinergic or 5-hydroxytrptaminergic transmission. All of the agents used, if they showed anaesthetic properties, behaved in the same way as Group 1 anaesthetics or behaved as logical Group 1 anti-anaesthetics in that they decreased the latency and increased the amplitudes of the first positive and first negative waves of the evoked cortical responses (Figure 2 (4)).

Changes seen in the evoked cortical responses can be a consequence of:

- A change in the responsiveness of cells located in the cortex.
- A change in the responsiveness of cells along the sensory projection pathway(s).
- A combination of both of these factors.

To answer the question 'where do the different groups of anaesthetics exert their action?' it is necessary to record from single cells along the projection pathway from the forepaw to the cerebral cortex. The fast pathway, i.e. that giving rise to the first 20 msec of the evoked cortical response after electrical or mechanical stimulation of the forepaw in the rat, is the dorsal column-medial lemniscal-thalamocortical pathway (Angel, Berridge & Unwin, 1973). Thus it is necessary to examine the behaviour of cells in the cuneate nucleus, the ventroposterolateral part of the ventrobasal thalamic nucleus and from cells in the 5 cellular layers of the cerebral cortex. (In the urethane anaesthetised rat cells in cortical layers II & III are difficult to excite so that only cells in cortical layers IV, V & VI have been examined). A necessary restriction for the cells studied is that each cellular population should be monosynaptically activated by its input fibres: cuneate cells from the dorsal column fibres, thalamic cells by cuneothalamic afferents and cortical cells by thalamocortical afferents.

That a reduction in somatosensory transmission to the cerebral cortex is involved in the process of anaesthesia is indicated for the following reasons:

- All anaesthetics, by definition, reduce aesthetic sensation
- The effects of anaesthetics on the evoked cerebral responses are reversed by high ambient pressures (Angel, Berridge, Halsey & Wardley-Smith, 1980)
- The potency of anaesthetics in changing the evoked cortical responses is directly related to their lipophilicity.

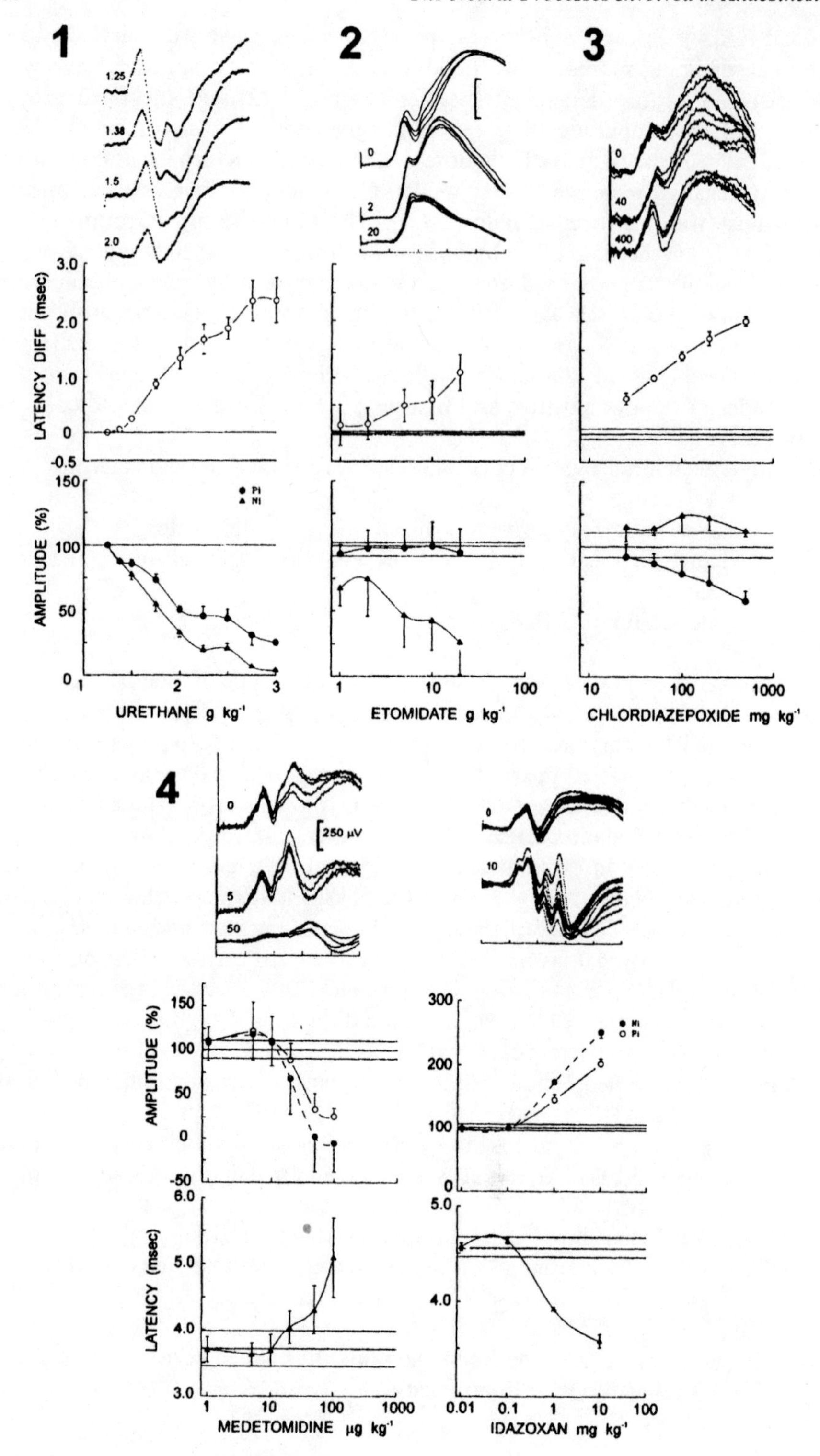
1
1.25
1.38
1.5
2.0
2
0
2
20
3
0
40
400
LATENCY DIFF (msec)
3.0
2.0
1.0
0
-0.5
AMPLITUDE (%)
150
100
50
0
Pi
Ni
URETHANE g kg⁻¹
ETOMIDATE g kg⁻¹
CHLORDIAZEPOXIDE mg kg⁻¹
4
0
250 μV
5
50
10
MEDETOMIDINE μg kg⁻¹
IDAZOXAN mg kg⁻¹
LATENCY (msec)
6.0
5.0
4.0
3.0
300
200

Effects of anaesthetic agents on cellular responses

Group 1.

No anaesthetics in this group had any large effects on the behaviour of single cells recorded from the cuneate nucleus in response to supramaximal electrical stimulation of the forepaw. Occasionally activation of cells in this nucleus produced complex and prolonged discharge patterns (see Figure 3 A) in which case the later parts of the response showed a small increase in latency, a greater degree of regularity in their firing and no change in probability of response. However, the restriction that thalamic cells be monosynaptically activated by the cuneate output precludes any thalamic effect of such late discharges. All thalamic cells investigated behaved in a completely different way to those in the cuneate nucleus (see Figure 3B & 3E). In contrast to cuneate cells thalamic cells responded with a variable latency and often failed to respond to a peripheral stimulus. Cuneate cells always gave one short latency response per stimulus. On average thalamic cells gave one response to around 80% of the presented stimuli. The effect of all Group 1 anaesthetics was to increase the latency and decrease the probability of response of VPL cells to supramaximal electrical stimulation of the forepaw. This same pattern of behaviour was also seen in cells in cortical layers IV and VI which appeared to mirror the effect of the thalamic cells from which they derive their input (Figure 3D). In contrast to these cortical cells those recorded from layer V appeared to be more susceptible to anesthetic action (see Figure 3F). A summary of the effects of

---

Figure 2 *This shows the effects of the four functional groups of anaesthetic agents (1-4) on evoked sensory cortical responses. Each block shows either single or superimposed averaged responses at various doses of anaesthetic and graphs showing the effect of the anaesthetic agent on the latency (ordinates) of the cortical response (expressed as changes from the starting level (**1-3**) or absolute measures (**4**) versus the log dose of the chemicals used (abscissae); and the amplitudes of the first positive deflection (Pi; measured from the baseline to the peak of the response) or first negative deflection (Ni; measured from the peak of Pi to the trough: ordinates) versus the log dose. Each point on the graph is the mean and the vertical lines represent the SEM. For the amplitude graphs the SEM of the control value of Pi is represented by the line above the 100% level the SEM of the Ni by the line below the 100% level. All the anaesthetic agents used were superimposed on a basal level of urethane (1.25 g kg^{-1}). The number of animals used were 12 for urethane and 5 for each of the other chemicals. Each mean value was obtained by averaging at least 5 averaged responses (N=64, to supramaximal electrical stimulation of the contralateral forepaw at a rate of 1s^{-1}) when the animal had reached a stable response level for each dose. The mean starting levels of latency for the first three groups were not statistically significantly different from the starting levels of those shown for the medetomidine experiments (the three horizontal lines represent the mean $\pm$ SEM). For the Idazoxan experiments the depth of urethane anaesthesia was increased to 1.5 g kg^{-1} before administration since it showed an anti-anaesthetic effect. For these experiments all the animals remained at surgical levels of anaesthesia , i.e. no reflex withdrawal of a hindpaw to a strong pinch, at the highest dose level used (10 mg kg^{-1}).*

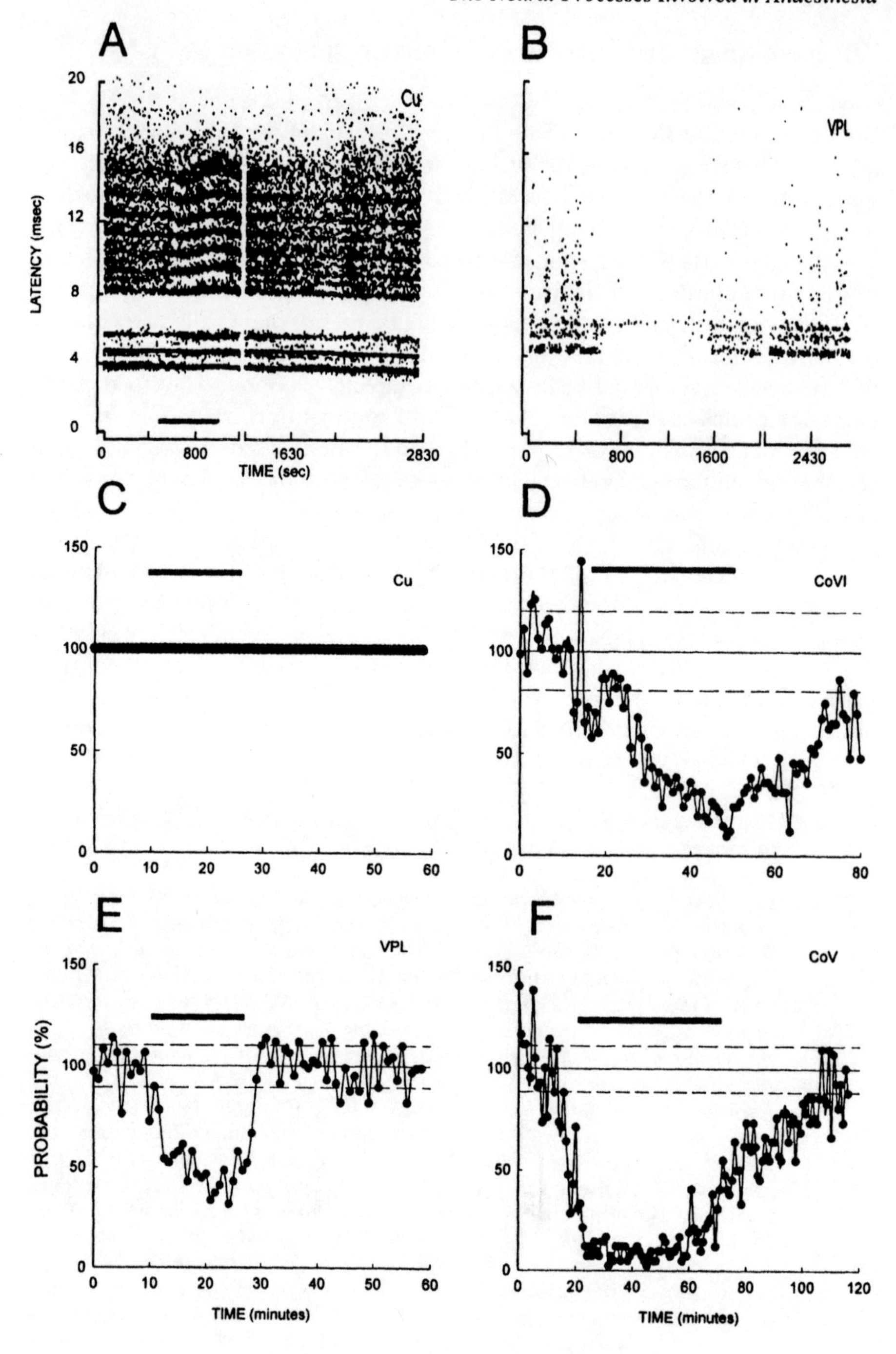
A
B
C
D
E
F
20
16
12
8
4
0
LATENCY (msec)
Cu
0
800
1630
2830
TIME (sec)
VPL
0
800
1600
2430
150
100
50
0
Cu
0
10
20
30
40
50
60
CoVI
0
20
40
60
80
VPL
PROBABILITY (%)
TIME (minutes)
CoV
0
20
40
60
80
100
120
TIME (minutes)

urethane on the probabilities of cellular responsiveness at each level of the somatosensory pathway is shown in Figure 4B. Note that the dose response curve for cells in cortical layer V is shifted to the left compared to those in cortical layer IV and the ventrobasal thalamus. Group 1 comprises the majority of anaesthetics and includes nitrous oxide, cyclopropane, di-ethyl ether, chloroform, Enflurane, Halothane, Methoxyflurane, the barbiturates, Althesin and urethane.

Group 2

Anaesthetics in this group are Propofol and Etomidate. Once again they have no effect on the probability or latency of response of cells in the cuneate nucleus. In contrast to the action of Group 1 anaesthetics, they show no effect on the probability of response of VPL cells (Figure 4A VPL) until very high dose levels are reached (e.g. for Etomidate >10 mg kg^{-1}). There is, however, a small dose dependent increase in the latency of the thalamic cell responses. Cells in cortical layers IV and VI show both a dose dependent increase in their latencies of response and a reduction in the probability that they will respond (Figure 4A CoIV). In common with the actions of Group 1 anaesthetics, Group 2 anaesthetics exert a greater effect on cells in cortical layer V, i.e. the dose response curve is shifted to the left (Figure 4B top).

Group 3

This group comprises the benzodiazepines. Again they have no effect on the response properties of cells in the cuneate nucleus. Cells in the VPL and cortical layers IV, V and VI show a reduced probability of response and an increase in latency. The latency changes depend upon the dose administered but the reduction in probability is achieved by small doses and is not further reduced by increasing doses. The reduction in probability of response appears to be the result of stabilizing the cells' discharges. Figure 5 shows the results obtained from a cortical (Figure 5A) and a VPL (Figure 5B) cell. In the control condition the cortical cell gave two action potentials per stimulus (between 6.5-7.5 msec

---

Figure 3 *This shows the effects of administering 3% forane to an animal with a basal level of urethane anaesthesia (1.25 g kg^{-1}) whilst recording from a cuneate cell (**A**) and, one hour later, a ventroposterolateral thalamic cell (**B**). The records are presented as serial dot raster response histograms in which an occurrence of an action potential is represented by a dot at its latency of occurrence (ordinate) versus the time of a supramaximal electrical stimulus applied to the forepaw (abscissa). The responses were obtained at a rate of 1s^{-1}. The period of forane administration is shown by the thick horizontal bars.*

*The remainder of the graphs show the effect of administering 1% Halothane on the probability of response of a cuneate cell (**C**), a ventroposterolateral thalamic cell (**E**), a cortical cell from layer VI (**D**) and a cortical cell from layer V (**F**) from four different animals with a basal level of urethane anaesthesia (1.25 g kg^{-1}). Each point on the graph shows the probability of response to 50 consecutive supramaximal electrical stimuli applied to the forepaw. The mean ± SD of the control period is shown by the horizontal lines, the control probability was scaled to 100% for each cell to allow direct comparison. The period of Halothane administration is shown by the thick horizontal bars.*

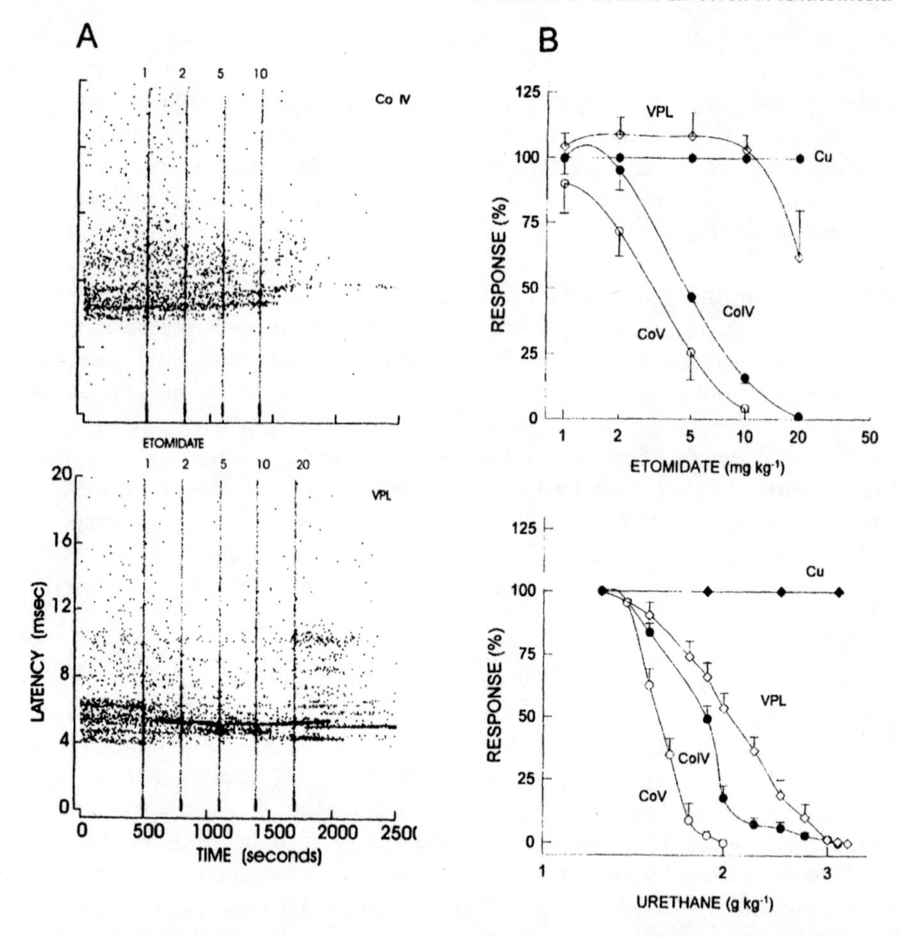

Figure 4 *A. Shows dot raster sequential histograms of the responses from a single cortical layer IV cell (top,* ***CoIV****) and a single thalamic ventroposterolateral cell (bottom,* ***VPL****) to increasing doses of etomidate, administered by intraperitoneal injection (dose given at the vertical lines, with the cumulative dose indicated for each administration in mg* kg^{-1}*). For each response to supramaximal electrical stimulation of the contralateral forepaw a cellular event is displayed as a dot at a particular latency (ordinate) versus the time of stimulation (abscissa).* ***B.*** *Shows a comparison of the effects of increasing anaesthetic depth with etomidate (top) and urethane (bottom) on the probability of response of cells along the somatosensory pathway in rats initially anaesthetised with urethane (1.25 g* kg^{-1}*). Each point on the graph shows the mean* $\pm$ *SEM for 5 cells of each type for the etomidate experiments and for 35 cuneate (****Cu****), 30 ventroposterolateral (****VPL****), 40 layer IV or VI cortical (****CoIV****) and 4 layer V cortical (****CoV****) cells for the urethane experiments.*

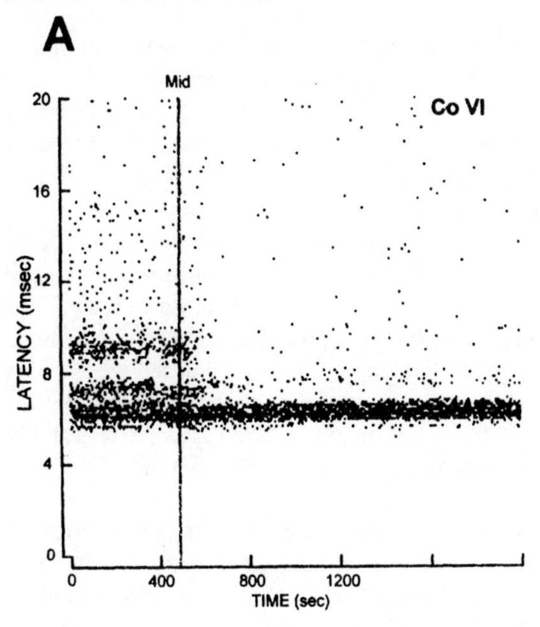

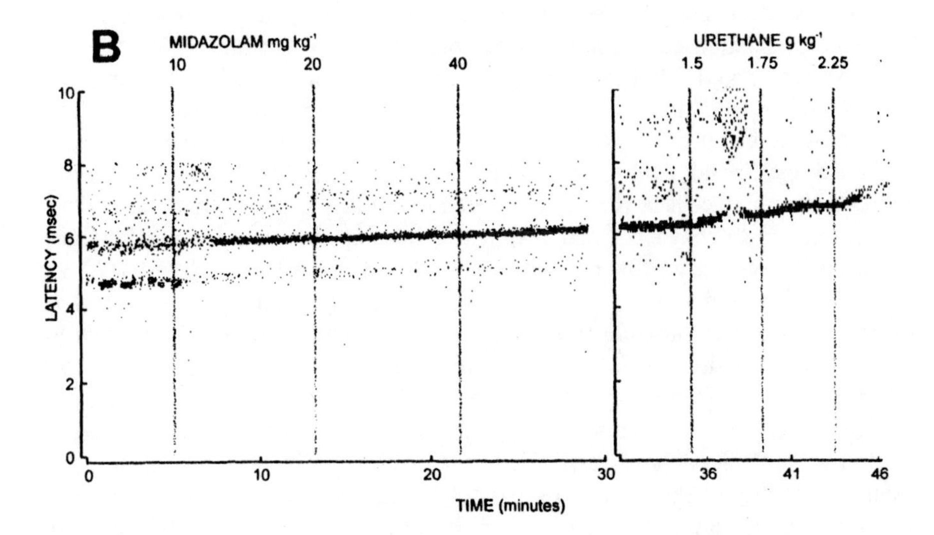

Figure 5 *This shows the effect of midazolam (vertical lines) on two single cells from two rats initially surgically anaesthetised with urethane (1.25 g kg^{-1}) to supramaximal electrical stimulation of the contralateral forepaw. The top graph shows the effect of a single dose (20 mg kg^{-1}) on a short latency cortical cell from layer VI. The bottom graph is from a ventroposterolateral cell to dose of 10, 20 and 40 mg kg^{-1} and, for comparison, to incremental doses of urethane at cumulative doses of 1.5, 1.75 and 2.25 g kg^{-1}.*

and 8-9 msec). After midazolam administration the cell tended to respond with only one action potential per stimulus between 6.7-7 msec. The thalamic cell responded, on average, 1.5 times per stimulus with a bimodal latency distribution at 4.6 and 5.75 msec. After midazolam it gave only one action potential per stimulus with a probability of 75.9, 76.3 and 82.3% (compared to the control of 100%) at modal latencies of 5.8, 5.95 and 6.0 msec for doses of midazolam of 10, 20 and 40 mg kg-1, respectively. The behaviour of the thalamic cell is contrasted with the effect of increasing doses of urethane anaesthesia in which situation there was both a dose dependent increase in latency and reduction in probability of discharge, showing a modal latency change from 6.0 msec to 6.3, 6.4 and 7.05 msec for urethane doses which increased the dose level to 1.5, 1.75 and 2.25 g kg^{-1} respectively. At the same time the probability of response reduced from 82.3% to 74.2%, 61.5% and 20.7%. There was no evidence that the benzodiazepines potentiated the effect of urethane.

Group 4

Reference to Figure 1 shows that there are diffuse projections from the brain stem reticular formation of different cell groups, to the thalamus and cerebral cortex, which use either noradrenaline, 5-hydroxytryptamine or acetylcholine as their neurotransmitters. Drugs which modulate the action of these three neurotransmitters can also exert effects on the level of anaesthesia seen in the urethane anaesthetised animal. Medetomidine and clonidin which are α_2-adrenoreceptor agonists both exert the same effect on cells in the somatosensory pathway. In common with all other drugs showing an anaesthetic action they show no effect on the cells in the dorsal column nuclei monosynaptically activated by dorsal column afferents (Figure 6 Cu). However, they have marked effects on cells in the VPL and in cortical layers IV-VI. In concert with their effects on evoked cortical responses these agents also show a biphasic effect on the response properties of these cells. At doses below 10 μg kg^{-1} for medetomidine and 100 μg kg^{-1} for clonidine the level of anaesthesia is decreased. For the thalamic and cortical cells this is seen as a small decrease in latency and a small increase in probability of discharge. At higher doses these drugs behave as Group 1 anaesthetics when the evoked cortical responses are examined. However, their effects on thalamic and cortical cells are more complex. The cells always responded with a marked increase in latency but their probability of response paradoxically increased. This is because the cells fire for a much longer period after a stimulus (see Figure 6). If the early discharge of the cell is quantified then the probability of the cell firing is decreased. Hence the ability of the cells to transmit a coherent 'sharp' signal to cells further along the sensory chain is markedly diminished. The effect of Idazoxan, an α_2-adrenoreceptor antagonist, is the reverse. At small doses ($<$100 μg kg^{-1}) it produces a small increase in anaesthetic depth but at higher doses it acts as an anti-anaesthetic producing a decrease in anaesthetic depth. This is shown by the restoration of the early, high probability discharge of thalamic and cortical cells (Figure 6). The effect of other noradrenergic drugs and those which affect 5-hydroxytryptaminergic and cholinergic transmission are summarized elsewhere (Angel, 1993).

Possible mechanisms of anaesthetic action

Anaesthetic agents can theoretically exert their effects via a whole series of possible mechanisms. These include:

- Reduction of transmitter release from the presynaptic terminal (see Richards, this volume).
- Modulation of the interaction between ligands and specific neurotransmitter receptor molecules (see Franks & Lieb; Lambert, this volume).
- General neuronal depression.
- Activation of inhibitory neurones to increase phasic and recurrent inhibition.
- Depression of excitatory neurones to decrease phasic and tonic excitation.
- Any or all of these effects acting in concert.

Examination of the known effects of some anaesthetic agents on specific receptors shows that agents with distinctive receptor profiles e.g. ketamine - an NMDA antagonist at glutamate receptors, and barbiturates which facilitate the action of γ-aminobutyric acid on $GABA_A$ chloride ionophores, both belong to the Group 1 class of γ-aminobutyric acid on $GABA_A$ chloride ionophores, have an action which is quite distinct from the barbiturates. Thus it is not possible to attribute, at the present time, a mechanism of anaesthesia which results from a simple single anaesthetic/receptor perturbation. It is even difficult to concatenate two or more anaesthetic/receptor interactions into a cohesive or viable theory of anaesthetic action. This is mainly because of the vast range of effects which anaesthetics induce in the whole animal. It is possible, however, to fractionate a particular anaesthetic/receptor interaction to explain a single feature of anaesthesia. The effect of Halothane on potentiating the effects of the inhibitory transmitter glycine may partly explain the loss of spinal motor neurone output seen with anaesthesia. Glycine plays an important role in the feedback inhibition of spinal motor neurones (Bradley, Easton & Eccles,1953; Eccles 1964).

That anaesthetics do not give a general depression of nerve cells is apparent from the data on somatosensory cell behaviour presented above. All anaesthetics block the monosynaptic stretch reflexes at doses which leave protective polysynaptic withdrawal reflexes relatively unaffected (DeJong, Robles, Corbiu & Nace; 1968). None of the anaesthetics tested so far have shown any influence on the monosynaptic transfer of information through the cuneate nucleus to supramaximal electrical stimulation of the forepaw. Group 1 anaesthetics impede the transfer of information at two main sites; the ventroposterolateral thalamus and to cells in cortical layer V. Group 2 anaesthetics exert their effects by impeding the transfer of information to cells in cortical layers IV and VI and have a greater effect on cells in layer V. Thus there are differential effects in the somatosensory and motor pathways. That there is a reduced efficiency with which information is transferred through the ventroposterolateral

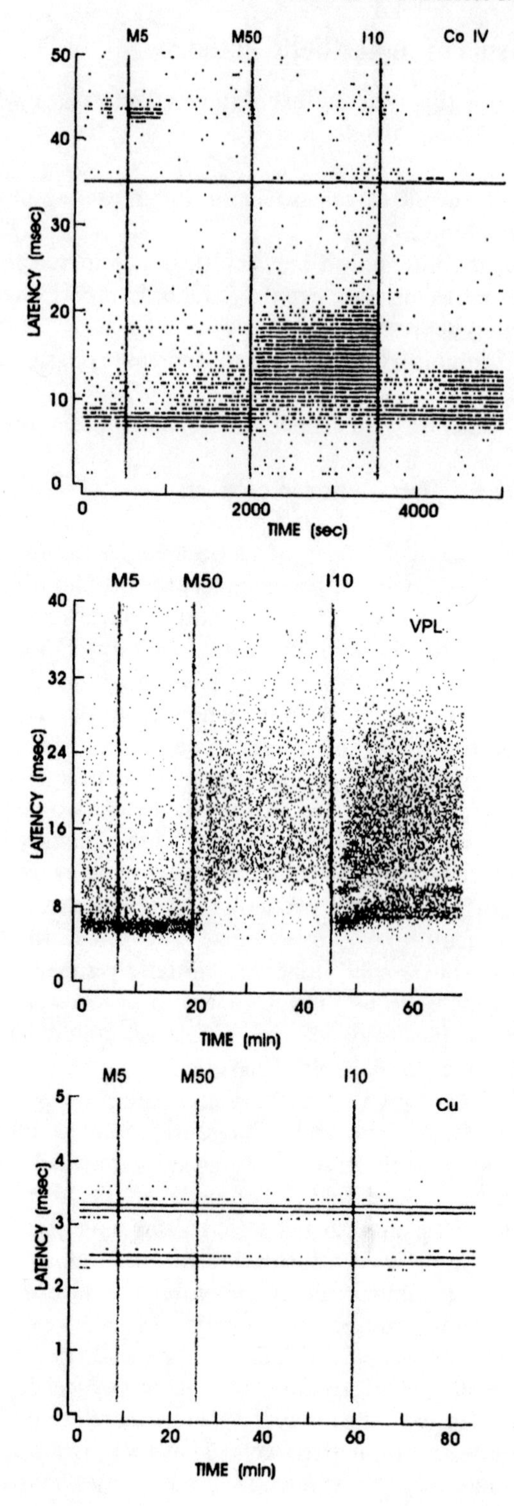
M5
M50
I10
Co IV
LATENCY (msec)
TIME (sec)
0
2000
4000
VPL
LATENCY (msec)
TIME (min)
Cu
LATENCY (msec)
TIME (min)

thalamus is evident from the fact that the responses of cells at this site are fairly labile. Reference to Figures 3B, 4A (VPL) and 5B shows that under control conditions the VPL cells show a marked variation of their firing pattern from moment to moment compared with the fixed pattern of cuneate cells (Figure 3A & 6 (Cu)). Thus the thalamic cells respond with a marked variability to a fixed input. One can thus pose the question "How does this occur?"

Electrode penetrations aimed at the lateral part of the ventroposterolateral thalamus can pass through the thalamic reticular nucleus and its boundary with the ventroposterolateral nucleus. Cells in the thalamic reticular nucleus and its boundary zone exhibit differences in their responsiveness to peripheral stimuli and in their spontaneous activity compared to VPL cells. Reticular/boundary cells are always spontaneously active and do not respond, at short latency, to supramaximal electrical stimulation of the forepaw. They do respond to potentially, or overtly, noxious stimuli applied anywhere on the body surface (pinching, cutting or heating the skin to above 45 °C). These sensory messages are conducted to the thalamic reticular/boundary cells via the spinothalamic tracts (Angel 1964). In contrast VPL cells respond with an early latency to electrical stimulation of the forepaw and respond to light touch, hair movement or gentle pressure from small (1-10 mm^2) skin fields on the contralateral hand, shoulder, arm and trunk. The majority of cells respond to tactile stimuli applied to the glabrous skin of the forepaw (Angel & Clarke, 1975). The responses of reticular/boundary cells reveal that there are two distinct populations of cell. One type decreases its frequency of discharge to noxious stimulation, the other increases its frequency of discharge. Similarly there are two distinct populations of cortical cell located in layers III, V and VI whose frequency of discharge is also modulated by noxious stimulation of the integument. The cortical cell populations can be further divided into two groups according to whether they also respond at short latency to tactile stimulation of the integument. Most (80%) cortical cells show both early tactile responses and also have their frequency of discharge modulated by noxious stimuli.

In the rat anaesthetised with urethane although there is a 'net' level of anaesthesia with any particular dose of anaesthetic this is usually not stable but spontaneously varies between two quite distinct states (Angel, Dodd & Grey 1976). Figures 7 & 8 show that the elctrocorticogram undergoes spontaneous, random transitions between high voltage low frequency activity (HVLF) and low voltage high frequency activity (LVHF). These transitions are seen whether or not the peripheral stimulus is turned on (compare the left hand [stimulus on] and right hand [stimulus off] electocortical activity records

---

Figure 6 *Shows three dot-raster sequential response histograms, to supramaximal electrical stimulation of the forepaw, for a cortical layer IV cell (top* ***CoIV****), a ventroposterolateral thalamic cell (middle,* ***VPL****) and a cuneate cell (bottom,* ***Cu****); from rats initially surgically anaesthetised with urethane (1.25 g kg^{-1}). The vertical lines show the time of intraperitoneal injections of medetomidine 5 μg kg^{-1} (M5) and 45 μg kg^{-1} (M50, cumulative dose equals 50 μg kg^{-1}) and to a dose of Idazoxan (10 mg kg^{-1})*

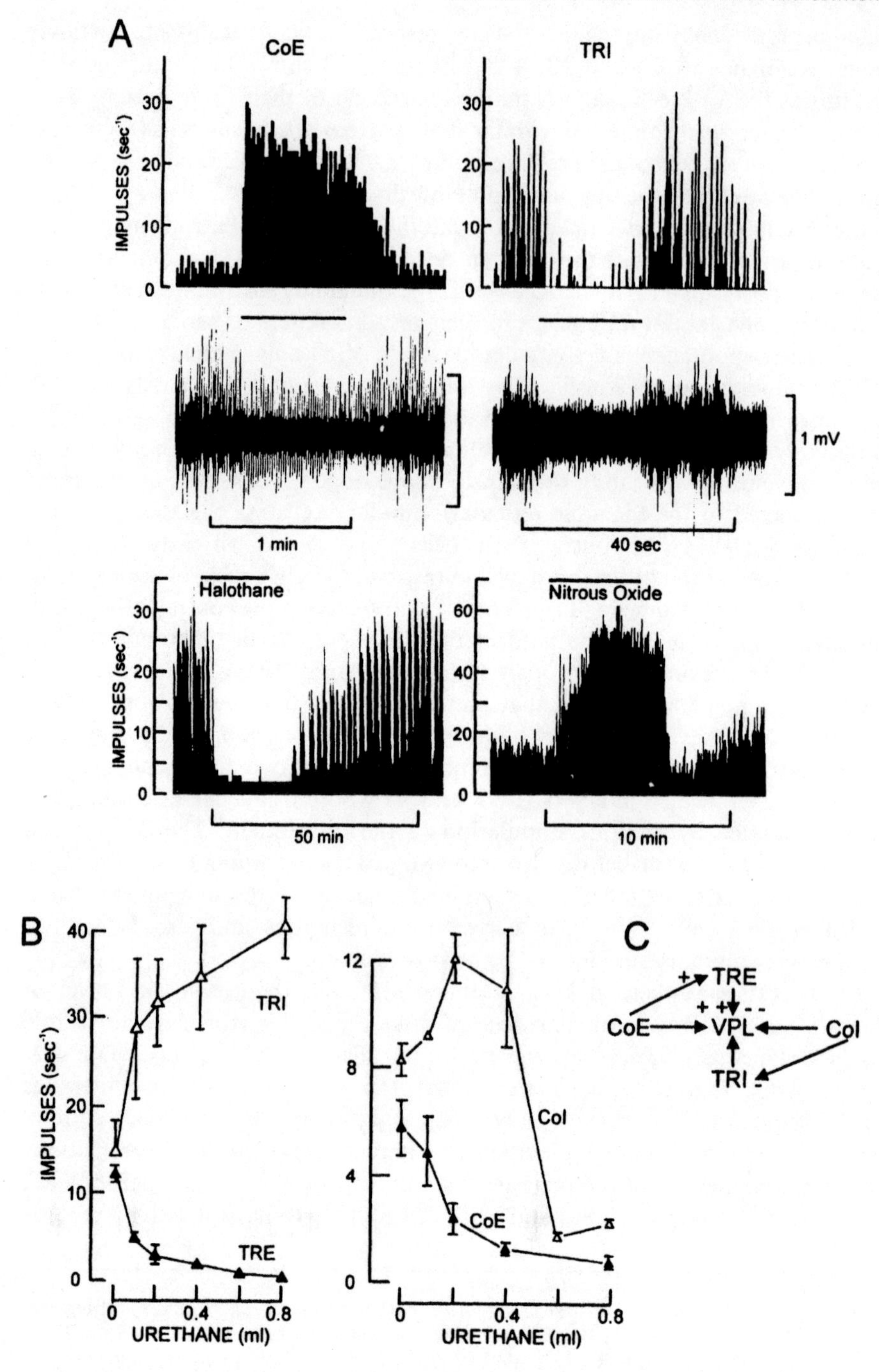
A
CoE
TRI
IMPULSES (sec⁻¹)
30
20
10
0
1 mV
1 min
40 sec
Halothane
Nitrous Oxide
60
40
50 min
10 min
B
TRI
TRE
CoI
CoE
URETHANE (ml)
0
0.4
0.8
12
8
4
C
TRE
VPL
CoE
CoI
TRI

in Figure 7). An analysis of the power spectra of the electrocorticogram in these two states is shown in Figure 8B and clearly shows a relative frequency shift from delta (0.5-3.5 Hz) to beta (15-30 Hz) activity when the amplitude changes from high voltage to low voltage. This electrocortical transition is called an 'arousal' reaction and can be mimicked by electrical stimulation anywhere in the brain stem reticular formation or by noxious stimuli applied to the skin.

When the electrocorticographic activity changes from HVLF to LVHF then the evoked cortical responses show a decreased latency of response and increased amplitudes of the first positive and negative waves (Figure 8C). At the same time transmission through the ventroposterolateral nucleus is increased. This is seen as an increased probability of response and a decreased latency of response (Figure 8D). At the same time, during HVLF to LVHF transitions, half the population of spontaneous cortical and thalamic reticular/boundary cells show an increase in their frequencies of discharge and the other half a reduction. This behaviour is shown for a cortical cell and a thalamic reticular cell in Figure 7A. Since an increase in thalamic transmission is seen when some thalamic reticular/boundary cells fire more often it can be hypothesized that these cells are involved in excitatory mechanisms. Similarly the other type, showing the reverse change in spontaneous activity, can be postulated to be involved in inhibitory mechanisms. If this is so then thalamic transmission can be presumed to be under both cortical excitatory and inhibitory and thalamic reticular inhibitory and boundary cell excitatory control. (Reference to the section on anatomical pathways above shows that this is precisely what one would predict from the anatomical arrangement of the cortex and thalamic reticular nucleus, except that functionally there

---

Figure 7 ***A.*** *The top two records of each column shows the changes in discharge frequency of a cortical potential excitatory cell (**CoE**) from cortical layer V and a thalamic reticular inhibitory cell (**TRI**) to a spontaneous change in the electrocorticogram (middle records) from HVLF activity to LVHF activity (the period of desynchronisation is marked by a horizontal bar) and to administration of 1% Halothane and 80% nitrous oxide for the same cell, recorded from rats anaesthetised with urethane (1.25 g kg^{-1}). The changes in cellular discharge are displayed as records of frequency of discharge in 1 sec epochs.*

B. *Shows the change in frequency of discharge (ordinates) of 6 thalamic reticular/boundary cells of each type either presumed inhibitory cells (**TRI**) or presumed excitatory cells (**TRE**) and for 8 presumed cortical excitatory cells (**CoE**) and 4 presumed cortical inhibitory cells (**CoI**) to increasing levels of urethane anesthesia (shown as ml of urethane added to the basal urethane level, 1.25 g kg^{-1} on the abscissae). Data are mean frequencies of discharge, the vertical bars represent the SD.*

C. *The simplest neuronal circuit linking cortical and thalamic reticular/boundary cells with the cortical excitatory cells (**CoE**) having a direct excitation (+) on ventroposterolateral cells (**VPL**) and an indirect excitation via the thalamic reticular excitatory cells (**TRE**) and cortical inhibitory cells (**CoI**) having a direct inhibition (−) on ventroposterolateral thalamic cells and an indirect inhibition via the thalamic reticular inhibitory cells (**TRI**).*

appears to be both excitatory and inhibitory balanced control). The functional pathway for this cortico-thalamic and cortico-reticulo-thalamic feedback control is shown in Figure 7C.

The importance of any simple hypothesis is that it is open to direct experimental question. The obvious corollary here is that, if this feedback loop is involved in the decrease in responsiveness of sensory transmission through the thalamus after administration of anaesthetic agents, then the putative excitatory cells at both thalamic and cortical levels should show a dose dependent decrease in their frequency of discharge as depth of anaesthesia is increased. The reverse, i.e. a dose dependent increase in discharge frequency should be seen for the putative inhibitory cells. For thalamic reticular inhibitory and boundary cell excitatory cells this is precisely what is seen with all group 1 anaesthetic agents (see Figure 7A & B). {*The distinction between thalamic reticular inhibitory and boundary cell excitatory activity is necessary from the observation that all cells in the thalamic reticular nucleus contain the neurotransmitter GABA (Yen, Conley, Hendry, & Jones, 1985*).] Cortical inhibitory cells, on the other hand at first increase their frequency of discharge as anaesthetic depth is increased and then suddenly decrease their discharge

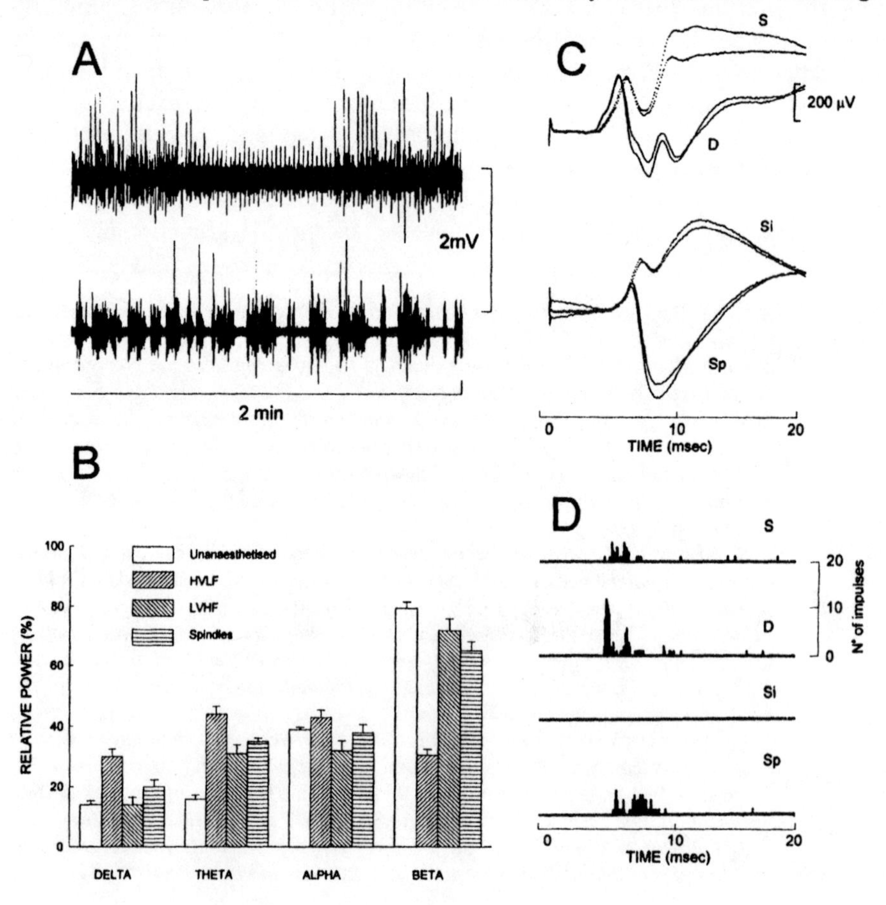

(Figure 7B). This apparent discrepancy in theoretical and actual behaviour is slightly less troublesome when one considers what happens at the transition from an increased discharge to the sudden decrease. At this level of anaesthesia the electrocortical activity undergoes a transition from HVLF activity to one in which the cortex starts to show spindling activity i.e. periods of electrocortical silence interrupted by bursts of high frequency activity (Figure 8A & B). If evoked cortical responses are obtained during the silent periods then they are of long latency with decreased amplitudes of the first positive and first negative waves (Figure 8C). At the same time, during periods of silence, transmission of information through VPL is severely reduced. During the spindles, however, thalamic transmission is enhanced to a value somewhere between that seen in the control HVLF and LVHF states (Figure 8D) and the evoked cortical responses show a large increase in the amplitude of the first negative wave. Power spectral analysis of spindle activity shows that the frequency content is between that seen in the control HVLF and LVHF states. Thus the effect of the anaesthetics on the cortical presumed inhibitory cells is

---

Figure 8 *A Shows the four states of electrocorticographic activity seen in the urethane (1.25 g kg^{-1}) anaesthetised rat. The top record shows a spontaneous transition from HVLF to LVHF back to HVLF activity. The bottom trace shows the electrocorticogram from the same rat after the dose of urethane had been increased to 1.75 g kg^{-1} and shows periods of electrocortical activity interspersed with bursts of activity ("spindling").*

***B** Shows the relative spectral content from 12 rats either unanaesthetised, during HVLF, LVHF (urethane 1.25g kg^{-1}) or spindling periods (urethane dose between 1.75 to 2.0g kg^{-1}) of electrocortical activity. The bars represent the mean activity and the vertical lines the SD. For this figure the electrocorticogram was digitized at 125 samples s^{-1}. Ten second epochs were then subjected to Fast Fourier Transform analysis (at least 50 epochs per animal per state). The power spectra were determined and the total power of the frequencies between 0.1 and 50 Hz during the epoch was given the value of 100%. The percentage relative powers of each of the international EEG wavebands: delta, 0.3-3.5 Hz; theta, 4-7 Hz; alpha, 8-13 Hz and beta, 13-30Hz were then determined. For the spindling activity the data was examined to see that each epoch was selected during the spindling phase and not during the silent phase.*

***C** Shows 2 averaged evoked cortical responses (N=64) to supramaximal electrical stimulation of the contralateral forepaw obtained in each of the four electrocorticographic activity states from a rat anaesthetised with urethane (1.25 g kg^{-1}) during HVLF activity (**S**), LVHF activity (**D**), spindling activity (**Sp**) and electrocortical silence (**Si**). (For these latter the dose of urethane was increased to 1.8 g kg^{-1}). The records are arranged to compare HVLF/LVHF states and silent/spindling states, the top and bottom pairs of responses respectively. The voltage calibration represents 100 μV for the Si and Sp records.*

***D** Shows the responses of a single ventropolsterolateral thalamic cell during the four electrocortical activity states (**S**, **D**, **Sp** & **Si** respectively) obtained at the same time as the evoked responses in part C above. Each graph shows a post-stimulus histogram expressed as number of responses (ordinate) at a particular latency (abscissa) to 200 supramaximal electrical stimuli delivered to the contralateral forepaw.*

such that inhibition is turned off during the spindling activity, which is reflected in an average decrease in the frequency of discharge of these cells. This is accompanied by transient decreases in thalamic reticular inhibitory cell discharge and transient increases in cortical and thalamic reticular boundary cell discharge during spindling activity.

Group 1 anaesthetics thus exert their effects on sensory transmission by increasing the inhibitory control, and decreasing the excitatory control of the thalamic feedback loops. A decrease in thalamic transmission means that less information reaches the cerebral cortex. This is amplified by an additional, as yet undefined decrease in the excitability of cortical layer V cells. Nonetheless the action of Group 1 anaesthetics loses some of its mysticism when one considers that they switch on a mechanism which could serve to allow attentional switching in the unanaesthetised animal.

The mechanism of action of Group 2 anaesthetics seems to be primarily at the cortical level and Group 3 anaesthetics exert their effects by increasing GABAergic inhibition at thalamic and cortical sites.

Group 4 anaesthetics pose a small problem. Their effects are peculiar when considering the noradrenergic system. The biphasic action of the α_2-agonists can perhaps be explained by assuming that the presynaptic autoreceptors have a higher affinity for noradrenaline than the postsynaptic receptors. If this is so autoreceptor activation by the agonists would decrease noradrenaline release and thus cause excitation and postsynaptic agonist action at higher doses would mimic the effect of noradrenaline and cause a depression of cellular responsiveness. Why the discharge of the somatosensory thalamic and cortical cells should be prolonged at the same time is not yet clear.

Cholinergic blockade with muscarinic centrally acting antagonists gives a clear increase in anaesthetic depth but even with huge doses of atropine one can never achieve more than a 20% increase in anaesthetic depth. Similarly centrally acting anticholinesterases (which prolong the effect of released acetylcholine) only give a moderate decrease in anaesthetic levels. Cholinergic antagonists do however, act in the same fashion as Group 1 anaesthetics. The position with 5-hydroxytryptamine is less clear because of the relative paucity of specific centrally acting agonists and antagonists. That these latter possess anaesthetic activity is clearly indicated by both their effects on cerebral evoked responses and also the warning on prescription drugs that they cause drowsiness.

Part 2

A comparison of the effects of the two Group 1 anaesthetics Halothane and nitrous oxide

Earlier work on the effects of nitrous oxide on somatosensory transmission in the urethane anaesthetized rat showed that it gave a dose dependent increase in latency and a decrease in the amplitudes of the first positive and negative waves.

There were, however, some observations which led to the suggestion that the effect of nitrous oxide was transient in nature (Angel, Berridge & Unwin, 1973). This peculiar behaviour of nitrous oxide has now been re-investigated and contrasted with that of Halothane in the same animals. The problem that nitrous oxide gives is illustrated in Figure 9. It is customary, when studying the effects of drugs, to derive a cumulative dose:response curve. With nitrous oxide this usually simple response paradigm proved to be very difficult. Examination of the top three and bottom three evoked cortical responses shows that nitrous oxide can behave like a Group 1 anaesthetic; the cortical responses show an increasing latency as the inspired concentration of nitrous oxide was increased from 20 to 80% and, at the same time, a reduction in the amplitudes of the first positive and first negative waves. However, examination of the responses of the cortical cell when attempting to derive a dose: response curve shows that the cellular responsiveness was restored part way through the administration of 40% and during the administration of 60% nitrous oxide. Thus the activity of nitrous oxide, as an anaesthetic, has to be assessed by giving periods of anaesthetic administration at a particular dose followed by a period of recovery before the next concentration was administered. With an experimental paradigm of alternating 500 seconds of administration of randomly arranged inspired nitrous oxide concentrations and recovery the animals showed that, for short periods of time, nitrous oxide behaved in exactly the same way as any other Group 1 anaesthetic. Thus its effect was seen first at the second synapse in the somatosensory pathway i.e. in VPL; the cells in cortical layers IV & VI mirrored this reduced cortical input. Cells in the thalamic reticular/boundary zone and spontaneously active cortical cells in cortical layers III, V & VI all displayed the appropriate changes in their levels of activity (see e.g. Figure 7A). Since this sort of transient effect has never been reported in investigations of the effect of anaesthetic/receptor interactions at the molecular level it is important to try to find out what is happening. This phenomenon, together with that of the spontaneous changes in level of anaesthesia seen with urethane (see above), could be important for understanding the effect of anaesthetics on the intact nervous system. Accordingly the effects of long term (50 minute) administrations have been studied. In all the experiments so far performed, whenever nitrous oxide was administered, regardless of inspired concentration, the initial effect was always that of a Group 1 anaesthetic. However, this effect was nearly always (92% of experiments) abruptly replaced by a different effect. Since this was always an abrupt change from one stable level to another, this response pattern has been referred to as a "snap". Figure 10 compares the effect of equipotent doses of Halothane and nitrous oxide. The "snap", which occurred about 15 minutes after the start of the nitrous oxide administration, affected both the spontaneous electrocortical pattern of activity and the only motor act present in these animals their respiratory effort. The records clearly show that, in the control condition, there were many spontaneous HVLF to LVHF transitions which were accompanied by changes in respiratory frequency. Nitrous oxide administration increased the power in the low frequency bands and decreased the

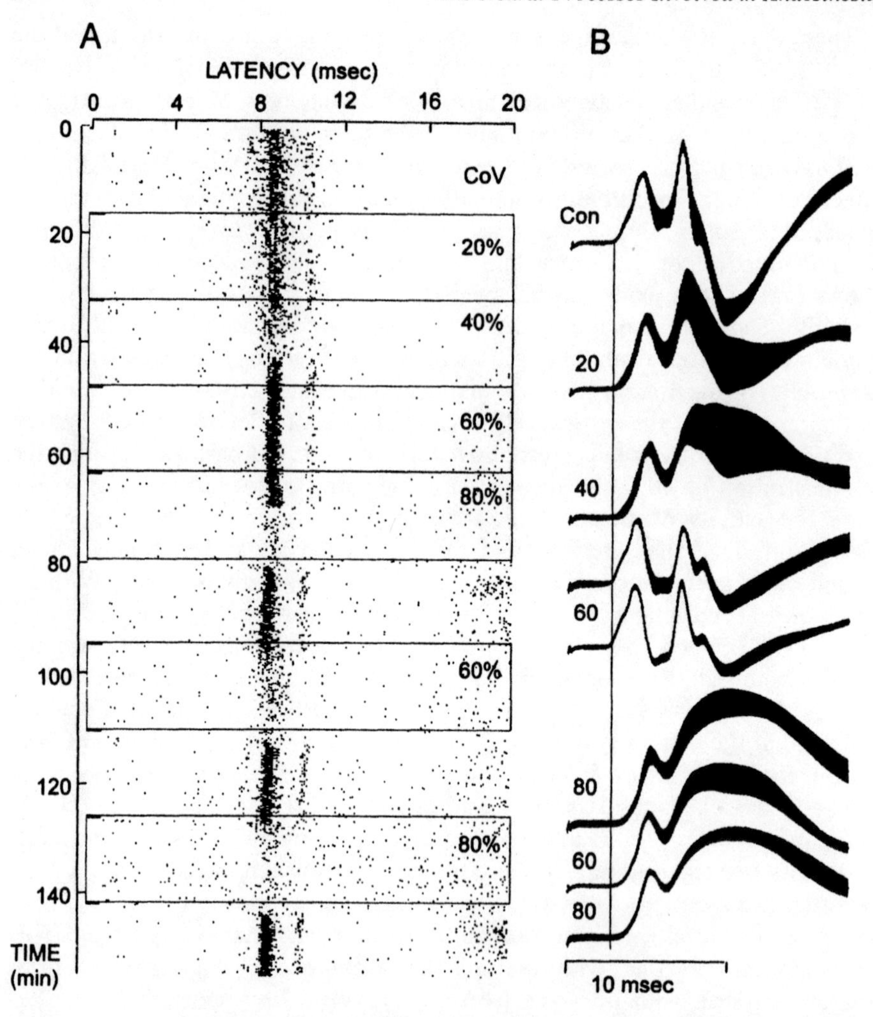

Figure 9 *Shows on the left, a dot-raster sequential histogram of a single cortical layer IV cell to supramaximal electrical stimulation of the contralateral forepaw (1 s^{-1}), in an animal initially anaesthetised with urethane (1.25 g kg^{-1}). The vertical lines represent periods of nitrous oxide administration at the doses shown (% nitrous oxide in an oxygen stream). Averages of 200 consecutive cortical evoked responses recorded at the same time are shown on the right. The vertical line is placed to indicate the mean latency of the response in the control condition. Each average shows the mean ± the SD. Note the 'peculiar' behaviour of the cell towards the end of the 40% through the 60% and towards the end of the 80% nitrous oxide administration and the concomitant evoked responses for the period of the 60% nitrous oxide administration.*

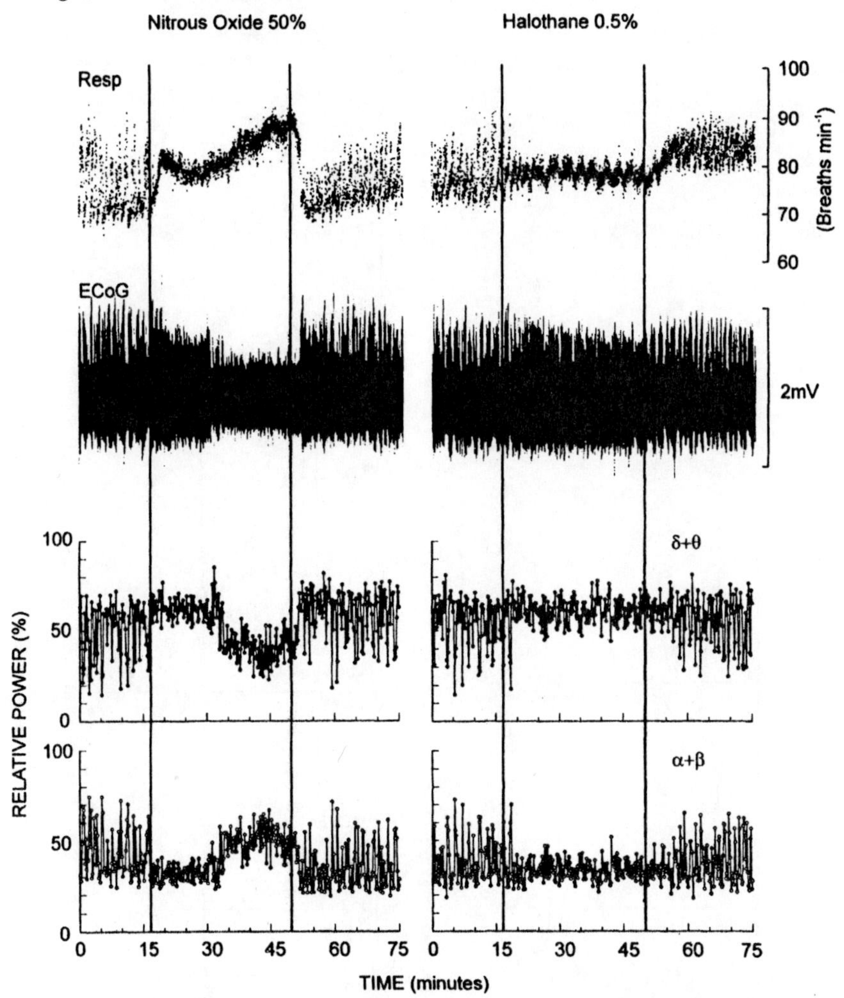

Figure 10 *Compares the effect of nitrous oxide and halothane on respiration (**RESP** measured as intra-tracheal pressure and expressed as instantaneous respiratory frequency; ordinate) and electrocortical activity (**ECoG**). These records were taken at the same time as those illustrated in Figure 11. The relative powers in the high frequency (alpha and beta, $\alpha + \beta$) and low frequency (delta and theta, $\delta + \theta$) bands are also shown. Periods of anaesthetic administration are indicated by the vertical lines.*

power high frequency bands of the electrocorticogram and also acted as a respiratory stimulant. This latter must be a central phenomenon since the inspired oxygen concentration was 50%. The snap decreased the power in the low frequency band and increased it in the high frequency band and caused a further increase in respiratory frequency. The changes in the power spectra of the electrocorticographic activity indicate that the level of anaesthesia had abruptly changed to a lesser degree of anaesthesia than the initial effect of

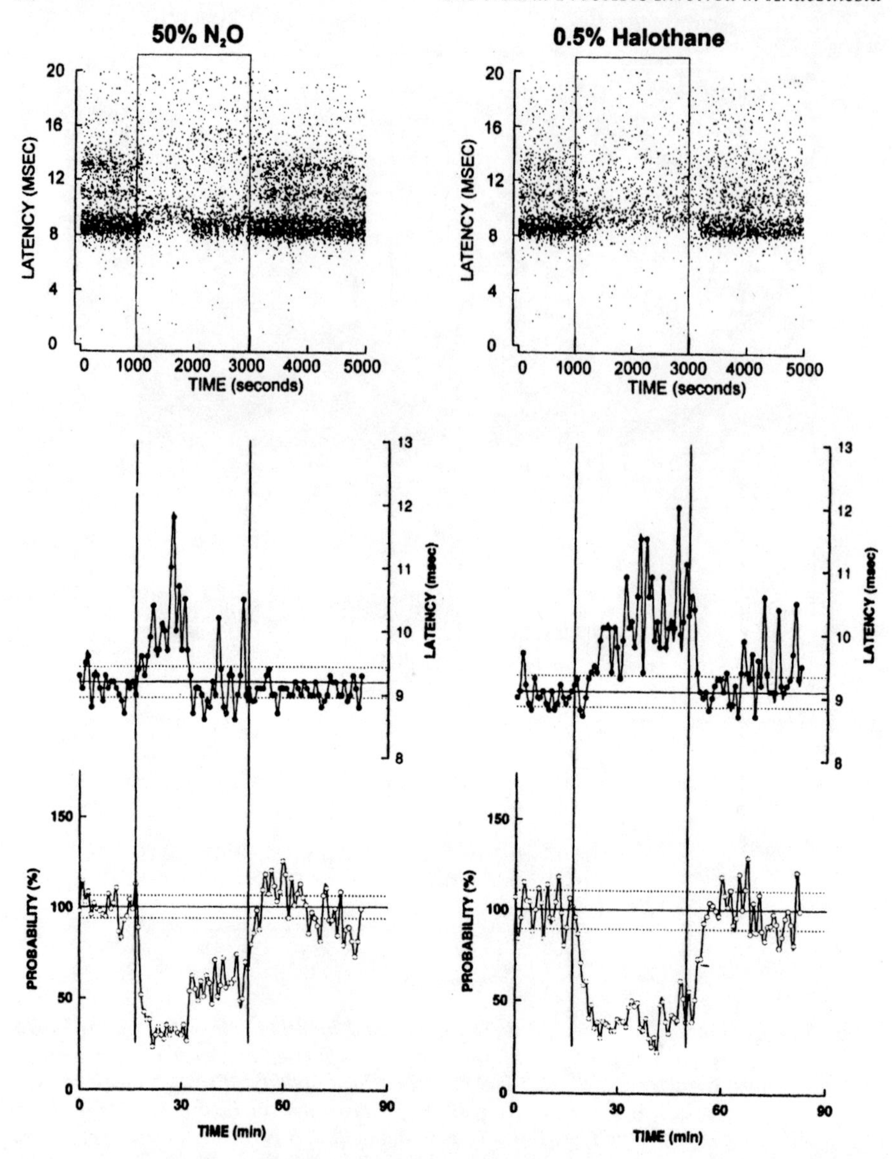

Figure 11 *Compares the effect of 50% nitrous oxide and 0.5% Halothane (approximately equipotent doses) on a ventroposterolateral thalamic cell. The top record shows the dot-raster sequential response histograms to supramaximal electrical stimulation of the contralateral forepaw in an animal initially surgically anaesthetised with urethane (1.25 g kg^{-1}). The periods of anaesthetic administration are indicated by the vertical lines. The bottom graph shows the median latency and probability of response of the cell in consecutive 50 sec periods. The interval between the end of the nitrous oxide and beginning of the Halothane administration was 60 minutes.*

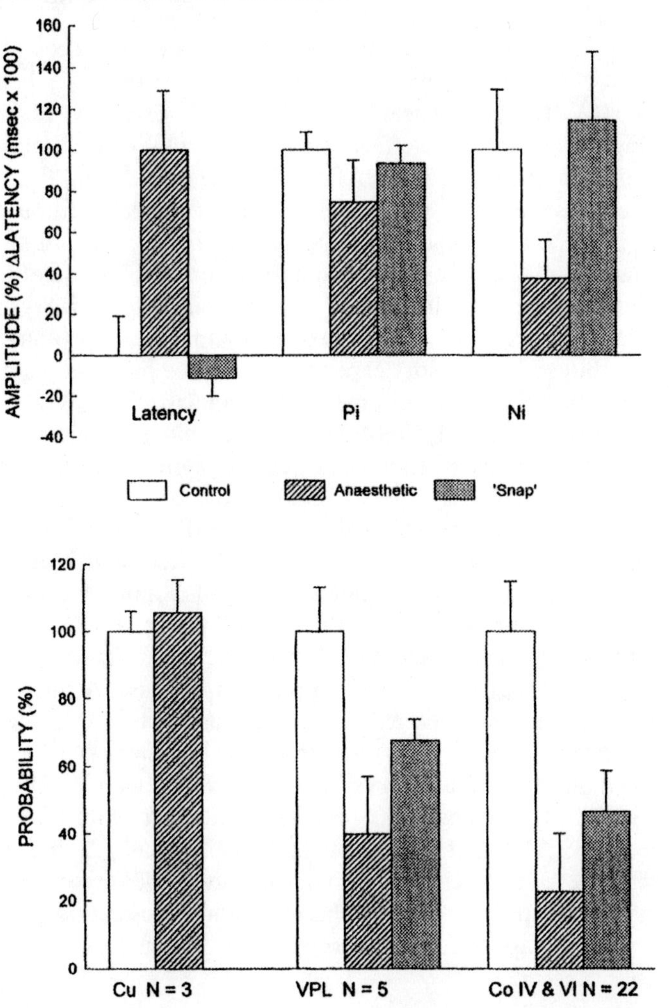

Figure 12 *This shows the effects of nitrous oxide (80%) on the evoked cortical responses for the initial anaesthetic effect and the subsequent "snap" for 22 animals (top) and the behaviour of single cells along the somatosensory pathway (bottom). Each vertical column represents the mean and the bars the SEM.*

nitrous oxide but that it still remained at a deeper level than the control condition.

The effect of Halothane in direct contrast was to give a maintained increase in anaesthetic depth as evidenced by the power spectral shift in the electrocorticographic activity and in the stabilisation of respiratory rate at a level similar to that seen in the control HVLF condition. The same behaviour was seen with the evoked cortical responses (Figures 9 & 12). In this case the "snap" gave cortical evoked responses which were statistically not significantly different to control values for their latencies and amplitudes of the first positive and negative waves. This is in contrast to the anaesthetic effect which gave statistically significant ($p<0.5$) increases in latency and decreases in the amplitudes of the first positive and first negative waves. The behaviour of single cells in VPL and cortical layers IV & VI also showed the "snap" response. Figure 11 shows the effect of equipotent doses of nitrous oxide and Halothane on a single cortical layer IV cell. Note that during the "snap" the latency of response returned to normal values but the cell, although showing an increase in response probability, remained depressed; but not so depressed as in the anaesthetic response. In direct contrast the effect of Halothane was to give a maintained increase in latency and decrease in response probability. Thus the data suggest that nitrous oxide would give a prolonged loss of consciousness but somatosensory experience would be partially restored within 5-15 minutes after the commencement of administration. Why the central nervous system can suddenly 'refuse to accept' that the animal is inspiring nitrous oxide at any concentration is, at present , somewhat of a mystery. The data show that all the mechanisms which are activated to produce the effect of a Group 1 anaesthetic are affected i.e. the cortico-thalamic and cortico-reticular-thalamic somatosensory control are all affected during the "snap" phase of the action of nitrous oxide. Prolonged administration of nitrous oxide, for up to 5 hours shows that this "snap" phase is not maintained but that the animal randomly and spontaneously varies its anaesthetic state between the initial anaesthetic effect and the "snap" phase. A similar effect has been found in man and termed 'acute tolerance' to both the analgesic (Ramsay, Brown & Woods, 1992), electrocortical changes (Avramov, Shingu & Mori, 1990) and loss of 'awareness' (Ruperht, Dworacek, Bonke & Dzoljic, 1985). A similar phenomenon has been reported in rodents (Smith, Winter, Smith & Eger, 1979) which may occur as a result of adaptation of dihydropyridine-sensitive calcium channels (Dolin & Little, 1989). However, none of these observations can be used to explain the random oscillations between the anaesthetic and "snap" effect of nitrous oxide which is seen on prolonged exposure to this agent.

References

Angel, A. (1964) The effect of peripheral stimulation on units located in the thalamic reticular nuclei. Journal of Physiology, 171, 42-60.

Angel, A. (1993) Central neuronal pathways and anaesthesia. British Journal of Anaesthesia, 71: 148-163.

Angel, A., Berridge, D. & Unwin, J. (1973) The effect of anaesthetic agents on primary evoked cortical responses. British Journal of Anaesthesia, 45: 824-836.

Angel, A. & Clarke, K.A. (1975) An analysis of the representation of the forelimb in the ventrobasal thalamic complex of the albino rat. Journal of Physiology, 249: 399-423.

Angel, A., Dodd, J. & the late J.D. Gray (1976) Fluctuating anaesthetic state in the rat anaesthetized with urethane. Journal of Physiology, 259, 11-12P.

Angel, A., Gratton, D.A., Halsey, M.J. & Wardley-Smith, B. (1980) Pressure reversal of the effect of urethane on the evoked cerebral cortical response in the rat. British Journal of Pharmacology, 70: 241-247.

Avramov, M.N., Shingu, K. & Mori, K. (1990) Progressive changes in electroencephalographic responses to nitrous oxide in humans: a possible acute drug tolerance. Anesthesia & Analgesia, 70: 369-374.

Bradley, K., Easton, D.M. & Eccles, J.C. (1953) An investigation of primary or direct inhibition. Journal of Physiology, 122: 474-488.

Brodal, A. (1957) The reticular formation of the brain stem. Anatomical Aspects and Functional Correlations London: Oliver & Boyd.

De Wildt, D.J., Hillen, F.C., Rauws, A.G. & Sangster, B. (1983) Etomidate-anaesthesia, with and without fentanyl, compared with urethane-anaesthesia in the rat. British Journal of Pharmacology, 79, 461-469.

DeJong, R.H., Robles, R., Corbiu, R.W. & Nace, R.A. (1968) Effect of inhalational anaesthetics on monosynaptic transmission in the spinal cord. Journal of Pharmacology and Experimental Therapeutics, 162: 326-330.

Dolin, S.J. & Little, H.J. (1989) Effects of "nitrendipine" on nitrous oxide tolerance and physical dependence. Anesthesiology, 70: 91-97.

Eccles, J. C. (1964) The Physiology of Synapses. Springer-Verlag, Berlin. P 204.

Magoun, H.W. & Rhines, R. (1946) An inhibitor mechanism in the bulbar reticular formation. Journal of Neurophysiology, 6: 165-171.

Moruzzi, G. & Magoun, H.W. (1949) Brain stem reticular formation and activation of the EEG. Electroencephalography and Clinical Neurophysiology, 1: 455-473.

Ramsay, D.S, Brown, A.C. & Woods, S.C. (1992) Acute tolerance to nitrous oxide in humans. Pain, 51: 367-373.

Rhines, R & Magoun, H.W. (1946) Brain stem facilitation of cortical motor responses. Journal of Neurophysiology 9: 219-229.

Ruperht, J., Dworacek, B., Bonke, B. & Dzoljic, M.R. van-Eijndhoven, J.H. & de-Vlieger, M. (1985) Tolerance to nitrous oxide in volunteers. Acta Anesthesiologica Scandinavica, 29: 635-638.

Smith, R.A., Winter, P.M., Smith, M. & Eger, E.I. (1979) Rapidly developing tolerance to acute exposures to anesthetic agents. Anesthesiology, 50: 496-500.

Yen, C.T., Conley, M., Hendry, S.H.C. & Jones, E.G. (1985) The morphology of physiologically identified GABAergic neurons in the somatic sensory part of the thalamic reticular nucleus of the cat. Journal of Neuroscience, 5: 2254-2268.

Anaesthetic Actions on Fast Synaptic Transmission

Chris D. Richards

DEPARTMENT OF PHYSIOLOGY, ROYAL FREE HOSPITAL SCHOOL OF MEDICINE, LONDON, UK

Introduction

Anaesthetics are administered during surgical procedures to render patients unconscious so that they do not experience pain. A secondary benefit is that an anaesthetised patient will not actively oppose the surgeon as he goes about his task. Indeed, it is now customary to administer neuromuscular blocking drugs during general anaesthesia to suppress reflex contractions that may otherwise occur in response to the traction and cutting of living tissue. As the brain is the organ concerned with the maintenance of consciousness and the perception of pain, this must mean that it is primary site at which general anaesthetics act. The nerve cells are the principal signalling elements of the brain and the connections between nerve cells, the synapses, are well known to be affected by general anaesthetics.

Since the original description of synaptic transmission by Sherrington in the early part of this century, the detailed mechanisms that underpin synaptic transmission have gradually become clearer. At a synapse, the axon of one neurone (the presynaptic neurone) becomes enlarged to form a nerve terminal which closely abuts part of a second neurone (the post-synaptic neurone) (see Figure 1). The principal events are as follows: A nerve impulse travelling in the axon of the presynaptic neurone arrives at the nerve terminal. This leads to the depolarisation and the entry of calcium ions into the nerve terminal via specific membrane proteins called voltage-gated calcium channels. The rise in calcium concentration triggers the release of a chemical messenger (a neurotransmitter) from the nerve terminal into the space between the nerve terminal and the post-synaptic neurone (the synaptic cleft). This process is called neurosecretion. The neurotransmitter binds to specific receptor molecules on the post-synaptic membrane and this triggers changes in the excitability of the post-synaptic neurone via specific ion channels. Activity in the presynaptic neurone may tend to increase the activity of the post-synaptic neurone (an excitatory synapse) or it may tend to reduce the excitability (an inhibitory synapse).

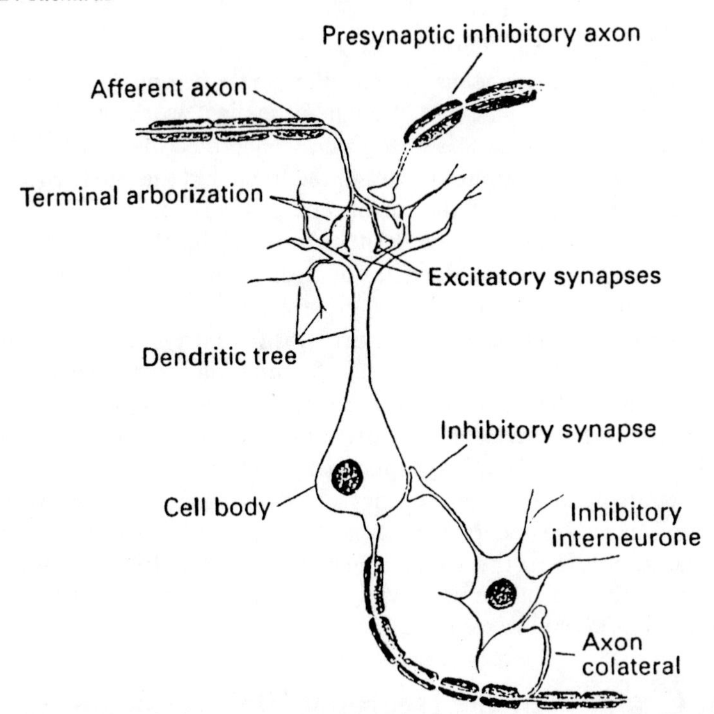

Figure 1 *A schematic drawing of the organsiation of a synaptic relay within the CNS. Action potentials arrive via the afferent axon and excite the post-synaptic cell. If the excitation is sufficiently intense the post-synaptic cell will discharge one or more action potentials which pass to other neurones via its axon. If this happens, small interneurones may be become activated and inhibit the first cell via an inhibitory synapse as shown. Action potentials in the afferent axon can be blocked by the activity of synapses between axons. This is known as presynaptic inhibition.*

It is well known that an astonishingly wide variety of substances can produce anaesthesia. These range from inert gases such as xenon and nitrogen to simple organic gases and vapours such as cyclopropane and diethyl ether to complex organic molecules typified by the barbiturates and steroids. In the late nineteenth century Mayer and Overton pointed out that anaesthetic potency of a given agent correlated closely with its oil-water partition coefficient. Following this realisation, the idea that anaesthetics might act by a common mechanism was born. The fundamental proposal was neatly summed up by Mullins (1954) who pointed out that anaesthetics will occupy a space in the membrane that depends on the concentration of anaesthetic in the plasma membrane of cells and on the volume occupied by each anaesthetic molecule. He then suggested that anaesthesia would occur when a critical volume of anaesthetic is achieved in the cell membrane. He further suggested that the membrane expansion that resulted from the presence of anaesthetics in the membrane would occlude the pores responsible for ion permeation and so disrupt neural processing. This is the critical volume hypothesis and it represented a major advance as it was the first attempt to describe the *mechanism* of anaesthetic action. Although subsequent work has shown that this simple model is incorrect, it is now generally accepted that anaesthetics do interact with relatively hydrophobic sites in the brain and that these sites reside in the hydrophobic regions of the membrane proteins responsible for regulating various aspects of neural excitability - the ion channels.

This article is concerned with the detailed mechanisms by which anaesthetics modulate synaptic transmission and will present evidence that anaesthetics act by modulating the properties of a number of different types of ion channel. From the brief description of the salient features of synaptic transmission given above, it should be clear that general anaesthetics could act at a number of sites. First, they could alter the properties of the presynaptic axon so that the action potential in the presynaptic nerve either fails reach the nerve terminal or is attenuated. Secondly, they could interfere with the process of neurosecretion in some way. Thirdly, anaesthetics could alter the response of post-synaptic receptors to the secreted neurotransmitter. Finally, they could modulate the properties of the post-synaptic neurone so that its response to a synaptic input was altered in some way.

The Ion Channels of the Neuronal Plasma Membrane

Ion channels can be visualised as water filled pores that provide an aqueous route for ions to traverse the cell membrane (see Aidley & Stanfield (1996) and Hille (1992) for detailed accounts of the properties of ion channels). Their function is to translocate ions across the plasma membrane and they can therefore be regarded as a specialised transport proteins which have properties analogous to enzymes, their substrate being a particular ionic species. The transport capacity of ion channels is very large some 10^6 - 10^8 ions being translocated each second. Despite this high transport capacity, ion channels are able to discriminate between ions of similar charge. For example, neuronal potassium channels achieve a selectivity of 100:1 for K^+ over Na^+. Ion channels are generally named according to their principal permeant ion species. Thus ion channels selective for Na^+, K^+, Ca^{2+}, Cl^- have been identified from electrophysiological studies and called sodium, potassium, calcium and chloride channels respectively although they may have a significant permeability to other ions. Some ion channels do not discriminate between ions of similar charge to any great degree; these are called non-selective cation or anion channels and such channels play a significant role in the process of excitatory synaptic transmission.

Ion channels may be opened by changes in the voltage across the cell membrane or by binding a specific substance such as a neurotransmitter. Those opened by voltage are called *voltage activated channels* and include the Na^+ channel responsible for action potential propagation, various K^+ channels which set the resting membrane potential and shape the response of neurones to synaptic inputs and Ca^{2+} channels which play an important role in cell signalling - including the initiation of neurosecretion. The channels opened by binding a neurotransmitter or other small molecule are known as *ligand operated channels*. These include the nicotinic acetylcholine receptor, various receptors to glutamate and receptors to serotonin ($5HT_3$ receptors) activation of which depolarise neurones via an excitatory postsynaptic potential or *epsp* which, as its name implies, lead to excitation. Activation of certain other ligand operated channels such as the $GABA_A$ receptor and the glycine receptor lead to hyperpolarisation via an inhibitory postsynaptic potential or *ipsp* resulting in inhibition of neural activity.

In addition to opening ion channels, many neurotransmitters bind to receptors that are coupled to second messenger systems, usually via a G-protein. These are often called metabotropic receptors to distinguish them from the ion channel linked receptors (which are also known as ionotropic receptors). Activation of metabotropic receptors often leads to slow changes in neural excitability. Examples are the muscarinic acetylcholine receptors, the metabotropic glutamate receptors and the $GABA_B$ receptor.

General Anaesthetics Both Depress Excitatory Synaptic Transmission and Enhance Inhibitory Synaptic Activity

Since general anaesthetics reduce the responsiveness of the nervous system to stimuli (especially noxious stimuli), the expectation is that anaesthetics will depress excitatory synaptic transmission and perhaps augment inhibitory synaptic transmission. Although not all synapses are affected equally, this expectation is broadly fulfilled. Although Sherrington (1906) in his classic book *The Integrative Action of the Nervous System* noted that synaptic transmission was relatively susceptible to the action of general anaesthetics, detailed analysis of the mechanism of anaesthetic action on synaptic function had to await the introduction of modern electrophysiological techniques. The first studies were carried out by Bonnet & Bremer (1948) and Bremer & Bonnet (1948) on the frog spinal cord and by Larrabee and Posternak (1952) in sympathetic ganglia. Both pairs of authors found that general anaesthetics depressed excitatory synaptic transmission without significantly depressing the propagation of action potentials in nerve axons. These early studies gave rise to the idea that anaesthetics exerted a selective depressant action the process of synaptic transmission.

Subsequent investigations established that a wide variety of general anaesthetics depress excitatory synaptic transmission in the spinal cord (Somjen & Gill, 1963; Somjen 1963, Weakly, 1969) and in the brain (el Beheiry & Puil, 1989; Richards, 1972, 1973; Richards, Russell & Smaje, 1975; Richards & Strupinski, 1986; Richards & White, 1975). Not all excitatory synapses, however, are easily blocked. While those of the olfactory cortex are depressed, the dendrodendritic excitatory synapses of the olfactory bulb are relatively resistant to depression by most anaesthetics (Nicol, 1972). The excitatory synapses of the primary afferent fibres in the cuneate nucleus have also been reported to be facilitated by clinically effective concentrations of general anaesthetics (Krnjevic & Morris, 1976, Morris, 1978).

In contrast to their predominant depressant effect on excitation, general anaesthetics have been found to enhance inhibitory synaptic transmission both in the spinal cord (Eccles, Schmidt & Willis, 1963;) and in the brain (Gage & Robertson, 1985, Nicol, 1972; Nicol, Eccles, Oshima & Rubia, 1975). Nevertheless, transmission at some inhibitory synapses is depressed (e.g. el Beheiry & Puil, 1989; Fujiwara, Higashi, Nishi, Shimoji & Yoshimura, 1988; Yoshimura, Higashi, Fujita & Shimoji, 1985).

This confusion concerning the effects of anaesthetics on synaptic activity is more apparent than real. The process of synaptic transmission is complex and differs from synapse to synapse. Some are concerned with reliable onward transmission of information (e.g. the synapses of the cuneate nucleus and those of primary afferents ending on spinal motoneurones) while others are concerned with more integrative activity where individual synaptic contacts may be relatively weak (e.g. those of the cerebral cortex and hippocampus). Nevertheless, one might reasonably expect that many (if not all) anaesthetics may act on a particular process in a broadly similar manner. This can only be established by a detailed study of the effects of a variety of anaesthetics on specific synaptic systems. In what follows, I shall explore the various mechanisms by which anaesthetics modulate fast excitatory and inhibitory synaptic transmission in the brain and spinal cord. Discussion of their effects on slow synaptic transmission is limited by a paucity of experimental data.

Presynaptic Effects of General Anaesthetics

The first convincing evidence that anaesthetics could depress the amount of transmitter secreted in response to nerve impulses was provided by Matthews & Quilliam (1964). They showed that the amount of acetylcholine secreted in response to stimulation of preganglionic sympathetic nerves was decreased by amylobarbitone and a number of other central depressants. Within the CNS such direct demonstrations are not easily achieved due to heterogeneity of neural connections. Instead, the effects of anaesthetics on chemically-evoked release of various neurotransmitters has been followed. Current evidence suggests that the secretion of neurotransmitters is depressed by a variety of anaesthetics. Thus the secretion of glutamate evoked by depolarisation with potassium is inhibited by pentobarbitone (Collins, 1980; Potashner, Lake, Langlois, Ploffe & Lecavalier, 1980). More recently, both Larsen, Grondahl, Haugstad & Langmoen (1994) and Schlame & Hemmings (1995) have shown that volatile anaesthetics depress the secretion of glutamate.

Some authors have reported that anaesthetic concentrations of barbiturates slightly increase the secretion of the inhibitory amino acid neurotransmitter GABA (Collins, 1980; Potashner et al. 1980) which could explain why barbiturates potentiate synaptically mediated inhibition. Other workers, however, have found that GABA secretion is inhibited (Jessell & Richards, 1977; Kendall & Minchin, 1982; Minchin, 1981). Since the fundamental steps of neurosecretion are not thought to be different for different neurotransmitters, the balance of the evidence must favour the idea that neurosecretion is inhibited by anaesthetics.

Some electrophysiological studies have been interpreted as evidence that barbiturates and other anaesthetics depress the secretion of neurotransmitter evoked by action potentials. This approach has been adopted by Weakly (1969) to study the effects of barbiturates, by Zorychta & Capek (1978) to investigate the effects of diethyl ether and by Kullman, Martin & Redman (1989) to study the effects of halothane and thiopentone. All three groups concluded that anaesthetics exert their actions by decreasing the quantal content of the excitatory post-synaptic potentials, which they interpret as indicating a decrease in the amount of neurotransmitter secreted in response to a nerve impulse. For reasons which will be discussed later, this interpretation is open to question.

What is the mechanism by which anaesthetics inhibit neurosecretion? The small size of most nerve endings and the heterogeneity of brain tissue makes this question difficult to answer for synapses in the brain. Instead it has proved more useful to study the effects of anaesthetics on the secretion of adrenaline and noradrenaline from adrenal medullary chromaffin cells. These endocrine cells are, like neurones, derived from neural crest tissue and are functionally homologous with and show many of the properties of sympathetic post-ganglionic neurones. Their neurosecretory properties have been intensively studied and correspond closely to those of nerve endings (see Figure 2).

Gothert and his colleagues (Gothert, Dorn & Loewnstein, 1976; Gothert & Wendt, 1977) using perfused adrenal glands, showed that anaesthetics depress the secretion of catecholamines evoked by nicotinic agonists and depolarisation with high potassium. Subsequently, I and my colleagues used isolated chromaffin cells to show that a wide variety of anaesthetics could inhibit catecholamine secretion evoked by direct depolarisation with high K^+ (Pocock & Richards, 1987,1988; Charlesworth, Pocock & Richards, 1994). For most clinically useful agents, the inhibition of secretion occurred within the range of concentrations expected during general anaesthesia (see Figure 3).

A number of mechanisms could account for this depression of secretion: Anaesthetics could decrease the amount of catecholamine in each secretory granule. They could disrupt the mechanism of exocytosis itself in some way, for example by inhibiting the docking of secretory granules at the plasma membrane. They could prevent the depolarisation of the secretory cell (or nerve ending) in some way – for example by increasing the membrane conductance to K^+ (this would not inhibit the secretion due to depolarisation with high extracellular K^+). They could inhibit the entry of Ca^{2+} in response to depolarisation by blockade of the voltage-activated Ca^{2+} channels. We have investigated each of these possibilities.

First, do anaesthetics decrease the catecholamine content of chromaffin granules or the transmitter content of synaptic vesicles? Evidence that this might happen was provided by Bangham & Mason (1980) who studied the efflux of catecholamines entrapped within liposomes. A subsequent thorough

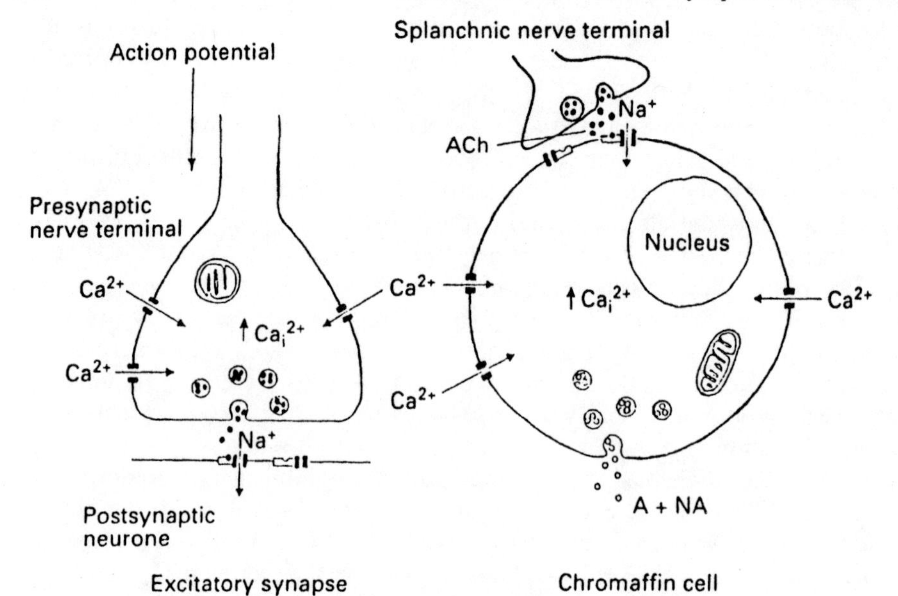

Figure 2 *Schematic diagram indicating the major steps involved in neurosecretion from a nerve ending (left) and an adrenal medullary chromaffin cell (right). Calcium ions are represented as moving through voltage-gated calcium channels. A - adrenaline, NA noradrenaline.*

study by Akeson & Deamer (1989) showed that anaesthetics did not deplete isolated chromaffin granules of their catecholamines. Equally, the content of neurotransmitter in brain tissue does not decline in anaesthesia. It is either unchanged or is increased, depending on the transmitter studied (Crossland & Merrick, 1954; Biebuyck, Dedrick & Scherer, 1975; Potashner et al., 1980). Furthermore, we have found that a variety of anaesthetics were without effect on the leakage of catecholamines from isolated chromaffin cells rendered leaky to small molecules by exposure to high voltage electrical discharge (electropermeabilised cells).

Second, do anaesthetics inhibit the intracellular events that lead to exocytosis? To investigate this possibility we examined the effects of three types of anaesthetic on the Ca^{2+} activation of secretion from electropermeabilised chromaffin cells. As Figure 4 shows the anaesthetics did not alter the basal secretion (10^{-8}M Ca^{2+}) or the secretion evoked by raising the free Ca^{2+} ($>10^{-6}$ M) (for further details see Pocock & Richards, 1987, 1988). Consistent with this, Schlame & Hemmings (1995) have recently reported that halothane did not inhibit the release of glutamate evoked by ionomycin from isolated synaptic terminals although it did inhibit the secretion evoked by 4-amino pyridine.

Since Ca^{2+} is an important second messenger which regulates neurosecretion and neuronal excitability in addition to a host of other important actions, the nature of the channels concerned with controlling Ca^{2+} entry into the cell

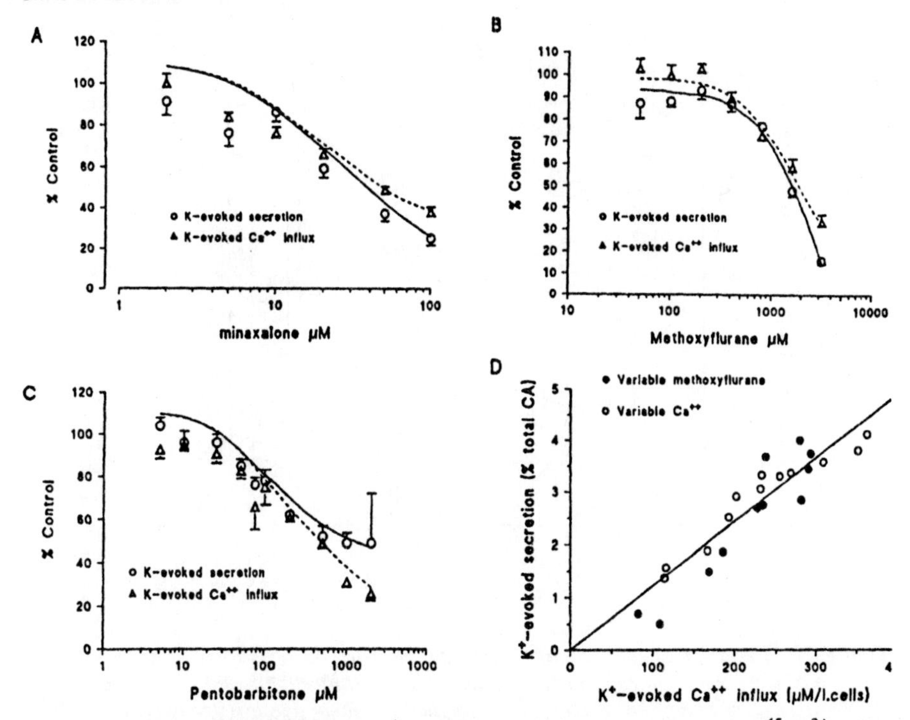

Figure 3 *The parallel inhibition of K^+-evoked catecholamine secretion and $^{45}Ca^{2+}$ influx by the water-soluble steroid anaesthetic minaxalone (A), methoxyflurane (B) and pentobarbitone (C). Panel D shows the relationship between Ca^{2+} influx and catecholamine secretion with varying concentrations of anaesthetic and with varying concentrations of Ca^{2+} in the bathing medium.*

has been the subject of intense interest. It is now thought that Ca^{2+} entry is controlled by six different types of voltage-gated channel which are known as the L, N, P, Q, R and T subtypes. The different subtypes can be distinguished by their sensitivity to various inhibitors and by their electrophysiological characteristics. The role of the different sub-types in regulating neural function is still being clarified but N-type and P-type channels have been implicated in the regulation of transmitter release at the neuromuscular junction of frogs and mammals (Uchitel, Protti, Sanchez, Cherksey, Sugimori & Llinas, 1992).

Do anaesthetics inhibit the activity of voltage-gated Ca^{2+} channels? Wertz & Macdonald (1985) found that pentobarbitone inhibited the voltage-dependent calcium currents in cultured dorsal root ganglion cells. Subsequently, Gross & Macdonald (1988) examined the action of pentobarbitone on specific subtypes of calcium current in these neurones. They found that pentobarbitone affected the high threshold N- or L- calcium currents but had little effect on the low threshold T current. The rate of inactivation of the high threshold currents was increased and the amplitude depressed. Herrington, Stern, Evers & Lingle (1991) have investigated the action of halothane on the calcium currents of GH3 cells and found that, as with the action of pentobarbitone

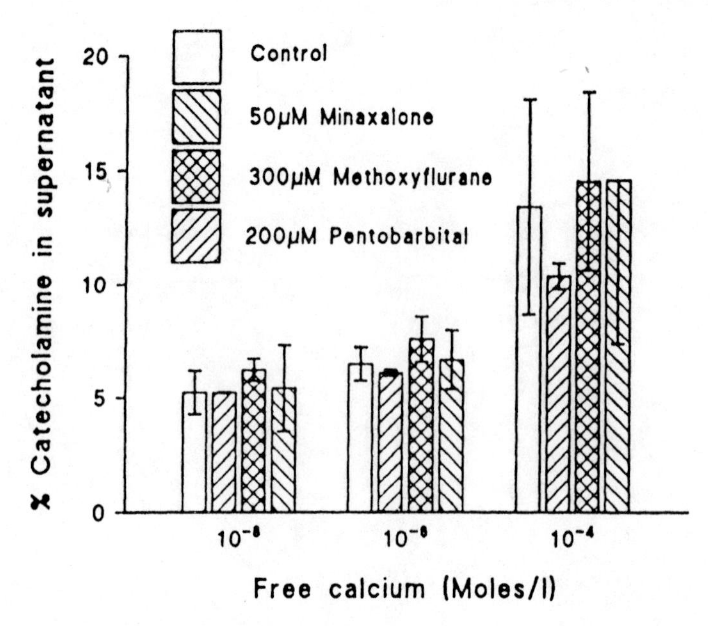

Figure 4 *The effect of anaesthetic agents on the calcium activation of catecholamine secretion from electropermeabilised chromaffin cells. the lowest concentrations shown reflect basal secretion plus the catecholamine lost from damaged cells. 10^{-6} M Ca^{2+} is close to threshold for activation of secretion and 10^{-4} M Ca^{2+} will elicit maximal secretion. The concentrations of anaesthetics used would exert a substantial inhibition of K^+-evoked secretion (30-50%) from intact cells.*

on dorsal root ganglion neurones, halothane had little effect on T currents but it inhibited the high threshold L currents with an IC_{50} of about 0.5 mM. In adrenal chromaffin cells Charlesworth, Pocock & Richards (1994) have shown that a variety of anaesthetics inhibit the currents due to activation of calcium channels of the N and L types although halothane had little inhibitory effect. There was no evidence to suggest a selective action of the other anaesthetics on either N or L channels. Hall, Leib & Franks (1994) have reported that the P-type calcium channels are relatively insensitive to a variety of anaesthetics.

Is the decrease in Ca^{2+} entry brought about by block of Ca^{2+} currents sufficient to account for the inhibition of secretion? To investigate this the $^{45}Ca^{2+}$ influx into freshly isolated cells was measured and compared with the amount of catecholamine secreted. We found that the depression of K^+-evoked secretion was closely paralleled with the inhibition of $^{45}Ca^{2+}$ influx (Figure 3D). This led us to the conclusion that the anaesthetic-induced depression of catecholamine secretion could be completely explained by inhibition of Ca^{2+} influx (Pocock & Richards, 1987, 1988, 1991). Nevertheless, it is important to recognise that not all anaesthetics inhibit neurosecretion. Halothane, ketamine and the local anaesthetic procaine have little effect on

K^+-evoked secretion of catecholamines at concentrations likely to be found in the brain during anaesthesia (Pocock & Richards, 1988; Takahara, Wada, Arita, Sumikawa & Izumi, 1986). Moreover, ketamine and halothane are poor inhibitors of K^+-evoked secretion of excitatory amino acids from brain tissue slices and synaptosomes (Oshima & Richards, 1988; Arai, Hatano & Mori, 1990; Schlame & Hemmings, 1995). Further work on more physiological intact systems is required before the various contradictions can be completely resolved.

Post-synaptic Effects of General Anaesthetics

From the data discussed in the previous section, it is clear that many anaesthetics depress neurosecretion at concentrations likely to be encountered in the brain during general anaesthesia. Nevertheless, it has long been known that anaesthetics modulate the responses of neurones to artificially applied neurotransmitters. The principal excitatory neurotransmitter is now generally thought to be glutamate. Although ionotropic glutamate receptors also serve a major role in fast synaptic transmission between CNS neurones, they are now known to show little structural homology to other neurotransmitter receptors and so belong to a distinct receptor family. Nevertheless, they also possess sequences of hydrophobic amino acids which presumably form the membrane spanning domains. As with the voltage-gated calcium channels, the glutamate receptor-ion channel complexes are now known to consist of several different types. The primary division is between those receptors that respond to kainate and AMPA (α-amino-3-hydroxy-5-methylisoxazole-4-propionic acid) and those that respond to NMDA (N-methyl D-aspartate). There are also G-protein linked glutamate receptors but little is known about their modulation by anaesthetic agents.

Early experiments by Crawford & Curtis (1966) showed that barbiturates decreased the ability of neurones in the cerebral cortex to respond to glutamate applied by micro-electrophoresis (ionophoresis). Halothane did not significantly affect the sensitivity to ionophoretically applied glutamate (Crawford, 1970). Subsequently Richards & Smaje (1976) and Richards, Russell & Smaje (1975) found that a number of anaesthetics including diethyl ether, methohexitone and the steroid anaesthetic alphaxalone all decreased the sensitivity of olfactory neurones to ionophoretically applied glutamate. Halothane had little effect. Following the recognition that there are two major types of ionotropic glutamate receptor (the NMDA and the kainate/AMPA receptors discussed above) attention has focused the action of anaesthetics on these receptor subtypes. Ketamine, phencyclidine, ethanol and diethyl ether have all been shown to depress responses of the NMDA subclass preferentially (Lodge & Anis, 1982; Lodge, Anis & Burton, 1982; Oshima & Richards, 1988; Weight, Lovinger, White & Peoples, 1991) whereas barbiturates depress the sensitivity of both receptor subtypes (Sawada & Yamamoto, 1985; Weight et al. 1991). In contrast, Perouansky, Baranov, Salman & Yaari (1995) concluded that halothane has little effect on either NMDA or on AMPA receptors. Oocytes

$$\text{closed} \underset{\alpha_1}{\overset{\beta_1}{\rightleftharpoons}} \text{open} \underset{\alpha_2}{\overset{\beta_2}{\leftrightharpoons}} \text{closed}$$

$$b \upharpoonleft\downharpoonright f[\text{c}]$$

$$\text{blocked} \underset{k_{-1}}{\overset{k_{+1}}{\rightleftharpoons}} (\text{closed} - \text{blocked})$$

Figure 5 *A simple kinetic scheme that can account for many of the effects of pentobarbitone on the NMDA channel. A similar (but not identical) scheme can account for the effects of a number of anaesthetics on nAChR channels. (From Charlesworth, Jacobson & Richards, 1995).*

injected with mRNA from rat brain express kainate receptors and the response of these cells to kainate is depresses by phenobarbitone (Daniels, Zhao, Inman, Price, Shelton & Smith, 1991).

The molecular cloning of the glutamate receptors has led to increased scrutiny of the role of the different subtypes of receptor. Experiments with recombinant receptors have shown that the sensitivity of receptors formed by different subunits show different sensitivities to anaesthetic agents (Harris, Mihic, Dildy-Mayfield & Machu, 1995; Dildy-Mayfield, Eger & Harris, 1996). Nevertheless, the recombinant AMPA receptors are, in general, rather insensitive to volatile anaesthetics.

The detailed mechanisms of anaesthetic block of glutamate ionotropic channels block are largely unknown but there have been some recent studies of the action of anaesthetics on NMDA channels. Pentobarbitone depresses the function of native NMDA receptors by blocking the open channel (see Figure 5 and Charlesworth, Jacobson & Richards, 1995). This results in a shortening of the lifetime of the channel open state. The conductance of the open channel was not affected so that the anaesthetic did not change the ionic selectivity of the channel. Low concentrations of isoflurane have been reported to decrease the probability of channel opening without a significant effect on the mean open time of channels gated by NMDA. Higher concentrations (>1mM) also reduce mean open time (Yang & Zorumski, 1991). Alkanols have similar effects (Weight et al. 1991). Dissociative anaesthetics such as ketamine appear to block open NMDA channels preferentially and this block is both voltage- and use-dependent (Halliwell, Peters & Lambert, 1989; Heutner & Bean, 1988; Macdonald & Nowak, 1991). In this case, the channel closes while the anaesthetic is bound and dissociation of the anaesthetic from the blocked state is very slow. Consequently, the number of blocked channels accumulates with repeatedly activation.

The best studied example of a fast signalling ligand-gated receptor is the nicotinic receptor of skeletal muscle. This is a member of a superfamily of ligand-gated ion channel receptors that includes other nicotinic acetylcholine receptors (nAChR) including the neuronal nicotinic receptor and the inhibitory $GABA_A$ and glycine receptors. Nicotinic ACh receptors have been widely used to investigate the actions of anaesthetics on ligand-gated ion channels. Initially this work concentrated on the receptors of the neuromuscular junction but more recently neuronal nicotinic receptors have received more attention particularly those of bovine chromaffin cells. Some years ago Adams (1976) showed that barbiturates and procaine depressed the second of a pair of closely spaced endplate potentials much more effectively than the first. He proposed that this occurred because the anaesthetics preferentially blocked ion channels when they were open. He went on to suggest that the blocked channel must then unblock and return to the open state before it can close. This simple scheme, which has guided investigation of anaesthetic action on ion channel kinetics since its introduction, is known as the sequential blocking model.

The sequential blocking model of anaesthetic receptor interactions makes a number of important predictions about the way anaesthetics may interact with ion channels which can readily be tested using patch-clamp recording techniques. These predictions are largely met when the acetylcholine receptor of the neuromuscular junction are exposed to low concentrations of the positively charged lidocaine derivative QX222 (Neher, 1983). With other anaesthetics the pattern of channel modulation is more complex than the model implies. Nevertheless, the principal effect observed when the nicotinic receptor is exposed to anaesthetics is a reduction in the time for which the channel opens after it has been activated by an agonist. Dilger & Brett (1991) and Dilger, Brett & Lesko (1992) have proposed a modification to the sequential blocking model which permits the blocked channel to enter a further long-lived blocked state. This extended block model accounts for much of the action of anaesthetics on the muscle nicotinic receptor including an apparent reduction in single channel amplitude. Dilger & Brett showed that this effect results from a rapid alternation between the open and blocked states that is too fast to resolve with present recording techniques.

The effect of anaesthetics on neuronal nicotinic channels has been studied in chromaffin cells. All anaesthetics studied inhibited the catecholamine secretion evoked by nicotinic agonists at lower concentrations that were required to inhibit K^+-evoked secretion. Detailed analysis showed that this effect was due to inhibition of the nAChRs themselves (Pocock & Richards, 1987, 1988; Charlesworth et al. 1992). Detailed study with the single channel patch-clamp technique allowed some conclusions regarding the mechanism of the blockade to be drawn. Pentobarbitone (Jacobson, Pocock & Richards, 1991), methohexitone, methoxyflurane and etomidate (Charlesworth & Richards, 1995) and procaine (Charlesworth, Jacobson, Pocock & Richards, 1992) all reduced the time for which the channels were open. However, neither the sequential blocking model nor the extended block model would adequately explain how the anaesthetic reduced the lifetime of the open channel. A similar conclusion has been reached by Wachtel (1995) for the effects of volatile anaesthetics on the nAChR of BC_3H1 cells.

One of the most dramatic and remarkable effects of many anaesthetics is their ability to potentiate the action of the neurotransmitter GABA at inhibitory synapses (see above). It is now clear that this potentiation results from a direct action of anaesthetics on GABA-activated channels themselves. Two studies have investigated the anaesthetic modulation of $GABA_A$ channel kinetics. Macdonald, Rogers & Twyman (1989) found evidence to suggest that $GABA_A$ channels have three different modes of opening and showed that pentobarbitone increased the relative frequency of the opening mode with the longest duration. This has the effect of increasing the mean open time of the channel and this would prolong the inhibitory effect of GABA. Halothane has been reported to show a similar action (Yeh, Quandt, Tanguy, Nakahiro, Narahashi & Brunner, 1991). Studies of recombinant $GABA_A$ receptors have revealed that the effects vary with the subunit composition (Harris et al. 1995) and recombinant expression of the ε-subunit of the $GABA_A$ receptors with the α- and β-subunits blocks the potentiating effect of anaesthetics on the response to GABA (Davies, Hanna, Hales & Kirkness. 1997).

In the spinal cord, glycine rather than GABA is the principal inhibitory transmitter. In contrast to the extensive literature for glutamate and GABA, there have been few studies of the effects of anaesthetics on glycine receptors. Volatile anaesthetics potentiate the response to glycine both in native and in recombinant receptors (Downie, Hall, Lieb & Franks, 1996). Although glycine-evoked currents are potentiated by propofol and chlormethiazole they are not potentiated by either pentobarbitone or alphaxalone (Hales & Lambert, 1991, 1992; Hill-Venning, Peters & Lambert, 1993).

Do Anaesthetics Preferentially Affect Presynaptic or Postsynaptic Events?

So far I have been treating the effects of general anaesthetics on presynaptic and post-synaptic mechanisms as though they were isolated topics. In an intact synapse an anaesthetic may act primarily on the mechanisms involved in neurosecretion or on the receptors of the postsynaptic membrane or on both. At a chemical synapse, it is generally considered that the secretory process consists of the exocytotic release of neurotransmitter from one or more vesicles. Therefore, at each synaptic contact the magnitude of the post-synaptic response is related to the number of vesicles released and the amount of transmitter in each vesicle. The presently available experimental evidence suggests that anaesthetics do not affect the quantity of neurotransmitter present in the synaptic vesicles. Thus, to depress synaptic transmission, an anaesthetic must decrease the probability that a given vesicle will release its contents (so that the total quantity of neurotransmitter secreted will be less than normal) or it must decrease the sensitivity of the post-synaptic membrane to the released transmitter (or both). To enhance synaptic transmission either more transmitter must be secreted or the effect of the released transmitter must be enhanced in some way.

Several attempts have been made to determine the relative contributions of presynaptic and post-synaptic actions to the anaesthetic modulation of excitatory synaptic transmission by examining the effect of anaesthetic on the fluctuations in the amplitude of the synaptic potentials. This type of analysis (quantal analysis) is based on the known behaviour of transmitter secretion by the nerve terminal of the neuromuscular junction where the steps in the amplitude of synaptic potentials are known to result from changes in the number of vesicles releasing their contents into the synaptic cleft. Weakly (1969), Zorychta & Capek (1978) and more recently Kullmann, Martin & Redman (1989) have produced evidence to show that the quantal content of epsps is reduced by general anaesthetics. They also found that quantal size was unaffected. This data would suggest that anaesthetics exert their effects primarily on the presynaptic events i.e. they act by decreasing the release of neurotransmitter in response to an action potential. The principal weakness of this work is the assumption that quantal size reflects the exocytotic release of neurotransmitter by a single vesicle. In the CNS, however, there is evidence that quantal size is determined by the number of available post-synaptic receptors (Edwards, Konnerth & Sackmann, 1990; Larkman, Stratford & Jack, 1991).

Consider the situation where the amount of transmitter secreted exceeds the number of available receptors by a considerable margin. When the synapse is activated, all the receptors will bind the neurotransmitter and participate in the generation of the synaptic potential (epsp or ipsp). If the anaesthetic were to affect both neurosecretion and the post-synaptic receptors equally, it would be the effect on the post-synaptic receptors that would dominate because the amount of transmitter secreted would have to be significantly reduced before it failed to saturate the available receptor population. If, however, the neurotransmitter did not saturate the available pool of receptors, a reduction in secretion would lead to a proportionate reduction in the amplitude of the epsp even though the receptors themselves were unaffected by the anaesthetic. Such considerations may go some way to explaining the differences in the susceptibility of different synapses to modulation by anaesthetic agents. A modest reduction in the secretion of neurotransmitter resulting from inhibition of calcium channel activity in the presynaptic terminal together with a decrease in open channel lifetime of the post-synaptic receptors would lead to depression of excitation. A similar reduction in neurosecretion at an inhibitory synapse could be offset by an increase in open channel lifetime (cf. the effects of barbiturates on the $GABA_A$ channels) and result in a prolongation of an ipsp.

Acknowledgement. This work has been supported by the Wellcome Trust.

References

Adams, P.R. (1976) Drug blockade of open end-plate channels. *J. Physiol. (Lond)* **260,** 531-552

Aidley, D.J. & Stanfield, P.R. (1996) *Ion channels : Molecules in Action.* Cambridge University Press, Cambridge.

Akeson, M.A. & Deamer, D.W. (1989) Steady-state catecholamine distribution in chromaffin granules. *Biochemistry*, 28, 5120-5127.

Arai, T., Hatano, Y. & Mori, K. (1990) Effects of halothane on the efflux of [^{3}H] D-aspartate from rat brain slices. *Acta Anaesth. Scand.* **34**, 267-270.

Bangham, A.D. & Mason, W.T. (1980) Anaesthetics may act by collapsing pH gradients. *Anaesthesiol.* 53, 135-141.

el-Beheiry, H. & Puil, E. (1989) Anaesthetic depression of excitatory synaptic transmission in neocortex. *Exp. Brain Res.* **77**, 87-93.

Biebuyck, J.F., Dedrick, D.F., Scherer, Y.D. (1975) Brain cyclic AMP and putative transmitter amino acids during anesthesia. pp 451-470 In : *Molecular Mechanisms of Anesthesia (Progress in Anesthesiology Vol 1.)* New York, Raven Press.

Bonnet, V. & Bremer, F. (1948) Analyse oscillographique des depressions fonctionnelles de la substance grise spinale. *Arch Int. Physiol.* 56, 97-99.

Bremer, F. & Bonnet,V. (1948) Action particuliere des barbituriques sur la transmission synaptique centrale. *Arch Int. Physiol.* 56, 100-102.

Charlesworth, P., Jacobson, I., Pocock, G. and Richards, C.D. (1992) The mechanism by which procaine inhibits catecholamine secretion from bovine chromaffin cells. *Br. J. Pharmac.* **106**, 802-812.

Charlesworth, P., Pocock, G. and Richards, C. D. (1992) The action of anaesthetics on stimulus-secretion coupling. *General Pharmacology,* **23**, 977-984.

Charlesworth, P., Pocock, G. and Richards, C.D. (1994) Calcium currents in bovine adrenal medullary cells and their modulation by anaesthetic agents.
J. Physiol. (Lond) **418**, 543-553.

Charlesworth, P. and Richards, C.D. (1995) Anaesthetic modulation of nicotinic channel kinetics. *Br. J. Pharmac* **114**, 909-917.

Charlesworth, P., Jacobson, I. and Richards, C.D. (1995) Pentobarbital modulation of NMDA channels in neurons isolated from the rat olfactory brain. *Br. J. Pharmac.* **116**, 3005-3013

Collins, G.C.C.S. (1980) Release of endogenous amino acid neurotransmitter candidates from rat olfactory cortex :possible regulatory mechanisms and effects of pentobarbitone. *Brain Research* **190**, 517-523.

Crawford, J.M. (1970) Anaesthetic agents and the chemical sensitivity of cortical neurones. *Neuropharmac.* **9**, 31-46.

Crawford, J.M. & Curtis, D.R. (1966) Pharmacological studies on feline Betz cells. *J. Physiol. (Lond.)* **186,** 121-138.

Crossland, J. & Merrick, A.J. (1954)The effect of anaesthesia on the acetylcholine content of brain. *J. Physiol. (Lond.)* **125**, 56-66.

Daniels, S., Zhao, D.M., Inman, N., Shelton, C.J. & Smith, E.B. (1991) Effects of general anaesthetics and pressure on mammalian excitatory receptors expressed in Xenopus oocytes. *Ann. N. Y. Acad. Sci.* **625**, 108-115.

Davies, P.A., Hanna, M.C., Hales, T.G. & Kirkness, E.F. (1997) Insensitivity to anaesthetic agents conferred by a class of $GABA_A$ receptor subunit. *Nature (Lond.)* **385**, 820-823.

Dildy-Mayfield, J.E., Eger, E.I. & Harris, R.A. (1996) Anaesthetics produce subunit selective actions on glutamate receptors. *J. Pharm. Exp. Ther.*, **276**, 1058-1065.

Dilger, J.P. & Brett, R.S. (1991) Actions of volatile anaesthetics and alcohols on cholinergic receptor channels. *Ann.N. Y. Acad. Sci.* **625**, 616-627.

Dilger, J.P., Brett, R.S. and Lesko, L.A. (1992). Effects of isoflurane on acetylcholine receptor channels. 1. Singlechannel currents. *Molec. Pharmac.* **41**, 127133.

Downie, D.L., Hall, A.C., Lieb, W.R. & Franks, N.P. (1996) Effects of inhalational general anaesthetics on native glycine receptors in rat medullary neurones and recombinant glycien receptors in Xenopus oocytes. *Br. J. Pharmac.* **118**, 493-501.

Eccles, J.C., Schmidt, R.W & Willis, W.D. (1963) Pharmacological studies on pre-synaptic inhibition. *J. Physiol. (Lond.)*, **168**, 500-530.

Edwards, F.A., Konnerth, A. & Sackmann, B. (1990) Quantal analysis of inhibitory synaptic transmission in the dentate gyrus of rat hippocampal slices : a patch-clamp study. *J. Physiol. (Lond.)* **430**, 213-249.

Fujiwara, N. Higashi, H, Nishi, S., Shimoji, K. Sugita, S., & Yoshamura, M. (1988) Changes in spontaneous firing patterns of rat hippocampal neurones induced by volatile anaesthetics. *J. Physiol. (Lond.)*, **402**, 155-175.

Gage, P.W. & Robertson, B. (1985) Prolongation of inhibitory currents by pentobarbitone, halothane and ketamine in CA1 pyramidal cells in rat hippocampus. *Br. J. Pharmac.* 85, 675-681.

Gothert, M., Dorn, W. & Loewenstein, I. (1976) Inhibition of catechaolamine release from the adrenal medulla by halothane : Site and mechanism of action. *Naunyn-Schmiedebergs Arch. Pharmac.* **294**, 239-249.

Gothert, M. & Wendt, J. (1977) Inhibition of adrenal medullary catecholamine secretion by enflurane: I. Investigations *in vitro*. *Anesthesiology,* **46**, 400-403.

Gross, R.A. & Macdonald, R.L. (1988) Differential actions of pentobarbitone on calcium current components of mouse sensory neurones in culture. *J.Physiol. (Lond.)* **405**, 187-203.

Hales, T.G. & Lambert, J.J. (1991) The actions of propofol on inhibitory amino acid receptors of bovine adrenomedullary chromaffin cells and cortical neurones. *Br. J. Pharmac.* **104**, 619-628.

Hales, T.G. & Lambert, J.J. (1992) Modulation of $GABA_A$ and glycine receptors by chlormethiazole. *Eur. J. Pharmac.* **210**, 239-246.

Hall, A.C., Lieb, W.R. & Franks, N.P. (1994) Steroselective and non-steroselective actions of isoflurane on the $GABA_A$ receptor. *Br. J. Pharmac.* **112**, 906-910.

Halliwell, R.F., Peters, J.A. & Lambert, J.J. (1989) The mechanism of action and pharmacological specificity of the antoconvulsant NMDA antagonist MK-801 : a voltage-clamp study on neuronal cells in culture. *Br. J. Pharmac.* **96**, 480-494.

Harris, R.A., Mihic, S.J., Dildy-Mayfield, J.E. & Machu, T.K. (1995) Actions of anaesthetics on ligand-gated ion channels : Role of receptor subunit composition. *FASEB J.*, **9**, 1454-1462.

Herrington, J., Stern, R.C., Evers, A.S. & Lingle, C.J. (1991) Halothane inhibits two components of calcium current in clonal (GH3) pituitary cells. *J. Neurosci.* **11**, 2226-2240.

Heutner, J.E. & Bean B.P. (1988) Block of N-methyl-D-aspartate activated current by the anticonvulsant MK-801 : selective binding to open channels. *Proc. Nat. Acad. Sci. USA.* **85,** 1307-1311.

Hille, B. (1992) *Ionic Channels of Excitable Membranes.* Second Edition. Sinauer, Sunderland, Mass.

Hill-Venning, C., Peters, J.A. & Lambert, J.J. (1993) The interaction of steriods with inhibitory and excitatory amino acid receptors. *Clin. Neuropharmac.* **15**, Suppl. 1. 683A-684A.

Jacobson, I., Pocock, G. and Richards, C.D. (1991) The effect of pentobarbitone on the properties of channels activated by nicotinic agonists in chromaffin cells. *Eur. J. Pharmac.* **202**, 331-339.

Jessell, T.M. & Richards, C.D. (1977) Barbiturate potentiation of hippocampal ipsps is not mediated by blockade of GABA uptake. *J. Physiol. Lond* **269**, 42P.

Kendall, T.G.J. & Minchin, M.C.W. (1982) thje effects of anaesthetics on the uptake and release of amino acid transmitters in thalamic slices. *Br.J. Pharmac.* **75**, 219-227.

Krjnevic, K. & Morris, M.E. (1976) Input-output relation of transmission thorugh the cuneate nucleus. *J. Physiol. (Lond.)* **257**, 791-815.

Kullman, D.M., Martin, R.L. & Redman, S.J. (1989) Reduction by general anaesthetics of group Ia excitatory post-synaptic potentials and currents in the cat spinal cord. *J. Physiol. (Lond.)* **412**, 277-296.

Larkman, A., Stratford, K. & Jack, J. (1991) Quantal analysis of excitatory synaptic action and depression in hippocampal slices. *Nature (Lond.)* **350**, 344-347.

Larrabee, M.G. & Posternak, J.M. (1952) Selective actions of anaesthetics on synapses and axons in mammalian sympathetic ganglia. *J. Neurophysiol.* **15**, 92-114.

Larsen, M., Grondahl, T.O., Haugstad, T.S. & Langmoen, I.A. (1994) The effect of the volatile anaesthetic isoflurane on Ca^{2+}-dependent glutamate release from rat cerebral cortex. *Brain Research* **663**, 335-337.

Lodge, D. & Anis, N.A. (1982) Effects of phencylidine on excitatory amino acid activation of spinal interneurones in the cat. *Eur. J. Pharmac.* **77**, 203-204.

Lodge, D., Anis, N.A. & Burton, N.R. (1982) Effects of optical isomers of ketamine on excitation of cat and rat spinal neurones by amino acids and acetylcholine. *Neurosci. Lett.* **29**, 281-286.

Macdonald, J.F & Novak, L.M. (1991) Mechanism of blockade of excitatory amino acid receptor channels. *Trends Pharmac. Sci.* **12**, 25-30.

Macdonald, R.L., Rogers, C.J. & Twyman, R.E. (1989) Barbiturate regulation of kinetic properties of the $GABA_A$ receptor channel of mounse spinal neurones in culture. *J. Physiol. (Lond.)* **417**, 483-500.

Matthews, E,K. & Quilliam, J.P. (1964) Effects of central depressant drugs uppon acetylcholine release. *Br. J., Pharmac.* **22**, 415-440.

Minchin, M.C.W. (1981) The effects of anaesthetics on the uptake and release of γ-aminobutyric acid and D-aspartate in rat brain slices. *Br.J. Pharmac.* **73**, 681-690.

Morris, M.E. (1978) Facilitation of synaptic transmission by general anesthetics. *J. Physiol. (Lond.)* **284,** 307-325.

Mullins, L.J. (1954) Some physical mechanisms in narcosis. *Chem. Rev.* **54**, 289-323.

Neher, E. (1983) The charge carried by single channel currents in cultured rat myotubes. *J. Physiol. (Lond.)* **339**, 663-678.

Nicol, R.A. (1972) The effect of anaesthetics on synaptic excitation and inhibition in the olfactory bulb. *J. Physiol. (Lond.)* **223**, 803-814.

Nicol, R.A., Eccles, J.C., Oshima, T. & Rubia, F. (1975) Prolongation of hippocampal inhibitory post-synaptic potentials by barbiturates. *Nature (Lond.)* **258**, 625-627.

Oshima, E. and Richards, C.D. (1988) An *in vitro* investigation of the action of ketamine on excitatory synaptic transmission in the hippocampus of the guinea-pig. *Eur. J. Pharmac.* **148**, 25-33.

Perouansky, M., Baranov, D., Salman, M. & Yaari, Y. (1995) Effects of halothane on glutamate receptor mediated excitatory post-synaptic currents : A patch-clamp study in adult mouse hippocampal slices. *Anaesthesiology*, **83**, 109-119.

Pocock, G. and Richards, C.D.(1987) The action of pentobarbitone on stimulus-secretion coupling in adrenal chromaffin cells. *British Journal of Pharmacology* **90**, 71-80.

Pocock, G. and Richards, C.D.(1988) The action of volatile anaesthetics on stimulus-secretion coupling in bovine adrenal chromaffin cells. *Br. J. Pharmac.* **95**, 209-217.

Pocock, G. and Richards, C.D.(1991) Anesthetic action on stimulus-secretion coupling. *Ann. N. Y. Acad. Sci.* **625**, 71-81.

Potashner, S,J., Lake, N., Langlois, E.A., Ploffe, L. & Lecavallier, D. (1980) Pentobarbital : differential effects on amino acid transmitter release. pp. 469-472 In :

Molecular Mechanisms of Anesthesia, Progress in Anesthesiology Vol. 2. Ed. B.R. Fink. Raven Press, New York.

Richards, C.D. (1972). On the mechanism of barbiturate anaesthesia. *J. Physiol. (Lond.)*, **227**, 749767.

Richards, C.D. (1972). On the mechanism of halothane anaesthesia. *J. Physiol. (Lond).*, **233**, 439456..

Richards, C.D., Russell W.J. & Smaje, J.C. (1975) The action of ether and methoxyflurane on synaptic transmission in isolated preparations of the mammalian cortex. *J. Physiol. (Lond.)*, **248**, 121-142.

Richards, C.D. and Smaje, J.C. (1976). Anaesthetics depress the sensitivity of cortical neurones to lglutamate. *Br. J. Pharmac.* **58**, 347357.

Richards C.D. & Strupinski, K. (1986) An analysis of the action of pentobarbitone on the excitatory post-synaptic potentials and membrane properties of neurones in the guinea-pig olfactory cortex. *Br. J. Pharmac.* **89**, 321-325.

Richards, C.D. and White, A.E.(1975) The actions of volatile anaesthetics on synaptic transmission in the dentate gyrus. *J. Physiol. (Lond.)* **252**, 241-257.

Sawada, S. and Yamamoto, C. (1985). Blocking action of pentobarbital on receptors for excitatory amino acids in the guineapig hippocampus. *Exp. Brain Res.* **59**, 226231.

Schlame, M. & Hemmings, H.C. (1995) Inhibition by volatile anaesthetics of endogenous glutamate release from synaptosomes by a presynaptic mechanism. *Anesthesiology*. **82**, 1406-1416.

Sherrington, C.S. (1906) *The Integrative Action of the Nervous System*. Yale University Press, New Haven.

Somjen, G.G. (1963) Effects of thiopental on spinal presynaptic terminals. *J.Pharm. Exp. Ther.* **140**, 396-402.

Somjen, G.G. & Gill, M. (1963) The mechanism of blockade of synaptic transmission in the mammalian spinal cord by diethyl ether and thiopental. *J.Pharm Exp. Ther.* **140**, 19-30.

Takahara, H., Wada, M., Arita, K. Sumikawa, K. & Izumi, F. (1986) Ketamine inhibits $^{45}Ca^{2+}$ influx and catecholamine secretion by inhibiting $^{22}Na^{+}$ influx in cultured bovine adrenal medullary cells. *Eur. J. Pharmac.* **125**, 217-224.

Uchitel, O.D., Protti, D.A., Sanchez, V. Chersky, B.D. Sugimori, M. & Llinas, R. (1992) Voltage-dependent calcium channel mediates presynaptic calcium influx and transmitter release in mammalian synapses. *Proc. Natl. Acad. Sci. USA.* **89**, 3330-3333.

Wachtel, R.E. (1995) Relative potencies of volatile anaesthetics in altering the kinetics of ion channels in BC3H1 cells. *J. Pharm. Exp. Ther.* **274**, 1355-1361.

Weakly, J.N. (1969) Effect of barbiturates on "quantal" synaptic transmission in spinal motoneurones. *J. Physiol. (Lond.)* **204**, 63-77.

Weight, F.F., Lovinger, D.M., White, G. and Peoples, R.W. (1991). Alcohol and anesthetic actions on excitatory amino acid activated ion channels. *Ann. N. Y. Acad. Sci.* **625**, 97107.

Werz, M.A. & Macdonald, R.L. (1985) Barbiturates decrease voltage-dependent calcium conductance of mouse neurones in dissociated cell culture. *Molec. Pharmac.* **28**, 269-277.

Yang, J. and Zorumski, C.F. (1991). Effects of isoflurane on NmethylDaspartate gated ion channels in cultured rat hippocampal neurons. *Ann. N. Y. Acad. Sci.* **625**, 287289.

Yeh, J.Z., Quandt, F.N., Tanguy, J., Nakahiro, M., Narahashi, T. & Brunner, E.A.

(1991) General anesthetic action on gamma-amino butyric acid-actiavted channels. *Ann N. Y. Acad. Sci.* **625**, 155-173.

Yoshimura, M., Higashi, H., Fujita, S. & Shimoji, K. (1985) Selective depression of hippocampal post-synaptic potentials and spontaneous firing by volatile anaesthetics. *Brain Research* **340**, 363-368.

Zorychta, E. & Capek, R. (1978) Depression of spinal monosynaptric transmission by diethyl ether : Quantal analysis of unitary synaptic potentials. *J. Pharm. Exp. Ther.* **207**, 825-836.

The Actions of Anaesthetics on Voltage-gated and Voltage-dependent Ion Channels

Ken T. Wann

WELSH SCHOOL OF PHARMACY, UNIVERSITY OF WALES
CARDIFF, UK

Introduction

The actions of all anaesthetics are expressed through their effects on the many membrane bound (integral) proteins of excitable cells, and general anaesthesia is clearly the consequence of the perturbation of a number of these integral proteins in key neurones in the central nervous system. Voltage-gated or voltage-dependent ion channel proteins clearly belong to the large family of proteins that are sensitive to a wide range of anaesthetic agents. What is certain is that the function of the major classes of these ion channels is modified by the many classes of anaesthetic agent and that this has profound implications for the operation of neural circuits. What is less clear is whether the interaction of the anaesthetic molecules with such target channel proteins is a direct, rather than an indirect action (e.g. through the membrane lipid), and whether the many effects on these ion channels are central to the production of anaesthesia. Indeed such issues remain hotly debated (Franks and Lieb, 1993; Elliott and Urban, 1995). In this chapter I consider firstly the nature of voltage-gated or voltage-dependent channels, and secondly some aspects of the action of anaesthetics on such ion channels. Different anaesthetics of course have different actions on these ion channels, but here an attempt is made to select a number of 'representative' effects for discussion.

Ion Channels

Voltage-gated or voltage-dependent ion channels are a diverse group of membrane proteins with an equally varied range of jobs to do. In the neurones of the central nervous system (CNS) each subtype of ion channel presumably has a different, and indeed important, role. These include controlling the speed of depolarisation of the action potential, shaping of the action potential and the after-hyperpolarisations following a single spike or spike train, regulating firing patterns, setting the resting membrane potential, and modulating neurotransmitter secretion. These channels are often distributed non-uniformly within different regions of the CNS,and even within separate membrane regions of the same cell. For example, channels can distribute preferentially in the nerve terminal region rather than on somatic or dendritic membrane. Thus in principle the surrounding lipid domains for various channel types could be subtly different, and this could have consequences for anaesthetic susceptibility. It would seem likely that there are diseases (eg the epilepsies, Alzheimer's, stroke) where either dysfunction of ion channels of the CNS occur, or where appropriate intervention at the level of such channels can have therapeutic benefit. Ion channels within the CNS therefore present potentially interesting targets for drugs (other than anaesthetics) with wide ranging clinical utility.

Our knowledge of voltage-gated and voltage-dependent ion channels has expanded rapidly in the last fifteen years or so, these advances stemming from the application of a range of new and diverse techniques. Thus in the field of electrophysiology the elegant and powerful patch-clamp technique was developed, in the area of molecular biology methods such as expression cloning have been introduced, and on the pharmacological side a range of selective toxins have proved to be potent analytical tools. The major classes of ion channel are first classified on the basis of the principal permeant species eg Na^+, K^+, or Ca^{2+} ions. It turns out that there are also many subclasses of each type of Na^+, K^+ and Ca^{2+} ion channel, so that it is now necessary to provide in each case some kind of channel taxonomy.

Na^+ Channels

Neuronal voltage-gated Na^+ channels permit the influx of Na^+ ions down a steep electrochemical gradient into the cell following depolarisation of a neurone to a 'threshold' level. The opening of Na^+ channels in response to depolarisation leads to the development of an inward membrane current which can be measured under voltage-clamp conditions (Barchi, 1988). To a first approximation the higher the density of Na^+ channels, the higher the peak depolarisation rate of the action potential, and the higher the conduction velocity (see Hodgkin, 1975). Any reduction in the number of functional Na^+ channels will result therefore in compromised nerve conduction.

Subtypes:

A number of independent pieces of evidence indicate that there is more than one subtype of Na^+channel. Firstly, there are clear differences in Na^+ channel properties in different tissues, and between immature and mature cells. This is evident from the wide variation in sensitivity of Na^+channels to the specific toxin tetrodotoxin (for references see Barchi, 1987). Secondly, and importantly, different subtypes of voltage-gated Na^+channel can co-exist in the CNS and indeed one neurone. These can be distinguished both on the basis of electrophysiological criteria, such as the voltage range over which the activation of channels occurs, or alternatively the rate of inactivation of the channels. The voltage-gated Na^+ channels which control the peak depolarisation rate open, or are activated, by relatively strong depolarisation and close (or inactivate) rapidly. However, there is another population of Na^+ channels which are activated by weak depolarisation and inactivate more slowly (eg French and Gage, 1985; Ogata and Tatebayashi,1990). The role of such Na^+ channels must be to contribute to the triggering of the action potential mechanism. Additionally, cloning experiments have identified in the rat CNS three Na^+ channel messenger RNA which exist simultaneously (Noda et al., 1986). These channel messages show strong (but not identical) sequence homology and the cDNA and the primary sequence has been obtained for each respective Na^+ channel. These three Na^+ channel subclasses therefore co-exist in the brain, and perhaps in the same neurone.

Subunit composition:
Early biochemical studies of the Na^+ channel relied on the high specificity and the high affinity of the ligand TTX, and a rich source of the channel (eg from the eel electroplaque). Equivalent purification of Na^+ channels was then achieved with mammalian systems. Such studies demonstrated that the principal subunit (α) is a glycoprotein Mwt 260 kDa with smaller associated (β) subunits (Agnew, 1984; Barchi,1988). If the α subunit is incorporated into a lipid bilayer membrane then functional Na^+ channels are observed (Rosenberg et al.,1984) thus indicating that the pore domain of the channel resides in this subunit. Advances in molecular biology mean that now we can make mRNA from the cDNA that codes for the α subunit of the Na^+ channel and inject it into an expression system (eg the oocytes of *Xenopus laevis*). It turns out that the Na^+ current recorded under voltage-clamp has characteristics which to a first approximation resemble that recorded from the native Na^+ channel (Stuhmer et al., 1987). This confirms that functional Na^+ channels can be formed from the α subunit alone.

What has become clearer in the last few years is the important role that associated subunits (e.g. the β subunits) have in the function of Na^+ and other voltage-gated channels (Isom et al., 1994). These subunits, when co-expressed with the principal α subunit, have profound effects on the activation or inactivation properties of the current generated by the α subunit, or indeed may modify the modulation of the current by pharmacological agents including anaesthetics (Makielski et al.,1996). In the case of the Na^+ channel there are two auxiliary subunits β_1 and β_2. The β_2 subunit is covalently attached by means of a disulphide linkage, the β_1 subunit is noncovalently associated (stoichiometry1:1:1). Co-expression of these subunits with the α subunit results in a Na^+ channel with a more rapid inactivation process.

The proposed topology of the amino acid chains of the α subunit and associated β subunits has been described in detail elsewhere (Noda et al., 1984; Catterall,1991; Hille,1992; Wann,1993; Isom et al., 1994). In as much as it is possible to assign functions to domains of amino acids within loops, and even to specific amino acids, it should eventually be possible to determine which residues are critical for the binding of anaesthetics and indeed experiments of this kind are already underway with the local anaesthetics (Ragsdale et al., 1996).

Actions of anaesthetics:
The targets for local anaesthetics are clearly the voltage-gated Na^+ channels which support nerve conduction and there are useful lessons to be learnt from the interaction between such agents and their receptor. The action of local anaesthetics has been discussed previously (eg Wann,1993; see also Strichartz,1973; Courtney, 1975; Schwartz et al., 1977) and is therefore only briefly discussed here. One important aspect of the block achieved by such agents is that in some cases access (from the cytoplasmic side) to the binding site within the Na^+ channel is via the open gate, ie a hydrophilic route. As a consequence of this, the degree of block is dependent on the frequency of opening of the channel. This is known as use-dependent or frequency-dependent block and is achieved most effectively when the drug does not unbind when the channel is closed. Is general anaesthesia a special case of local anaesthesia ? This question is implicit in many of the investigations of anaesthetics with voltage-gated Na^+ channels, and a corollary of it would be that in principle it should be possible to induce general anaesthesia with a local anaesthetic - say lidocaine. The detailed effects of anaesthetics on voltage-gated Na^+ channels have largely been studied in peripheral (and in many cases invertebrate) nerve axons (Bean et al., 1981; Haydon and Urban,1983a,b,c; for references see Wann and Macdonald, 1988; Wann,1993). There is a notable exception to this in a study of Na^+ channels from human cortex (Frenkel and Urban, 1991). On the basis of the axon studies it has long been recognised that nerve conduction is somewhat insensitive to anaesthetics, in comparison to the process of synaptic transmission (see also Larrabee and Posternak, 1952), and that for this reason it should be ignored when considering the phenomenon of general anaesthesia. It is worth noting that this is not a universally held view however (Elliott and Urban,1995). It could be argued that in the central nervous system of a mammalian species Na^+channels have a different susceptibility, but at present there is little evidence to pursue this further (although see below). The argument that synaptic transmission is only sensitive by virtue of nerve conduction in the fine nerve terminals being more sensitive to anaesthetics is also improbable. Thus Richards (1982) showed that in the guinea-pig olfactory tract the block achieved by pentobarbitone, at a concentration which affects synaptic transmission only, is dissimilar to that achieved by a local anaesthetic or tetrodotoxin, which almost by definition affect only voltage-gated Na^+ channels. In general then we must assume for the present that voltage-gated Na^+ channels are only affected at concentrations of anaesthetic in excess of that required to induce general anaesthesia (but see Elliott and Urban,1995).

We will now consider some of the principal effects of anaesthetics on the Na^+ current. All classes of anaesthetics (at some concentration) depress reversibly the macroscopic currents carried by voltage-gated Na^+ channels. However, this depression is not always achieved in the same way. There are clear differences in the mechanisms of action of the various classes of agent. This is best illustrated by summarising a systematic series of experiments from one laboratory. For this reason we will now consider data from the squid giant axon (Haydon and Urban,1983a,b,c; Urban,1993; Elliott and Elliott,1994). In brief, these studies showed that anaesthetics shift the activation of the Na^+ conductance to more positive potentials, and reduce the maximum value attained. They also shift the voltage range over which inactivation is switched on to more negative values, and speed up the rate of switching on of inactivation. The resultant effect is to reduce the macroscopic Na^+ current but what is more interesting is that hydrocarbons (apolar agents) such as cyclopropane predominantly affect the relationship between inactivation and voltage, and the maximum Na^+ conductance. In contrast, the alcohols and other polar agents have comparatively little effect on the maximum Na^+ conductance and preferentially shift the activation curve for the Na^+ conductance. The inhalation anaesthetics produced effects (depending on the agent) which were similar to that of alcohols (eg relatively small effects on maximum Na^+ conductance), or the hydrocarbons (eg relatively large shift in the inactivation curve). These multiple actions have been ascribed to the tendency of the apolar agents (hydrocarbons) to partition in the membrane interior and the polar agents (alcohols) to locate preferentially at a surface level. It is argued that the inhalation agents have intermediate physico-chemical properties. The proposed changes brought about following introduction of the anaesthetic are membrane thickening (eg hydrocarbons) or changes in surface potential (eg alcohols). Where increases in the rate of membrane processes occur (eg the speed of inactivation), it is proposed that fluidising actions are responsible (see Urban,1993). Whatever the mechanism, the conclusion must be that different anaesthetic classes affect different factors controlling the amplitude and time course of the macroscopic Na^+ current. Turning to the question of the concentrations required to induce such changes, it would seem that these investigations used concentrations of inhalation agents ranging from more than twice 'clinical' levels to values approaching an order of magnitude higher than the EC_{50} for general anaesthesia (see also Bean et al., 1981). In contrast in a more recent study with a reconstituted human Na^+ channel preparation propofol was effective at concentrations close to those required for general anaesthesia (Frenkel and Urban, 1991).

As indicated previously there is more than one type of Na^+ channel in the mammalian CNS, and thus far the actions of anaesthetics have been studied only on the channel subtype which supports peripheral nerve conduction (although see Frenkel et al., 1993; Fujiwara et al., 1988). Neuronal signalling in the CNS depends also on the operation of threshold Na^+ channels which contribute to the 'excitability state' of neurones. If such channels exist at postsynaptic membrane sites (somatic or dendritic membranes), then they would have a crucial role in determining postsynaptic excitability and the integration of incoming depolarising influences. Their behaviour would dictate, at least in part, the input-output relationship at synapses. In this sense they would be involved in the 'decision making of the cell' - to fire or not to fire (an action potential). Their role is therefore unlike that of voltage-gated Na^+ channel in the nerve trunks which determines solely the rate of delivery of a neuronal action potential. In other words 'The axon doesn't think. It only ax' (Bishop,1965). Such central mammalian voltage-gated Na^+ channel targets might be worth focusing on in future studies. At present however we can conclude that there are no compelling reasons to implicate modulation of Na^+ channels in the production of general anaesthesia.

Ca^{2+} Channels

The functions of neuronal Ca^{2+} channels are 1) to carry inward depolarising current and 2) to deliver Ca^{2+} to the intracellular fluid where it has a number of important roles which include the the triggering of Ca^{2+} release from internal stores, activating secretion of transmitters, activating ion channels and modulating enzyme function (Henzi and MacDermott,1992; Li and Hatton, 1997). The downside of Ca^{2+} entry is that an overload of intracellular Ca^{2+} results in cytotoxicity and cell death. The action of general anaesthetics on intracellular Ca^+ regulation has been reviewed by Kress and Tas (1993). The dynamics of intracellular Ca^{2+} concentration changes can be followed using quantitative fluorescent probes (Tsien, 1989) and optical imaging methods (see eg Denk et al., 1996) .

Subtypes:
A number of important subtypes of Ca^{2+} channel exist to deliver $Ca^{2+}$to the intracellular fluid. We can distinguish these on the basis of voltage sensitivity, pharmacological criteria such as the sensitivity to dihydropyridines and various neurotoxins, and finally subunit composition (Reuter, 1996). The principal forms are the L (non-inactivating), P (originally found in Purkinje cells), N (intermediate voltage sensitivity), T (carrying inactivating 'transient' or 'threshold' current), Q and R subtypes. These can presumably co-exist in the same neurone and mapping studies have been conducted with either patch-clamp methods, where the surface of the neuronal membrane is explored with a patch pipette, or using antipeptide antibodies to define the localisation of the channel subtypes. Such studies indicate, for example, that the L-type resides predominantly in the soma and proximal dendrites. In contrast N, P and Q type channels may 'live' predominantly in the membranes of nerve terminals which release excitatory transmitter. Pharmacological studies on exocytosis are consistent with this view. Thus transmitter release is inhibited by agents, such as ω-Conotoxin GVIA and ω-Agatoxin IVA, which are specific for these channel subtypes (for references see Reuter,1996; Smith and Cunnane,1997). The present view is that the transmitter release process in a number of central excitatory synapses depends on the normal operation of N-, P- and Q-type Ca^{2+} channels. In contrast a recent study shows that GABA release at a central inhibitory nerve terminal is dependent on either N or P alone (Poncer et al., 1997).

Subunit composition:
The Ca^{2+} channel pore resides, as is the case for the Na^+ channel, in the α_1 subunit. The neuronal Ca^{2+} channel α_1 subunits are thought to be the product of at least six genes, of which five have been cloned. The amino acids are

thought to be arranged in four repeats each of six transmembrane domains as in the case of the Na^+channels. What makes the pore Ca^{2+} selective is the presence of glutamate residues at positions occupied by lysine and alanine in the case of the Na^+ channel (Heinemann et al., 1992). Ca^{2+} channels are heteromeric proteins consisting of additional β and $\alpha_2\delta$ subunits. Here again the auxiliary subunits have regulatory functions and when these subunits are co-expressed with the α_1 subunit then the inactivation of the macroscopic current is speeded up (Isom etal., 1994).

Actions of anaesthetics:
Neuronal voltage-gated Ca^{2+} channels seem more sensitive to anaesthetics than Na^+channels although surprisingly there are considerably fewer studies available for comparison. Herrington et al. (1991) have shown that in a pituitary cell line both the L- and T-type Ca^{2+} channels are depressed reversibly by halothane at a concentration that does not affect the Na^+current in these cells. The T-type current was much less sensitive than the L-type (1.3 versus 0.8 mM). There was no obvious change in the voltage-dependence of the activation or inactivation of either current subtype, but in the case of the T-type there were changes in the kinetics of the current, ie both the rates of activation and inactivation were accelerated. The conclusion from this study is that halothane at least can depress the L-type Ca^{2+} channel at near clinically relevant concentrations. ffrench-Mullen et al. (1993) reported that in guinea pig hippocampal neurones (-)-pentobarbitone was as effective (IC_{50} of 3.5 μM) in reducing the inward Ca^{2+} current (which consisted of more than one component) as in enhancing the GABA-Cl^- current. Notably the peak and late components of the inward current were almost equally affected by the barbiturate showing that the early component of Ca^{2+} entry (which is the relevant fraction for triggering transmitter release) is exquisitely sensitive (Figure 1). Charlesworth et al.(1994) studied a fuller range of anaesthetics on macroscopic Ca^{2+} currents in chromaffin cells. Pharmacological dissection shows that such currents are probably carried by both L- and N-type Ca^{2+}channels. All anaesthetics studied (halothane, methoxyflurane, etomidate and methohexitone) decreased the macroscopic Ca^{2+}current peak amplitude and enhanced its rate of decay. These agents were not however equieffective. The conclusion was that clinical concentrations of halothane and etomidate (0.4 mM and 0.08 mM respectively) had little effect on the macroscopic Ca^{2+}current, but that methoxyflurane and methohexitone (0.4 mM and 0.08 mM respectively) would reduce the (total inward) Ca^{2+} current by about 20-30%. From the above studies using barbiturates we might tentatively conclude that such agents have mild (methohexitone) to dramatic (pentobarbtione) depressant actions on voltage-gated Ca^{2+} channels at clinical concentrations. Many anaesthetics depress the release of transmitter and it would be reasonable to propose that this is due to a reduction of Ca^{2+}entry through voltage-gated Ca^{2+}channels (for arguments see eg Pocock and Richards,1993). As indicated above, these presynaptic Ca^{2+}channels are likely to be of the N, P and Q subtypes. Hall et al (1994) have examined the action of both gaseous

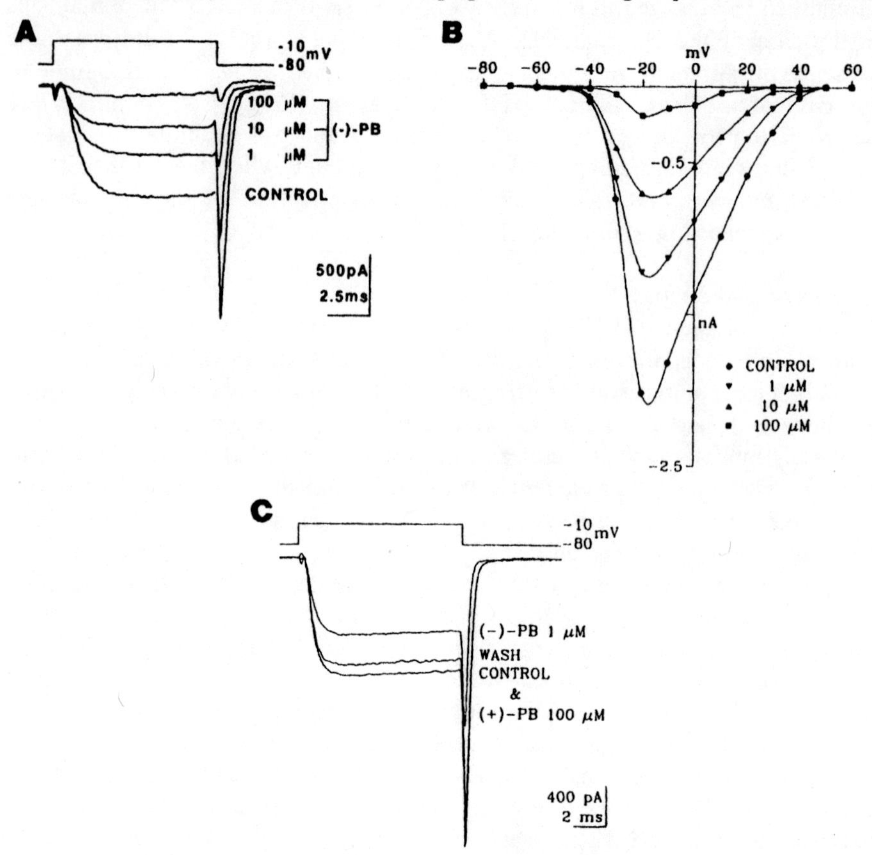

Figure 1 *The action of (-)-pentobarbitone on whole cell Ba^{2+} currents through Ca^{2+} channels recorded from an acutely dissociated guinea pig hippocampal neurone. A, The whole cell inward current elicited by depolarising from -80 mV to -10 mV (top trace) in control medium and in the presence of 1, 10 and 100 μM pentobarbitone. B, The current-voltage relationship for the peak inward currents in control medium and in the presence of increasing concentrations of pentobarbitone. C, A comparison of the effects of (+)- and (-)-pentobarbitone on the inward Ba^{2+} current. Note the concentration difference. Temperature 19-25 °C. (Reproduced by permission from ffrench-Mullen et al., 1993).*

(eg halothane and isoflurane) and intravenous (eg thiopentone, pentobarbitone and propofol) agents on P-type macroscopic currents in rat cerebellar Purkinje neurones. This study showed that at relevant clinical concentrations these agents (including a barbiturate) had only small effects (<10%) on the characteristics of the P-type current, and the authors concluded that these channels, unlike ligand-gated channels, were therefore not likely to be involved in producing general anaesthesia. In summary, the evidence for the involvement of voltage-gated Ca^{2+}channels in general anaesthesia remains unconvincing for most agents tested, the possible exception to this being some barbiturates.

K^+ Channels

The neuronal K^+ channels are a superfamily of ion channels and are the most diverse in terms of both structure and function. Their functions include setting the resting membrane potential, shaping the action potential by determining the rate of repolarisation, controlling the duration and amplitude of after-hyperpolarisations following an action potential or train of spikes, regulating firing frequency and the pattern of adaptation of action potential discharge. As a consequence of their varied functions, we would expect that modulation of K^+ channels by anaesthetics would have a number of implications for neuronal activity. The net effect of the anaesthetic will depend on the location of the target K^+ channel in the CNS. Because of the detailed sequence information now available for many K^+ channels, it is possible to raise antibodies against K^+ channel sequences and also to conduct *in situ* hybridisation experiments. In this way we can map the distribution of K^+ channel subclasses in the CNS. It is clear that there are interesting regional and cellular variations in the distribution of K^+ channel proteins. For example, K^+ channels are non-uniformly expressed within different membrane domains of the one neurone. Experiments have investigated the distribution in a number of brain regions (eg cortex, hippocampus, cerebellum) of members of the Shaker (eg Kv1.4 and Kv1.2), Shab (eg DRK1), Shaw (eg Kv3.4), Shal subfamilies (eg Kv4.2), and *slo poke* (Sheng et al., 1992; Sheng et al.,1993;Hwang et al.,1993; Roeper and Pongs, 1996). Such studies conclude, for example, that A-type channels, which carry a transient outward current, are expressed preferentially in nerve terminals and axons of the hippocampus (Sheng et al.,1992). However, this may not be the case for all nerve terminals, the precise complement of K^+channel subunits expressed presynaptically being dependent on which brain area is being studied (see Roeper and Pongs,1996).

Subtypes:

Some members of the family are voltage-gated, alternatively others are voltage-dependent being activated by other means eg a rise in intracellular Ca^{2+}. Many members of the family have now been cloned and on the basis of the molecular biology, we can divide these channels into three principal classes. These classes have respectively six, two and one membrane spanning domains. The K^+ channels with α subunits posessing six transmembrane segments belong to what is referred to as the S4 superfamily and in this category are included the classical delayed rectifier, A-type K^+ channel and the Ca^{2+}-activated K^+ channels. The K^+ channel proteins of the S4 superfamily resemble one of the four repeats of the Na^+ or Ca^{2+} voltage-gated channels and the current view is that at least *in vitro* a functional K^+ channel is a tetramer. In the case of the principal subunit of the inward rectifier K^+channel and the ATP-inhibited type of K^+ channel (K_{ATP}) there are only two transmembrane hydrophobic regions flanking the pore region, the latter being equivalent to the SS1-SS2 of the Na^+ and other classes of K^+ channels. The members of this K^+channel class therefore have a simpler membrane topology.

Voltage- and K^+-dependent gating in inward rectifiers depends on positively charged cytoplasmic factors such as Mg^{2+} ions (Matsuda, 1988: Stanfield et al., 1994) and polyamines (Lopatin et al., 1994). Additionally, the M2 spanning region carries an aspartate residue which is thought to be involved in the intrinsic gating mechanism of inward rectifiers. Much is still yet to be learned about the members of this K^+ channel family, including the precise channel subunit stoichiometry. The third class of channel possessing only one membrane spanning domain is referred to as the 'minK' (minimal K^+ channel). When expressed in oocytes it carries a slowly activating sustained outward current (Zorn et al. 1993). Members of the S4 and the inward rectifier families can be modulated by a number of important classes of neurotransmitter.

Subunit composition:

As in the case of the voltage-gated Na^+ and K^+ channels the α subunit possesses the SS1-SS2 pore region. An elegant set of experiments showed that the location of the fast inactivation machinery (N-type) in the Shaker K^+ channel (a clone originally from Drosophila) is in the chain of 20 amino acids at the NH_2 terminus (Hoshi et al., 1990; Zagotta et al., 1990). Here there are positively charged lysine and arginine residues which are critical determinants of N-type inactivation. Removal of these slows the fast inactivation process. The receptor for the terminus resides in an intracellular loop (acidic residues) between S4 and S5. The ball region can essentially be thought of as an open channel blocker. In addition to the N-type of inactivation machinery residing in the NH_2 terminus there is a C-type inactivation (see Latorre and Labarca,1996) which is a slower process dependent on residues in the S5, pore and S6 domains. Interestingly, anaesthetics such as halothane and ketamine appear to accelerate inactivation of recombinant K^+ channels by interacting with the amino acids in this cytoplasmic domain (Kulkarni et al., 1996). An intracellular auxiliary β subunit has an additional role in modulating the inactivation process of K^+ channels. A dramatic increase in inactivation rate is observed when α and β subunits are co-expressed (Rettig et al.,1994; Isom et al., 1994).These subunits can even it seems convert non-inactivating into rapidly inactivating K^+ currents.

Actions of anaesthetics:

The effects of anaesthetics have been studied principally on the S4 class of K^+ channels. There is one study on the action of halothane on the minK channel which shows that this channel is extremely sensitive to inhibition ($IC_{50} < 0.34$ mM, Zorn et al.,1993) The implications of this result for neuronal function however remain unclear. Neuronal K^+ channels exhibit a range of sensitivities to anaesthetics. Axonal delayed rectifier K^+ are even less sensitive to general anaesthetics than their voltage-gated Na^+ channel counterparts (Haydon and Urban,1986). In contrast, the non-inactivating K^+current of clonal pituitary cells is about five times more sensitive to halothane than the 'classical' delayed rectifier, being more sensitive than the voltage-gated Na^+ current of the same preparation (Herrington et al., 1991). Recombinant Kv2.1

channels and the K^+ channels controlling the slow AHP and adaptation of spike discharge seem to be the most sensitive studied to date (Fujiwara et al., 1988; Southan and Wann, 1992; Pearce, 1996; Kulkarni et al., 1997). The slow AHP and the accommodation of action potential discharge in mammalian central neurones depend on the operation of at least two classes of K^+ channel. One is the voltage-gated M-type K^+ channel (its hallmark is that is a non-inactivating outward current shut down by muscarinic receptor agonists). The other is a voltage-dependent, Ca^{2+}-activated K^+ channel of small conductance, ie the SK channel. A number of studies have shown that both the AHP and action potential accommodation in hippocampal neurones are decreased by halothane, isoflurane and enflurane at clinical concentrations, showing that the current through these K^+ channel subclasses is reduced by these anaesthetics (see Fujiwara et al., 1988; Southan and Wann, 1989; Pearce, 1996). As a consequence the firing pattern of these cells is altered from a phasic to tonic discharge, the number of spikes in the discharge continuing to increase in response to stronger depolarising inputs (Figure 2). In the presence of the anaesthetics neurones therefore more faithfully encode any depolarising inputs. Since these anaesthetics also hyperpolarise central neurones (see Southan and Wann,1989) the result is that the response of such neurones to weak depolarising inputs is *depressed.* In contrast since spike accommodation is reduced in the presence of the anaesthetics their response to strong inputs is *enhanced.* This tendency to ignore weak inputs and to amplify stronger inputs is induced in these same neurones by transmitters such as noradrenaline (Madison and Nicoll,1982) and is analagous to increasing the signal/noise ratio. The actions of one of these agents, ie enflurane, on the non-inactivating M-current has been studied under voltage-clamp conditions (Figure 3). A small depression was observed at close to the MAC level, ie 2.5% (Wann and Southan,1992). The reduction in the AHP would occur if the Ca^{2+} current in hippocampal neurones were depressed by the anaesthetics. However, the consensus is that the I_{AHP}, the current through the SK channel, is itself sensitive to the anaesthetics, there being little action on the Ca^{2+} influx which triggers the opening of this channel (see eg Pearce, 1996). In conclusion then it seems that the classes of K^+ channel that control adaptation and the AHP of central neurones are the most sensitive to some of the inhalational anaesthetic agents. It is curious that these same channels are also the most sensitive of the non ligand-gated channels to high pressure (Southan and Wann, 1996). In both cases the K^+ channels are blocked by these challenges, excitant effects resulting. It seems difficult to reconcile these excitant effects *in vitro* to the CNS depression we associate with general anaesthesia. It is more likely therefore that these effects on K^+ channel function contribute to the proconvulsant actions of some of these agents (Stevens et al., 1984; Reilly,1998 this volume).

Concluding Remarks

As a general conclusion, it is clear that a number of voltage-gated or voltage-dependent ion channels are affected by general anaesthetics. However, equally

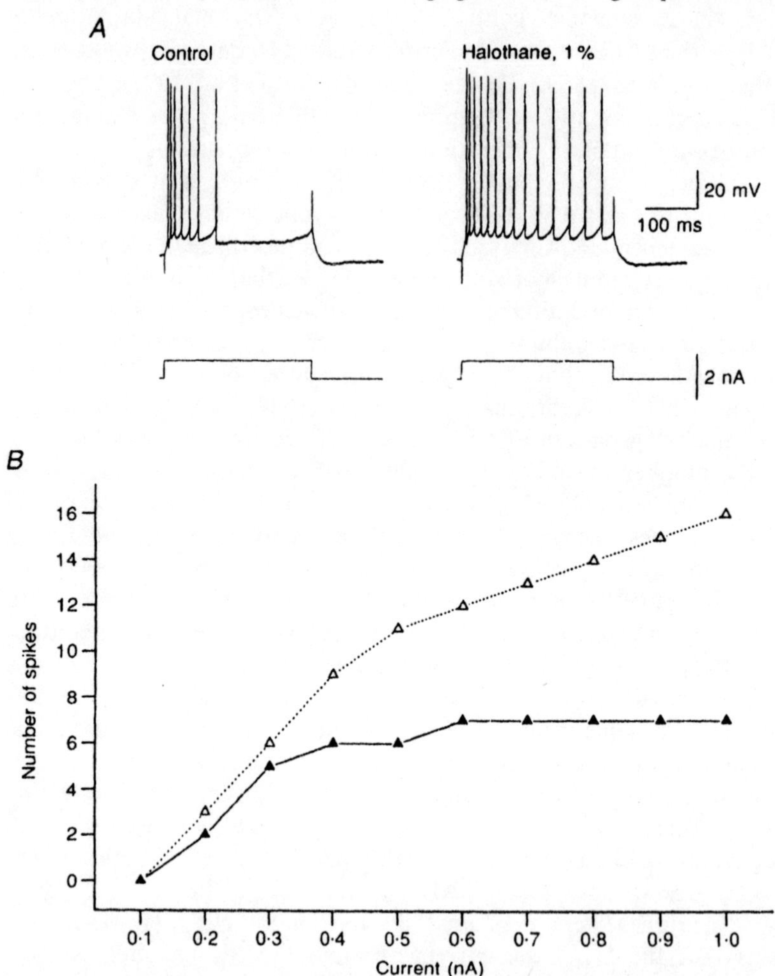

Figure 2 *The effect of 1% halothane on the accommodation of action potential discharge in a rat CA1 pyramidal neurone in a hippocampal brain slice. A, The response of a CA1 pyramidal neurone is shown to a depolarising current pulse of 0.8 nA in control (left panel) and in the presence of 1% halothane (right panel). B, The relationship between the number of action potentials in the train and the intensity of the depolarising current in control (Ù) and the presence (Δ) of 1% halothane. Temperature 36 °C. (Reproduced by permission from Pearce, 1996).*

evident is that only a few classes of ion channels are sensitive at, or near clinical concentrations. There are therefore no strong reasons at present to implicate many of these types of ion channels in the production of anaesthesia it seeming probable that their role is not a crucial one. In particular modulation of voltage-gated Na^+ channels would appear to contribute little to the induction of anaesthesia. Modulation of voltage-gated Ca^{2+} channels may contribute to the depression of transmitter by anaesthetics, and if this is relevant to the production of anaesthesia, then we would have to consider a role for subtypes

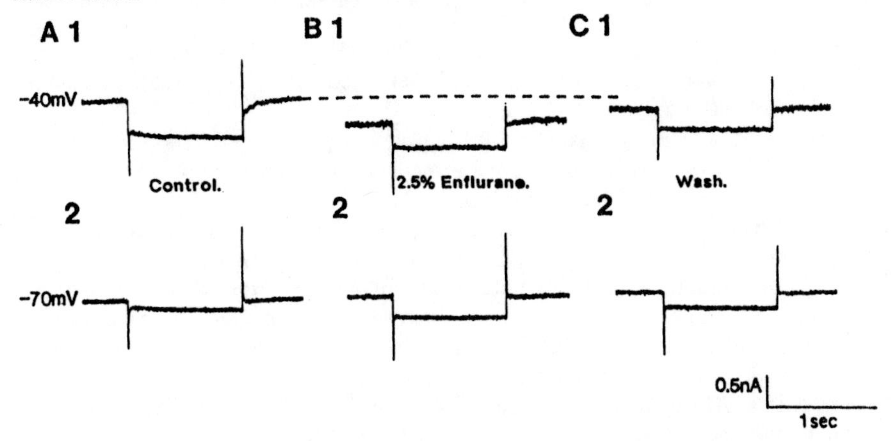

Figure 3 *The effect of enflurane (2.5%) on the M-current recorded in a rat hippocampal CA1 neurone in a brain slice. The M-current is a non-inactivating, voltage-gated K^+ current which is elicited by changing the membrane potential in control medium from -40 mV to - 55 mV (A1). An inward current relaxation is observed which is absent when stepping from -70 mV to -85 mV (A2). A1, control; B1, 2.5% enflurane; C1, recovery. Note the shift in steady state outward current (reduction in M-current) and the decline in the inward current relaxation in the presence of enflurane. A2,B2 and C2 show the corresponding data when stepping from -70 mV to -85 mV. Temperature 36°C. (Reproduced by permission from Wann and Southan, 1992).*

of these channels in generating the anaesthetic state. Although some subclasses of K^+ channel (either voltage-gated or voltage-dependent) are sensitive to some anaesthetics at relevant concentrations, an obvious role in anaesthesia remains unclear.

A clear role for any of the channels discussed here in anaesthesia remains to be found, but one conclusion is clear. The generation of anaesthesia is unlikely to be brought about by modulation of voltage-gated and voltage-dependent channels alone, ligand-gated channels being in many cases more sensitive to the anaesthetic challenge.

References

Agnew, W.S. (1984) Voltage-regulated sodium channel molecules. Ann. Rev. Physiol., 46, pp 517-530.

Barchi, R.L. (1988) Sodium channel diversity: subtle variations on a complex theme. TINS., 10, pp 221-223.

Barchi, R.L. (1988) Probing the molecular structure of the voltage dependent sodium channel. Ann. Rev. Neurosci.,11, pp 455-495.

Bean, B.P., Shrager, P. and Goldstein, D.A. (1981) Modification of sodium and potassium channel gating kinetics by ether and halothane. J. gen. Physiol.,77, pp233-253.

Bishop, G. H. (1965) My life among the axons. Ann. Rev. Physiol., 27, pp 1-8.

Catterall, W.A. (1991) Structure and function of voltage-gated sodium and calcium channels. Current Opinion in Neurobiology,1, pp5-13.

Charlesworth, P., Pocock, G. and Richards, C.D. (1994) Calcium channel currents in bovine adrenal chromaffin cells and their modulation by anaesthetic agents. J. Physiol.,481.3, pp 543-553.

Courtney, K.R. (1975) Mechanism of frequency dependent inhibition of sodium currents in frog myelinated nerve by the lidocaine derivative GEA 968. J. of Pharmacol. and Exp. Therap., 195, pp 225-236.

Denk, W.,Yuste, R., Svoboda, K.,Tank, D.W. (1996) Imaging calcium dynamics in dendritic spines. Current Opinion in Neurobiology, 6, pp 372-378.

Elliot,J.R. and Elliott, A.A. (1994) The effects of alcohols and other surface-active compounds on neuronal sodium channels. Prog. in Neurobiology, 42, pp 611-683.

Elliott, J.R. and Urban, B.W. (1995) Integrative effects of general anaesthetics: why nerve axons should not be ignored. Eur. J. Anaes.., 12, pp 41-50.

Franks, N.P. and Lieb, W. R. (1993) Selective actions of volatile general anaesthetics at molecular and cellular levels. B. J. Anaes., 71, pp 65-76.

French, C.R., Gage, P.W. (1985) A threshold sodium current in pyramidal cells in rat hippocampus. Neurosci. Letters., 56, pp 289-293.

Frenkel, C., Duch, D.S. and Urban, B.W. (1993) Effects of I.V. anaesthetics on human brain sodium channels. B. J. Anaes., 71, pp 15-24.

Frenkel, C., Urban, B.W. (1991) Human brain sodium channels as one of the molecular target sites for the new intravenous anaesthetic propofol (2,6-diisopropylphenol). Eur.J. Pharmacol. - Molecular Pharmacol. Section., 208, pp 75-79.

Fujiwara, N., Higashi, H., Shimoji, K., Sugita, S. and Yoshimura, M. (1988) Changes in spontaneous firing patterns of rat hippocampal neurones induced by volatile anaesthetics. J. Physiol., 402, pp 155-175.

Hall, A.C., Lieb, W.R., Franks, N. P. (1994) Insensitivity of P-type calcium channels to inhalational and intravenous general anaesthetics. Anesthesiology, 81, pp 117-123.

Haydon, D.A., Urban, B.W. (1983a) The action of hydrocarbons and carbon tetrachloride on the sodium current of the squid giant axon. J. Physiol., 338, pp 435-450.

Haydon, D.A., Urban, B.W. (1983b) The action of alcohols and other non-ionic surface active substances on the sodium current of the squid giant axon. J. Physiol., 341, pp 411-427.

Haydon, D.A., Urban, B.W. (1983c) The effects of some inhalation anaesthetics on the sodium current of the squid giant axon. J. Physiol., 341, pp 429-439.

Haydon, D.A., Urban, B.W. (1986) The actions of some general anaesthetics on the potassium current of the squid giant axon. J.Physiol., 373, pp 311-327.

Heinemann, S.H.,Terlau, H., Stuhmer, W., Imoto, K. and Numa, S. (1992) Calcium channel characteristics conferred on the sodium channel by single mutations. Nature., 356, pp 441-443.

Henzi, V. and MacDermott, A. B. (1992) Characteristics and function of Ca^{2+} - and inositol 1,4,5-triphosphate-releasable stores of Ca^{2+} in neurons. Neurosci., 46, pp 251-273.

Herrington, J., Stern, R.C., Evers, A.S. and Lingle, C.J. (1991) Halothane inhibits two components of calcium current in clonal (GH_3) pituitary cells. J.Neurosci., 11(7), pp 2226-2240.

Hille, B. (1992) Ionic Channels of Excitable Membranes. 2nd Edn. Sinauer Associates Inc.

Hodgkin, Sir A. (1975) The optimum density of sodium channels in an unmyelinated nerve. Phil. Trans. R. Soc. Lon. B., 270, pp297-300.,

Hoshi, T., Zagotta, W.N. and Aldrich, R.W. (1990) Biophysical and molecular mechanisms of Shaker potassium channel inactivation. Science, 250, pp 533-538.

Hwang, P. M., Fotuhi, M., Bredt, D.S., Cunningham, A.M. and Snyder, S.H. (1993)

Contrasting immunohistochemical localizations in rat brain of two novel K^+ channels of the *Shab* subfamily. J. Neuroscience, 13(4), pp 1569-1576.

Isom, L.L., De Jongh, K.S. and Catterall, W.A. (1994) Auxiliary subunits of voltage-gated ion channels. Neuron., 12, pp 1183-1194.

Jarlath, M.H. ffrench-Mullen, Barker, J.L. and Rogawski, M.A. (1993) Calcium current block by (-)-Pentobarbital, Phenobarbital, and CHEB but not (+)-Pentobarbital in acutely isolated hippocampal CA1 neurons:Comparison with effects on GABA-activated Cl^- current. J. Neurosci.,13(8), pp 3211-3221.

Kress,H.G. and Tas, P.W.L. (1993) Effects of volatile anaesthetics on second messenger Ca^{2+} in neurons and non-muscular cells. B. J. Anaes., 71, pp 47-58.

Kulkarni, R.S., Zorn, L.J., Anatharam, V., Bayley, H. and Treistman, S.N. (1996) Inhibitory effects of Ketamine and halothane on recombinant potassium channels from mammalian brain. Anesthesiology,84(4), pp 900-9.

Larrabee, M.G. and Posternak, J.M. (1952) Selective action of anaesthetics on synapses and axons in mammalian sympathetic ganglia. J.Neurophysiol., 15, pp 92-114.

Latorre, R. and Labarca, P. (1996) Potassium channels: Diversity, Assembly and Differential Expression. In 'Potassium channels and their modulators', chapter 6, pp 123-156. Taylor and Francis, eds Evans, J.M., Hamilton, T.C, Longman, S.D. and Stemp, G.

Li, Z. and Hatton, G.I. (1997) Ca^{2+} release from internal stores: role in generating depolarizing after-potentials in rat supraoptic neurons. J. Physiol., 498, pp 339-350.

Lopatin, A. N., Makhina, E.N. and Nichols, C.G. (1994) Potassium channel block by cytoplasmic polyamines as the mechanism of intrinsic rectification. Nature., 372, pp 366-369.

Madison, D.V. and Nicoll, R.A. (1982) Noradrenaline blocks accomodation of pyramidal cell discharge in the hippocampus. Nature., 299, pp 636-638.

Makielski, J.C., Limberis, J.T., Chang, S.Y., Fan, Z. and Kyle, J.W. (1996) Coexpression of $\beta 1$ with cardiac sodium channel α subunits in oocytes decreases lidocaine block. Molecular Pharmacol., 49. pp 30-39.

Matsuda, H. (1988) Open-state substructure of inwardly rectifying potassium channels revealed by magnesium block in guinea-pig heart cells. J. Physiol., 397, pp 237-258.

Noda, M., Ikeda, T., Kayano, T., Suzuki, H. and Takeshima, H. (1986) Existence of distinct sodium channel messenger RNAs in rat brain. Nature,320, pp 188-192.

Noda, M., Shimizu, S., Tanabe, T., Takai, T. and Kayano, T. (1984) Primary structure of *Electrophorus electricus* sodium channel deduced from cDNA sequence. Nature, 312, pp 121-127.

Ogata, N, and Tatebayashi, H. (1990) Sodium current kinetics in freshly isolated neostriatal neurones of the adult guinea pig. Pflugers Archiv., 416, pp 594- 603.

Pearce, R.A. (1996) Volatile anaesthetic enhancement of paired-pulse depression investigated in the rat hippocampus *in vitro.* J. Physiol.,492.3, pp 823-840.

Pocock, G. and Richards, C.D. (1993) Excitatory and inhibitory synaptic mechanisms in anaesthesia. B. J. Anaes., 71, pp 134-147.

Poncer, J.C., McKinney, R.A.,Gahwiler, B.H. and Thompson S.M. (1997) Either N- or P- type calcium channels mediate GABA release at distinct hippocampal inhibitory synapses. Neuron., 18, pp 463-472.

Ragsdale, D.S., McPhee, J.C., Scheuer, T. and Catterall, W.A. (1996) Common molecular determinants of local anaesthetic, antiarrhythmic, and anticonvulsant block of voltage-gated Na^+ channels. Proceedings Nat. Acad. Sci., 93, pp 9270-5.

Reuter, H. (1996) Diversity and function of presynaptic calcium channels in the brain. Current Opinion in Neurobiology, 6, pp 331- 337.

Richards C.D. (1982) The actions of pentobarbitone, procaine and tetrodotoxin on synaptic transmission in the olfactory cortex of the guinea-pig. B. J. Pharmacol., 75, pp 639-646.

Rettig, J., Heinemann, S.H., Wunder, F., Lorra, C., Parcej, D.N., Dolly, J.O. and Pongs, O. (1994) Inactivation properties of voltage-gated K^+ channels altered by presence of β-subunit. Nature.,369, pp 289-294.

Roeper, J. and Pongs, O. (1996) Presynaptic potassium channels. Current Opinion in Neurobiology,6, pp 338-341.

Rosenburg,R.L., Tomiko, S.A. and Agnew, W.S. (1984) Reconstitution of neurotoxin-modulated ion transport by the voltage-regulated sodium channel isolated from the electroplax of *Electrophorus electricus.* Proceedings Nat. Acad. Sci., 81, pp 1239-1243.

Schwarz, W., Palade, P.T. and Hille, B. (1977) Local anaesthetics: effects of pH on use-dependent block of sodium channels in frog muscle. Biophys. J., 20, pp 343-368.

Sheng, M., Liao, Y.J., Jan, Y.N. and Jan, L.Y. (1993) Presynaptic A-current based on heteromultimeric K^+ channels detected *in vivo*. Nature.,365, pp 72-75.

Sheng, M., Tsaur, M.L., Jan, Y.N. and Jan, L.Y. (1992) Subcellular segregation of two A-type K^+ channel proteins in rat central neurons. Neuron., 9, pp 271-284.

Smith, A.B. and Cunnane, T.C. (1997) Multiple calcium channels control neurotransmitter release from rat postganglionic sympathetic nerve terminals. J. Physiol., 499.2, pp 341-349.

Southan, A.P. and Wann, K.T. (1989) Inhalation anaesthetics block accommodation of pyramidal cell discharge in the rat hippocampus. B.J.Anaesth., 63, pp 581-586.

Southan, A.P. and Wann, K.T. (1996) Effects of high helium pressure on intracellular and field potential responses in the CA1 region of the *in vitro* rat hippocampus. Eur. J. Neurosci.,8, pp 2571-2581.

Stanfield, P.R., Davis, N.W., Shelton, P.A., Sutcliffe, M.J., Khan, I.A., Brammer, W.J. and Conley, E.C. (1994) A single aspartate residue is involved in both intrinsic gating and blockage by Mg^{2+} of the inward rectifier, IRK1. J. Physiol. 478.1, pp 1-6.

Stevens, J.E., Fujinaga, M,. Oshima, E. and Mori, K. (1984) The biphasic pattern of the convulsive property of enflurane in cats. B. J. Anaes.,56, pp 395-403.

Stuhmur, W., Methfessel, C., Sakmann, B., Noda, M. and Numa, S. (1987) Patch clamp characterisation of sodium channels expressed from rat brain cDNA. Europ. Biophys. J., 14, pp 131-138.

Tsien, R.Y., (1989) Fluorescent probes of cell signalling. Annu. Rev. Neurosci., 12, pp 227-253.

Urban, B. W. (1993) Differential effects of gaseous and volatile anaesthetics on sodium and potassium channels. B. J. Anaes., 71, pp 25-38.

Wann, K. T. (1993) Neuronal sodium and potassium channels: Structure and function. B. J. Anaes., 71, pp 2-14.

Wann, K. T. and Macdonald, A. G. (1988) Actions and interactions of high pressure and general anaesthetics. Prog. Neurobiol., 30, pp 271-307.

Wann, K. T. and Southan, A. P. (1992) The action of anaesthetics and high pressure on neuronal discharge patterns. Gen. Pharmacol., 23, pp 993-1004.

Zagotta, W. N., Hoshi, T. and Aldrich, R.W. (1990) Restoration of inactivation in mutants of *Shaker* potassium channels by a peptide derived from ShB. Science., 250, pp 568-571.

Zorn, L., Kulkarni, R., Anantharam, V., Bayley, H. and Treistman, S.N. (1993) Halothane acts on many potassium channels, including a minimal potassium channel. Neurosci. Lett., 161(1), pp 81-4.

The GABA$_A$ Receptor: An Important Locus for Intravenous Anaesthetic Action

Jeremy J. Lambert, Delia Belelli, Susan Shepherd, Anna-Lisa Muntoni, Marco Pistis and John A. Peters

NEUROSCIENCES INSTITUTE, DEPARTMANT OF PHARMACOLOGY AND CLINICAL PHARMACOLOGY, NINEWELLS HOSPITAL AND MEDICAL SCHOOL, THE UNIVERSITY OF DUNDEE, SCOTLAND, UK

Introduction

Despite intensive research, the molecular events that underlie the state of general anaesthesia remain uncertain. The early observations of Meyer and Overton, which established a strong correlation between anaesthetic potency and solubility in fatty solvents, suggest a hydrophobic site of anaesthetic action which has traditionally been equated with the lipid component of the plasma membrane of neurones (reviewed by Little, 1996). Theories with various degrees of refinement postulate that an anaesthetic induced membrane perturbation (*e.g.* changes in volume, thickness or phase-behaviour) are transmitted to sensitive membrane proteins that affect neuronal function (Franks and Lieb, 1987; Little, 1996). The identity of such proteins, and how their modulation might plausibly result in anaesthesia, is often neglected in this approach. Lipid theories currently face several unresolved problems. Firstly, the structural changes induced by clinically relevant concentrations of anaesthetics are small and can be mimicked by modest changes in temperature that are well within the range of diurnal variation in man and other mammals (Franks and Lieb, 1987). Secondly, the cut-off phemonenon, in which a loss of anaesthetic activity is observed for higher, yet still lipid soluble, members of homologous series of anaesthetics such as the *n*-alcohols and *n*-alkanes is not accommodated by the Meyer-Overton rule (Franks and Lieb, 1987). Furthermore, the validity of the latter has been undermined by recent studies with certain polyhalogenated and perfluorinated compounds which did not demonstrate the anaesthetic activity predicted by the Meyer-Overton rule (Koblin *et al.*, 1994; Liu *et al.*, 1994; Raines and Miller, 1994).

Against this background, the focus of mechanistic research has shifted towards the role of cellular proteins in anaesthetic action (Franks and Lieb, 1994; Harris *et al.*, 1995). A precedent for a direct interaction between a wide range of anaesthetic structures and a lipid-free model protein, with a rank order of effect paralleling anaesthetic potency, was set by Franks and Lieb (1987) in their studies of the firefly luciferase enzyme. Although such elegant evidence remains to be garnered for any neuronally expressed protein that might theoretically be suspected to participate in anaesthesia, a strong case can be made for the involvement of certain transmitter gated ion channels (Franks and Lieb, 1994; Harris *et al.*, 1995). Potentiation of inhibitory- and/or suppression of excitatory-neurotransmission has the simplistic appeal of providing a logical explanation for the depressant activity of anaesthetics. Indeed, it is now clear that several transmitter gated ion channels, including the $GABA_A$, glutamate, glycine, nicotinic and 5-HT_3 receptor subtypes, are susceptible to allosteric regulation by a range of anaesthetic structures (Tanelian *et al.*, 1993; Franks and Lieb, 1994; Harris *et al.*, 1995). What is less certain is whether the modulation of such candidates is essential, or merely incidental, to anaesthetic action. A further complication arises from the heterogeneous nature of the receptors in question. With the possible exception of the 5-HT_3 receptor, the aforementioned ligand-gated ion channels are multimeric assemblies of subunits that may be drawn from several structural classes (Ortells and Lunt, 1995). Anaesthetic activity may thus depend upon the precise receptor isoform examined. As elaborated below, this is known to be the case for certain anaesthetic-receptor interactions. Such specificity, which might eventually be exploited clinically, reinforces the contention that at least some anaesthetics act directly upon proteins.

A number of criteria might can applied in identifying plausible targets of anaesthetic action (Franks and Lieb, 1994). If any semblance of a unitary theory is to be retained, the rank order of activity at the putative site should parallel anaesthetic potency. Moreover, the action should be physiologically relevant and occur at anaesthetic concentrations encountered in clinical practice. Importantly, agents demonstrating stereoselectively, particularly enantioselecivity, in anaesthetic potency (*e.g.* isoflurane, barbiturates, etomidate, ketamine and pregnane steroids) should exhibit appropriate selectivity at their molecular target (Franks and Lieb, 1994). Applying these criteria, the $GABA_A$ receptor emerges as a particularly promising candidate. Some of the evidence supporting this view is summarised in this article where we focus upon the structurally diverse intravenous general anaesthetics, alphaxalone, pentobarbitone, propofol and etomidate.

$GABA_A$ receptors probably exist as a pentameric arrangement of subunits drawn from five families, termed α, β, γ, δ and ε whose members may co-assemble to create receptors with distinct pharmacological properties (Sieghart, 1995; Smith and Olsen, 1995). Furthermore, a number of isoforms of some of these subunits (*e.g.* α_{1-6}, β_{1-3}, γ_{1-3}) have now been identified which also impart distinct properties to the receptor (*e.g.* Sieghart, 1995; Smith and Olsen, 1995). Much of the work presented below utilizes the *Xenopus laevis* oocyte expression model. In this system, oocytes are pre-injected with cRNA transcribed from human $GABA_A$ receptor subunit cDNAs, and 24-48 hours later, are used in conjunction with the two electrode voltage-clamp technique to record the current response mediated by the $GABA_A$ receptor-gated increase in membrane chloride conductance (*e.g.* Hill-Venning *et al.*, 1997). To compare the GABA-modulatory actions of the anaesthetics, human recombinant $GABA_A$ receptors composed of α_1, β_2 and γ_2 subunits are utilized. This subunit combination is highly expressed in the mammalian CNS (Wisden *et al.*, 1992). In some experiments, the whole cell recording mode of the patch clamp technique was used to record GABA-evoked currents from a mouse epithelial cell line (L(tk-)) stably transfected with the $\alpha_6\beta_3\gamma_{2S}$ subunit combination (Hadingham *et al.*, 1996). In this study, particular attention has been given to the clinically utilized anaesthetic etomidate. Work presented below will demonstrate etomidate to be highly selective for the $GABA_A$ receptor. Furthermore, this interaction is stereoselective and highly dependent upon the subunit composition of the $GABA_A$ receptor, a specificity which involves a single amino acid. Collectively, these properties are inconsistent with etomidate acting in a non-specific manner to perturb the structure of the neuronal cell membrane. Instead, they support a specific interaction with the $GABA_A$ receptor protein.

Alphaxalone

The positive allosteric action of the synthetic steroidal anaesthetic alphaxalone (5α-pregnan-3α-ol-11,20-dione) at both native and recombinant $GABA_A$ receptors is well established (Harrison and Simmonds, 1984; Barker *et al.*, 1987; Cottrell *et al.*, 1987, see Lambert *et al.*, 1995 for a review). For example, alphaxalone produced a concentration-dependent (3 nM - 1 μM) enhancement of the current evoked by GABA recorded from *Xenopus laevis* oocytes expressing human $GABA_A$ receptors formed from α_1 , β_2 and γ_{2L} subunits (Figures 1 and 2). The concentration of anaesthetic producing a half maximal effect (EC_{50}) was 2.2 ± 0.3 μM and maximal enhancement (E_{max}) (to 78 ± 3% of the GABA maximum) occurred with 10 μM alphaxalone. To place this value in context, the minimum infusion rate of Althesin steroids that suppresses movement in response to a surgical stimulus in 50% of opioid premedicated patients receiving 67% N_2O, gives rise to a plasma concentration of alphaxalone of approximately 1.9 μg/ml (Sear *et al.*, 1983). Correcting for protein binding of 37.7% (Child *et al.*, 1972), the free aqueous concentration of alphaxalone can be estimated to be 3.6 μM, a value very similar to the EC_{50} for potentiation of GABA (Figure 2). However, it must be emphasised that the relationship between the plasma and brain concentrations of alphaxalone is unknown.

A diastereomer of alphaxalone, betaxalone, which differs only in the orientation (*i.e.* β-configuration) of the 3-hydroxyl group possesses neither anaesthetic potency, nor activity at the $GABA_A$ receptor. Although indicative of pronounced structural specificity, stereoisomerism of this type cannot distinguish between a direct effect of alphaxalone at the $GABA_A$ receptor, or an indirect effect subsequent to lipid perturbation, because diastereomers can have differing effects within an achiral environment. However, compelling

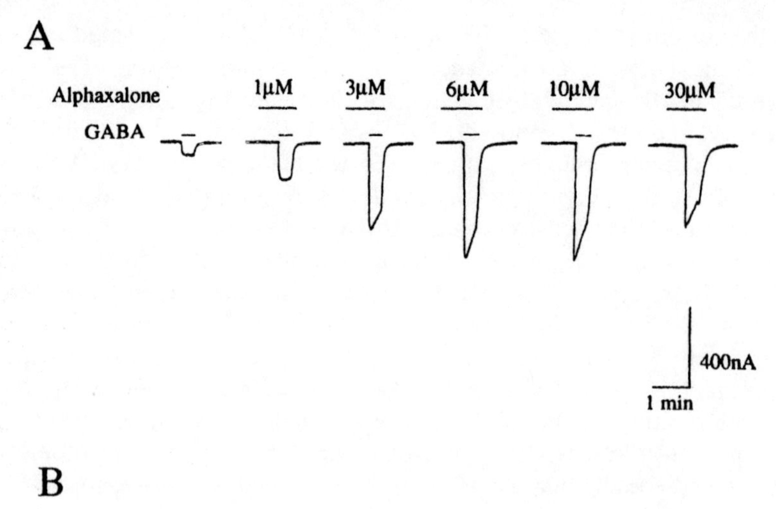

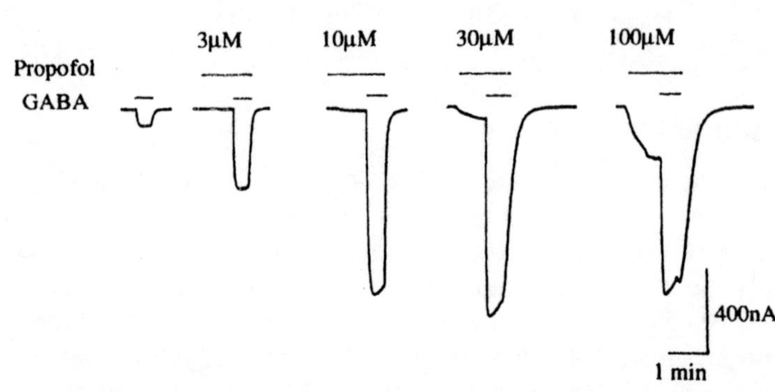

Figure 1 *The intravenous anaesthetic agents alphaxalone and propofol potentiate GABA-evoked currents mediated by human $GABA_A$ receptors with the subunit composition $\alpha_1\beta_2\gamma_{2L}$ expressed in* Xenopus *oocytes. A. Illustrated are inward current responses elicited by the bath application of GABA and their modulation by the co-application of alphaxalone (1 μM-30 μM). The concentration of GABA utilised produced a response of 10% of the GABA maximum (EC_{10}). Note the concentration-dependent enhancement of the GABA-evoked response by the anaesthetic (1-10 μM), although for greater concentrations (30 μM) the enhancement is reduced giving rise to a "bell-shaped" concentration response relationship. B. Propofol (3-100 μM) produced a concentration-dependent enhancement of the GABA(EC_{10})-evoked response. Note that in contrast to alphaxalone, the pre-application of relatively high concentrations of propofol produce an inward current response (GABA-mimetic) in the absence of GABA. All recordings are made from oocytes voltage-clamped at -60 mV.*

evidence for a direct interaction between certain pregnane steroids and the $GABA_A$ receptor has recently been adduced. The endogenous neurosteroid (+)-5α-pregnan-3α-ol-20-one (which differs from alphaxalone in that C11 is unsubstituted) is an anaesthetic and acts as a positive allosteric modulator of the $GABA_A$ receptor (Majewska *et al.*, 1986; Harrison *et al.*, 1987; Peters *et al.*, 1988). Significantly, the enantiomer (-)-5α-pregnan-3α-ol-20-one demon-

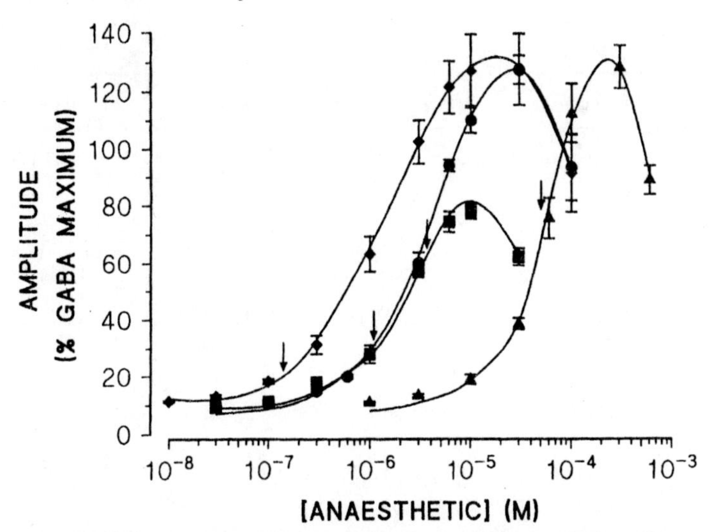

Figure 2 *Potentiation of GABA-evoked responses by structurally diverse intravenous anaesthetic agents. Illustrated are the concentration-response relationships for the enhancement of GABA(EC_{10})-evoked currents (recorded from oocytes expressing human $\alpha_1\beta_2\gamma_{2L}$ receptors) by etomidate (◆), propofol (●), alphaxalone (■) and pentobarbitone (▲). These data were analysed to give the concentration of anaesthetic which enhances the GABA response to 50% of the maximal effect for that anaesthetic (EC_{50}) and the maximal enhancement produced by the anaesthetic (E_{max}). In descending order of potency are etomidate (EC_{50} = 1.2 ± 0.1 μM; E_{max} = 127 ± 12%), alphaxalone (EC_{50} 2.2 ± 0.3 μM; E_{max} = 78 ± 3%), propofol (EC_{50} = 3.8 ± 0.2 μM; E_{max} = 127 ± 5%) and pentobarbitone (EC_{50} = 55 ± 4 μM; E_{max} = 128 ± 7%). The approximate free aqueous concentrations of these agents occurring during clinical anaesthesia in man (except pentobarbitone which is estimated for rat) is indicated by ↓. Oocytes were voltage-clamped at -60 mV. Each symbol represents the mean ±* S.E.M. of data obtained from at least four oocytes.

strates greatly reduced GABA-modulatory and anaesthetic potency in tadpoles and mice (Wittmer *et al.*, 1996). In this instance, the compounds act dissimilarly only in a chiral (*e.g.* protein) environment. A similar correlation between GABA-modulatory and anaesthetic potency has been established for androstane enantiomers bearing a 17β-carbonitrile substituent (Wittmer *et al.*, 1996). These observations provide a cogent argument for direct interactions between pregnane steroid anaesthetics and the $GABA_A$ receptor and reinforce earlier observations suggesting alphaxalone to be an effective modulator of GABA only when applied extracellularly (Lambert *et al.*, 1990).

The selectivity of action of alphaxalone has been examined by comparing its effects at several transmitter gated ion channels. Relatively high concentrations (*i.e.* 10 μM) of this anaesthetic have no effect on the strychnine-sensitive glycine receptor of mouse spinal neurones (Hill-Venning *et al.*, 1991), or recombinant glycine receptors composed of α_1 and β subunits expressed in *Xenopus laevis* oocytes (Table 1). Similarly, the excitatory amino acid

Table 1

	Alphaxalone	*Pentobarbitone*	*Propofol*	*Etomidate*
GABA$_A$ (α_1,β_2,γ_{2L},EC$_{50}$) (127±12%)	2.2±0.3 μM	55±4 μM (78±3%)	3.8±0.2 μM (128±7%)	1.2±0.1 μM (127±5%)
Glycine (α_1,β)	60 μM* No effect	757±35 μM (51±10%)	27±2 μM (98±6%)	Max=300 μM (29±2%)
Kainate (rat brain mRNA, IC$_{50}$)	100 μM No effect	173±20 μM	100 μM No effect	100 μM No effect
Nicotinic (α_4,β_2, IC$_{50}$)	5±1 μM	16±1 μM	20±3 μM	57±6 μM
5-HT$_3$ (h5-HT$_3$R-A$_s$)	~30 μM**	600±100 μM	46±1 μM	152±14 μM

The selectivity of action of intravenous anaesthetic agents. All experiments were performed on *Xenopus laevis* oocytes voltage-clamped at a holding potential of -60 mV. The sources of receptor were: GABA$_A$- human $\alpha_1\beta_2\gamma_{2L}$; glycine - human human α_1, rat β; neuronal nicotinic - rat $\alpha_4\beta_2$; 5-HT$_3$ - human 5-HT$_3$R-A$_S$ and kainate - rat brain mRNA. All experiments upon GABA$_A$ and glycine receptors utilized the EC$_{10}$ concentration of the natural agonist. Experiments investigating the inhibitory actions of anaesthetics on kainate and 5-HT$_3$ receptors utilized the agonist EC$_{50}$ (kainate = 7- 10 μM; 5-HT = 2 μM). The control concentration of nicotine for $\alpha_4\beta_2$ receptors was 1 μM. For experiments performed with GABA$_A$ and glycine receptors, the EC$_{50}$ and the maximum modulation produced, in parenthesis, are given. For kainate, nicotinic and 5-HT$_3$ receptors, the IC$_{50}$ values are listed. Where quantified, all data are the mean ± S.E.M. of observations made from 3 to 5 oocytes.

receptors, which *in vivo* are activated by glutamate, are insensitive to alphaxalone. Hence, kainate-evoked whole cell currents recorded from rat hippocampal neurones (Lambert *et al.*, 1990) or from oocytes pre-injected with a preparation of rat brain mRNA (Table 1) are unaffected by alphaxalone within the concentration range 10 to 100 μM. Similarly, these supra-anaesthetic concentrations had no effect on NMDA-evoked currents recorded from rat hippocampal neurones (Lambert *et al.*, 1990). In apparent contradiction, alphaxalone (10 μM) suppressed the depolarizing response of hippocampal neurones to ionophoretically applied glutamate (Lambert *et al.*, 1990). However, at these relatively high concentrations, alphaxalone directly activates the GABA$_A$ receptor (Cottrell *et al.*, 1987; Lambert *et al.*, 1995). Hence, the dissimilar influence of alphaxalone on glutamate receptor-mediated currents and depolarizations, recorded under voltage- and current-clamp respectively, is a product of the GABA-mimetic action of the anaesthetic which decreases the input resistance of the neurone and as a consequence reduces the glutamate-evoked depolarization.

The 5-HT$_3$ receptor is relatively insensitive to alphaxalone. Hence, only supra-anaesthetic concentrations of the steroid (IC$_{50}$ = 30 μM, see Table 1) produced a significant inhibition of the 5-HT-evoked current recorded from oocytes expressing homo-oligomeric receptors formed from the human 5-HT_3 R-A_S subunit (Belelli *et al.*, 1995). Potential stereoselective actions of alphaxalone at glycine, glutamate and 5-HT_3 receptor subtypes have not been investigated.

Alphaxalone produced a concentration-dependent (0.3-60 μM) and relatively potent (IC_{50} = 5 ± 1 μM) inhibition of the current induced by the bath application of 1 μM nicotine to oocytes expressing neuronal nicotinic receptors composed of α_4 and β_2 subunits (see Table 1). This value approximates to the aqueous concentration of alphaxalone at the anaesthetic ED_{50} and it is conceivable that an effect at neuronal nicotinic receptors could contribute to the central depressant actions of this steroid. However, the non-anaesthetic epimer, betaxalone, is equipotent with alphaxalone in blocking the nicotinic receptors of bovine chromaffin cells (Cottrell *et al.*, 1987). Therefore, it is unlikely that neuronal nicotinic receptors are the major locus of action for the steroidal anaesthetics.

Pentobarbitone

Numerous electrophysiological studies have demonstrated anaesthetic concentrations of pentobarbitone to enhance the interaction of GABA with the $GABA_A$ receptor, and somewhat higher concentrations to directly activate the $GABA_A$ receptor (Barker and Ransom, 1978; Huang and Barker, 1980; Akaike *et al.*, 1985). Similarly, for oocytes expressing human $GABA_A$ receptors constructed from α_1, β_2 and γ_{2L} subunits, pentobarbitone (3 μM-30 μM) produced a concentration-dependent enhancement of currents evoked by GABA (EC_{50} = 55 ± 4 μM; E_{max} = 128 ± 7%; see Figure 2 and Table 1). Much greater concentrations of the anaesthetic were required to directly activate the $GABA_A$ receptor complex (EC_{50} = 604 ± 85 μM; E_{max} = 17 ± 1%). In the rat, the aqueous concentration of pentobarbitone associated suppression of movement in response to a painful stimulus has been calculated to be 50 μM (Franks and Lieb, 1994).

Pentobarbitone exists as S(-) and R(+) enantiomers due to a chiral centre in its methyl butyl side chain (Andrews and Mark, 1982). The S(-) enantiomer exceeds the R(+) form in anaesthetic potency by a factor of approximately two (Christensen and Lee, 1973). Differential metabolism, adsorption or distribution do not account for this difference and, indeed, would tend to oppose the stereoselectivity that is observed (Andrews and Mark, 1982). In numerous assays of positive allosteric regulation of $GABA_A$ receptor function, including facilitation of the binding of [^{3}H]GABA or [^{3}H]diazepam, displacement of [^{35}S]TBPS, and enhancement and activation of receptor mediated chloride conductances in spinal neurones, the relative potencies of the enantiomers of pentobarbitone are similar to those observed *in vivo* (Huang and Barker, 1980; Olsen *et al.*, 1986; Ticku and Rastogi, 1986).

The strychnine-sensitive glycine receptor is relatively insensitive to barbiturates (Barker & Ransom, 1978; Akaike *et al.*, 1985). In agreement, in the present study, only relatively high concentrations of pentobarbitone (EC_{50} =

757 ± 3 μM; E_{max} = 51 ± 10%) produced an enhancement of glycine-evoked currents recorded from oocytes expressing α_1 and β glycine receptor subunits (Table 1). We have previously demonstrated that pentobarbitone inhibits kainate-evoked currents recorded from rat hippocampal neurones with an IC_{50} of 69 μM (Lambert *et al.*, 1991). This anaesthetic is similarly effective in rat cortical neurones (Marszalec and Narahashi, 1993). Here, slightly higher concentrations of pentobarbitone (IC_{50} = 173 ± 20 μM) were required to inhibit kainate-evoked currents recorded from oocytes preinjected with rat brain mRNA. Relatively low concentrations of pentobarbitone inhibited nicotine-induced currents recorded from oocytes expressing the neuronal nicotinic receptor subtype $\alpha_4\beta_2$ (IC_{50} = 16 ± 1 μM). By contrast 5-HT (2 μM)-evoked currents mediated by 5-HT_3 receptors were relatively insensitive (IC_{50} = 600 ± 100 μM) to this anaesthetic (Table 1).

Propofol

Propofol (0.1 - 10 μM) produced a concentration-dependent enhancement of GABA-evoked currents recorded from oocytes expressing human $\alpha_1\beta_2\gamma_{2L}$ $GABA_A$ receptors with a calculated EC_{50} of 3.8 ± 0.2 μM and a maximal enhancement (for 10 μM propofol) of 127 ± 5% of the GABA maximum (Table 1, Figures 1 and 2). Much greater concentrations of this anaesthetic were required to directly activate the $GABA_A$ receptor complex (EC_{50} = 70 ± 1 μM; E_{max} = 48 ± 8%). The plasma concentration of propofol necessary to maintain total intravenous anaesthesia (TIVA) in man is within the range 6 - 9 μg/ml (Shafer, 1993). Allowing for the extensive protein binding of propofol ± 97.8%; Servin *et al.*, 1988) suggests the free aqueous concentration of the anaesthetic to lie in a concentration range (0.75 to 1.1 μM) associated with modest potentiation of GABA-evoked currents. However, it should be noted that this value may be an underestimate, since in rats, the concentration of propofol determined in the brain and spinal cord during anaesthesia greatly exceeds that measured in plasma (Shyr *et al.*, 1995). Propofol does not contain a chiral centre.

Propofol also produced a concentration-dependent enhancement of currents evoked by the bath application of glycine to oocytes expressing $\alpha_1\beta$ glycine receptors, although concentrations approximately ten fold greater were required compared with those effective in modulating the $GABA_A$ receptor (EC_{50} = 27 ± 2 μM; E_{max} = 98 ± 6%; Table 1). Currents elicited by the bath application of kainate to oocytes preinjected with rat brain mRNA were unaffected by supra-anaesthetic concentrations (100 μM) of propofol. Furthermore, only high concentrations of this anaesthetic are reported to inhibit NMDA receptors (Orser *et al.* 1995).

Currents evoked by the application of nicotine to oocytes expressing $\alpha_4\beta_2$ neuronal nicotinic receptors were inhibited by propofol (IC_{50} = 20 ± 3 μM). Approximately two fold higher concentrations (IC_{50} = 46 ± 1 μM) of the anaesthetic were required to inhibit 5-HT-evoked currents recorded from oocytes expressing the human 5-HT_3R-A_S (Table 1).

Etomidate

Etomidate produced a potent (EC_{50} = 1.2 ± 0.1 μM) and large (E_{max} = 127 ± 12% for 10 μM etomidate) enhancement of GABA-evoked currents recorded from oocytes expressing human $\alpha_1\beta_2\gamma_{2L}$ $GABA_A$ receptors (Table 1). At much higher concentrations (EC_{50} = 83 ± 34 μM; E_{max} = 19 ± 2%), this anaesthetic directly activated the $GABA_A$ receptor complex. In patients premedicated with diazepam and fentanyl, the plasma concentration of etomidate required to maintain anaesthesia was found to be 0.5 μg/ml (Doenicke *et al.*, 1982). Allowing for protein binding of 93.7% (Meuldermanns and Heykants, 1976) the free aqueous concentration of etomidate may be estimated to be 0.14 μM. However, this value probably represents a lower bound, because following a bolus injection to experimental animals, the concentration of etomidate in the brain exceeds that measured in plasma (Heykants *et al.*, 1975).

Etomidate (300 μM) produced only a modest enhancement of glycine-evoked currents recorded from oocytes expressing $\alpha_1\beta$ glycine receptors. Similarly, 100 μM of this anaesthetic had no effect on kainate-evoked currents (Table 1) and relatively high concentrations were required to inhibit both neuronal nicotinic receptors (IC_{50} = 57 ± 6 μM) and 5-HT_3 receptors (IC_{50} = 152 ± 14 μM). Hence, etomidate is selective for the $GABA_A$ receptor.

The Interaction of Etomidate with the $GABA_A$ Receptor Is Stereoselective

Etomidate is optically active. The central depressant effect of this anaesthetic exhibit a clear preference for the (R)-(+) isomer (Doenicke and Ostwald, 1997). In a recent study in mice, (R)-(+) etomidate was found to be approximately ten times more potent as an hypnotic than the (S)-(-) enantiomer (D. Gemmell, personal communication). A similar enantioselective interaction occurs between etomidate and the $GABA_A$ receptor. Hence, in radioligand binding assays assessing the facilitation of the binding of [^{3}H]GABA or [^{3}H]diazepam, displacement of [^{35}S]TBPS, and the stimulation of [$^{36}Cl^-$]-efflux from preloaded brain slices, the potency of (R)-(+) etomidate is at least ten-fold higher than the (S)-(-) enantiomer (Olsen *et al.*, 1986). In the present study, we compared the actions of the enantiomers of etomidate upon $GABA_A$ receptors composed of α_6, β_3 and γ_{2L} subunits expressed in *Xenopus* oocytes (Figure 3). Inspection of the figure reveals the superior GABA-modulatory and GABA-mimetic effects of (R)-(+) *versus* (S)-(-)-etomidate. These data, together with those demonstrating the selectivity of etomidate, identify the $GABA_A$ receptor as an important molecular locus of action of this anaesthetic.

The Interaction of Etomidate with the $GABA_A$ Receptor Is Subunit Dependent

We have recently investigated the interaction of etomidate with the $GABA_A$ receptor in more detail, giving particular emphasis to the influence of subunit

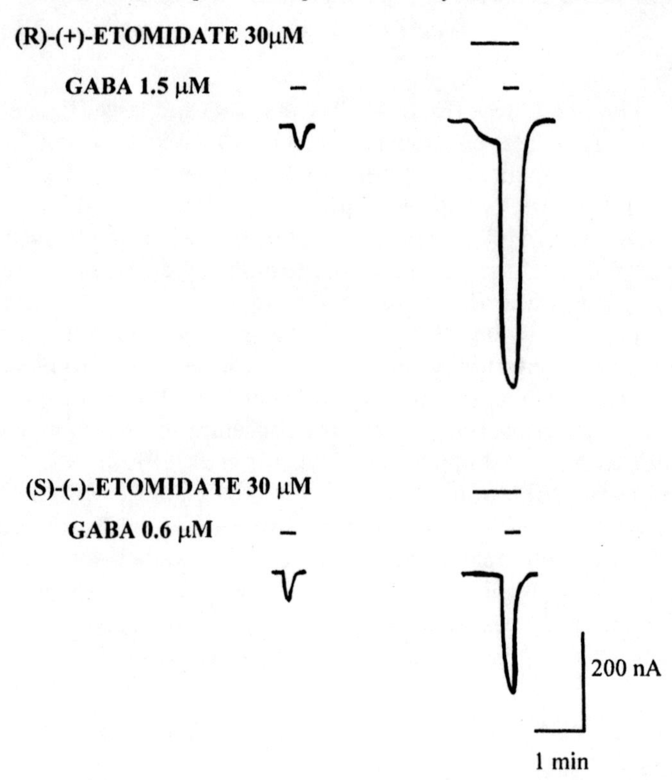

Figure 3 *The interaction of etomidate with the $GABA_A$ receptor is stereoselective. The traces ilustrate the inward current response resulting from the bath application of GABA (EC_{10}) to oocytes expressing human $\alpha_6\beta_3\gamma_{2L}$ receptors and their modulation by the stereoisomers of etomidate. Note that 30 μM of (R)-(+)-etomidate produces a much larger enhancement of the GABA-evoked response than 30 μM of (S)-(-)-etomidate. Additionally, a clear inward current (GABA-mimetic) response to R-(+)-etomidate is evident whereas only higher concentrations (not shown) of (S)-(-) etomidate produce modest inward currents. In these experiments the oocytes were voltage-clamped at -60 mV.*

composition. In oocytes, etomidate enhanced GABA-evoked currents mediated by all the human recombinant $GABA_A$ receptors tested ($\alpha_x\beta_y\gamma_{2L}$; where x = 1, 2, 3, and 6, y = 1, 2 and 3), although the properties of this interaction were dependent on both the β and α subunit isoforms (Hill-Venning *et al.*, 1997, Table 2). For all the β_2-containing receptors, the calculated EC_{50} for etomidate varied little (0.6 - 1.2 μM) with the α subunit isoform (see Table 2). However, the substitution of the β_2 for the β_1 subunit produced a 9-12 fold increase of the etomidate EC_{50} (~6-11 μM, see Table 2) for all α isoforms tested. Furthermore, for α_1, α_2 and α_6 (but not α_3) the maximal enhancement produced by the anaesthetic was much greater for β_2 *versus* β_1 containing receptors. Although the β_3 subunit was only investigated for the α_6 β_3 γ_{2L} receptor combination, comparison with the appropriate α_6 β_y

Table 2

	Etomidate			
Subunit Combination	*Modulation EC_{50}*	*Modulation E_{max}*	*Agonist EC_{50}*	*Agonist E_{max}*
$\alpha_1\beta_2\gamma_2$	1.2±0.1 μM	127±12%	83±34 μM	19±2%
*$\alpha_1\beta_1\gamma_2$	10.8±1.1 μM	79±2%	N.D.	4±1% (1mM)
*$\alpha_2\beta_2\gamma_2$	0.75±0.1 μM	108±1%	55±24 μM	26±6%
*$\alpha_2\beta_1\gamma_2$	6.3±0.3 μM	65±3%	N.D.	5±1% (1mM)
*$\alpha_3\beta_2\gamma_2$	1.0±0.1 μM	88±6%	108±4 μM	9±1%
*$\alpha_3\beta_1\gamma_2$	8.1±0.9 μM	75±8%	N.D.	<2% (1mM)
*$\alpha_6\beta_2\gamma_2$	0.6±0.04 μM	169±14%	22±1 μM	51±15%
*$\alpha_6\beta_1\gamma_2$	±0.6 μM	28±2%	N.D.	5±2% (1mM)
*$\alpha_6\beta_3\gamma_2$	0.7±0.06 μM	135±7%	23±2.4	96±24
*$\alpha_6\beta_{1\text{-}2}\gamma_2$	1.3±0.1 μM	226±47%	34 (n=2)	38 (n+2)
*$\alpha_6\beta_{2\text{-}1}\gamma_2$	7.7±1.0 μM	23±2%	N.D.	4.4±1.2 (0.6mM)
*$\alpha_6\beta_{1S290N}\gamma_2$	1.6±0.3μM	150±13%	79±6	45±13
*$\alpha_6\beta_{3\ N289S}\gamma_2$	5.7±2.6 μM	36±1%	55±6	14±3
*$\alpha_6\beta_{3\ N289M}\gamma_2$	N.D.	11.8±0.6%	N.D.	3.5±0.4 (0.6 mM)

$GABA_A$ receptor subunit specificity of etomidate. The numbers in paraenthesis give the maximum concentration of etomidate tested as an agonist for the β_1 subunit-containing receptors. The E_{max} is expressed as a percentage of the maximum response to GABA. ND = not determined due to the small magnitude of the current induced by etomidate. All data were obtained from oocytes voltage-clamped at a holding potential of -60 mV. Where quantified, all data are the mean ± S.E.M. of observations made from 3 to 5 oocytes. *Data extracted from Hill-Venning *et al.* (1997).

γ_{2L} receptors (where y = 1 or 2) reveals the interaction of etomidate to approximate to that of the β_2 containing receptors but to be distinct from receptors incorporating the β_1 subunit. Hence, the potency of etomidate is clearly influenced by the β subunit isoform, but the magnitude of the maximal effect is dictated by both the subtype of the β and α subunit. The GABA modulatory actions of pentobarbitone, propofol (Hill-Venning *et al.*, 1997) and anaesthetic steroids (Hadingham *et al.*, 1993) are not dependent upon the isoform of the β subunit and hence this feature distinguishes etomidate from these intravenous anaesthetics. The β subunit isoform also greatly influences the GABA-mimetic properties of etomidate.

For receptors containing either β_2 or β_3 subunits clear agonist effects of etomidate were present, whereas these were limited or absent for the β_1 containing receptors (Table 2). Such direct effects (for $\alpha_1\beta_2\gamma_{2L}$ receptors) were enhanced in amplitude by the coapplication of flunitrazepam (0.3 μM) and inhibited by picrotoxin (30 μM) observations which confirm the GABA-mimetic action of this anaesthetic (Hill-Venning *et al.*, 1997). For receptors incorporating the β_2 subunit both the EC_{50} and the maximal amplitude of the GABA-mimetic effect of etomidate was influenced by the α isoform (EC_{50} range = 22-130 μM; E_{max} = 9-51%).

The β Isoform Selectivity of Etomidate Resides with a Single Amino Acid

The discriminatory influence of the β-subunit isoform was further investigated by utilizing chimeric β_1 and β_2 subunits. For oocytes expressing receptors composed of α_6 and γ_{2L} subunits and a $\beta_{1\text{-}2}$ chimeric subunit (consisting of the N terminal extracellular domain of the β_1 subunit and the transmembrane domains, connecting loops and carboxy terminal region of the β_2 subunit) etomidate produced a potent ($EC_{50} = 1.3 \pm 0.1$ μM) and large potentiation ($E_{max} = 226 \pm 47\%$) of the GABA-evoked current (Table 2). In addition, this anaesthetic exhibited clear GABA-mimetic actions with an EC_{50} of 34 μM and a maximal effect of 38% (Table 2). Comparison of these values with those obtained for receptors containing wild type β subunits reveals they most closely resemble those observed for β_2 or β_3 subunit containing receptors. By contrast to receptors incorporating the $\beta_{1\text{-}2}$ chimera, those containing the N-terminal domain of the β_2-subunit (i.e. $\beta_{2\text{-}1}$) exhibited a reduced potency ($EC_{50} = 7.7 \pm 1$ μM) and maximal effect ($E_{max} = 23 \pm 2\%$, n = 3) of the GABA-enhancing effects of etomidate (Table 2). Additionally, only a limited GABA-mimetic effect was evident. These properties are similar to those reported for receptors which incorporate a β_1 subunit. Hence, these data suggest that a key region which confers the enhanced activity of etomidate is located distal to the N-terminal extracellular portion of the receptor protein.

The anticonvulsant loreclezole exhibits a β subunit selectivity similar to that of etomidate (Wafford *et al.*, 1994). Indeed, the structure of etomidate is similar to that of loreclezole (Hill-Venning *et al.* 1997). For loreclezole, mutagenesis studies have highlighted the importance of an asparagine residue (for β_2 and β_3 subunits) and a serine residue (for the β_1 subunit) located towards the extracellular side of the chloride ion channel (M2) (Wingrove *et al.*, 1994); see Figure 4. Mutation of the the β_1 serine to an asparagine residue ($\beta_{1\ S290N}$) greatly enhanced both the GABA modulatory and GABA-mimetic actions of etomidate for receptors that also contained α_6 and α_2 subunits (Table 2). By contrast, the reciprocal mutation of the β_3 asparagine to serine residue ($\beta_{3\ N289S}$) greatly reduced these parameters (see Table 2). We have previously demonstrated that the function of a GABA-gated recombinant receptor isolated from *Drosophila melanogaster* (*Rdl*) is enhanced by both pentobarbitone and propofol, but is insensitive to etomidate (Chen *et al.*, 1994; Belelli *et al.*, 1996). For this invertebrate receptor, a methionine residue occupies the equivalent position to the asparagine (β_2 and β_3) and serine (β_1) residue. Interestingly, the co-expression of α_6 and γ_2 subunits with a β_3 subunit carrying an asparagine to a methionine ($\beta_{3\ N289M}$) mutation, results in a receptor which, like the invertebrate receptor, is etomidate insensitive (Table 2).

Intracellular Etomidate Is Inert

The data presented above demonstrate etomidate to selectively modulate the GABA$_A$ receptor, an interaction which is greatly influenced by a single amino

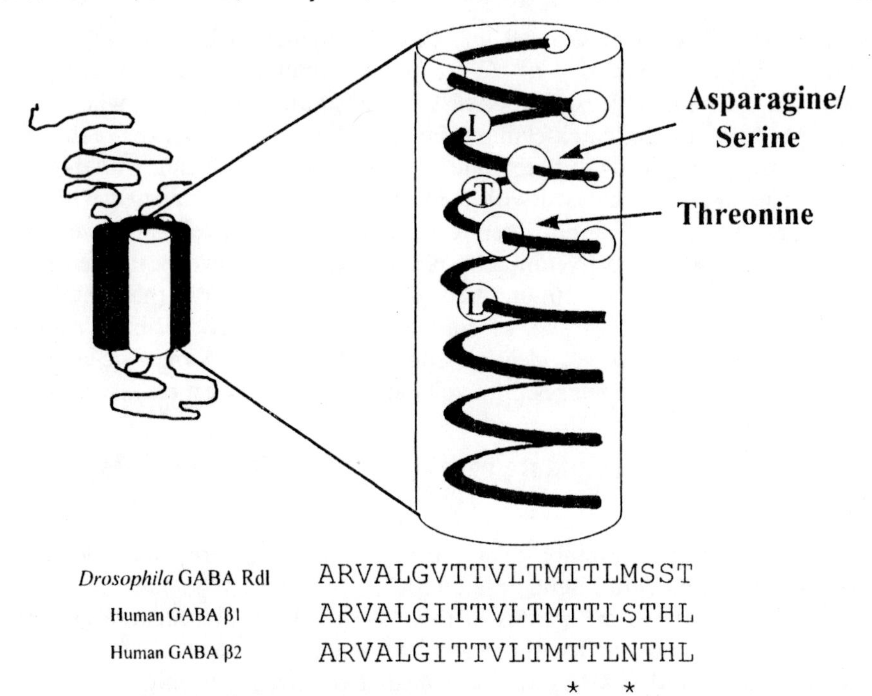

Figure 4 *A diagrammatic representation of a $GABA_A$ receptor subunit with the putative chloride ion channel region M2 shown in expanded format. The M2 region is presumed to be an α helix (with 3.2 amino acids per turn). The amino acids isoleucine (I) and threonine (T) indicated are thought to face the ion channel lumen (see text and Xu and Akabas, 1993) as is the conserved leucine (L). The critical asparagine (β_2 or β_3) or serine (β_1) residue for etomidate actions (based on the GABA α_1 sequence) does not line the ion channel pore (Xu and Akabas, 1993). Interestingly, a threonine (T) residue (located one helical turn away from the asparagine/serine residue) when mutated to a glutamine residue abolishes the GABA modulatory and GABA-mimetic effects of pentobarbitone (Birnir* et al., *1997).*

The alignment of the primary amino acid sequence (single letter code) of the putative M2 region of related amino acid subunits is also shown. The amino acids critical for etomidate and pentobarbitone effects are indicated *.

acid located within the transmembrane ion channel. By systematically mutating amino acids in the M2 region to cysteine residues and determining their subsequent susceptibility to covalent modification by charged, channel blocking, sulfhydryl reagents, inferences have been drawn concerning the amino acids which face the aqueous pore of the channel (Xu and Akabas, 1993). Such an analysis, made for the $GABA_A$ α_1 subunit, places the α_1 serine residue, which is equivalent to the β_1 serine or β_2 and β_3 asparagine away from the channel lumen and projecting either into the protein interior or the lipid bilayer. Should this orientation be common to the β subunits, etomidate could feasibly access this site *via* the membrane. In view of this, we determined whether intracellular etomidate can influence the $GABA_A$ receptor. The experiments utilized a mouse cell line (L(tk-)) engineered to express human $GABA_A$ receptors composed of α_6, β_3 and γ_2 subunits (Hadingham *et al.*, 1996) and currents evoked by GABA were studied using the whole-cell recording mode of the patch-clamp technique. The ability of extracellularly applied etomidate (1 μM) to potentiate GABA-evoked currents recorded from control cells and cells dialysed with a pipette solution supplemented with 10 μM etomidate was compared. Despite being present at 10-fold molar excess intracellularly, extracellular etomidate potentiated GABA-evoked currents to 184 $\pm$ 14% of control (n = 6) an effect indistinguishable from that obtained from control cells (211 $\pm$ 29% of control; n=5). Furthermore, both anaesthetic-loaded and control cells clearly responded to the GABA-mimetic activity of etomidate. These data suggest intracellular etomidate to be inert.

The Interaction of Etomidate with the $GABA_A$ Receptor May Be Neurone Specific

The differential susceptibility of $GABA_A$ receptor isoforms to allosteric regulation by etomidate might be anticipated to impart regional, or indeed neuronal, selectivity to the actions of this anaesthetic. A precedent is give by recent observations made for the etomidate analogue, loreclezole. This anticonvulsant enhanced $GABA_A$ receptor mediated currents in only 50% of rat hippocampal dentate granule cells (Kapur and MacDonald, 1996). In all other aspects examined (*e.g.* benzodiazepine and GABA pharmacology), this neuronal population was homogeneous. Given the β-isoform selectivity of loreclezole, it might be postulated that β_1 subunits are expressed by the insensitive cells, whereas the β_2 or β_3 subunit may predominate on responsive neurones, a proposal consistent with the β-subunit distribution with the dentate gyrus (Wisden *et al.*, 1992). In our preliminary studies of rat cerebellar granule cells, the local application of etomidate (30 μM) produced a large inward current on all cells tested, suggesting the dominant expression of $GABA_A$ receptors containing β_2 or β_3 subunits. These subunit mRNAs are well represented in this cell type, whereas the β_1 subunit mRNA is limited (Wisden *et al.*, 1992).

Concluding Remarks

Although general anaesthetic agents are structurally diverse, they may share a common molecular target, the $GABA_A$ receptor. Theoretically, such an action could result from a non-specific perturbation of the neuronal membrane surrounding the $GABA_A$ receptor protein. However, the evidence reported here strongly suggests the presence of distinct binding sites on the $GABA_A$ receptor protein for these anaesthetics. A number of laboratories are now applying the powerful techniques of domain exchange and site-directed mutagenesis to characterise the nature and location of the anaesthetic binding sites on the receptor protein and to determine amino acid residues that are essential for the perturbation of receptor channel kinetics that these agents produce. In this, the sesquicentenary of the first clinical use of chloroform

(Rushman *et al.*, 1996), perhaps we are beginning to understand the molecular mechanisms that underlie the remarkable behavioural effects that occur seconds after the administration of a general anaesthetic.

Acknowledgements

D.B. and M.P. are supported by the M.R.C., S.S. by Glaxo-Wellcome and A.-L.M. by Organon Teknika. This work was partly supported by an equipment grant from the Royal College of Anaesthetists. We are grateful to Dr. Paul Whiting for providing wild-type and mutant $GABA_A$ receptor cDNAs and L(tk-) cell line expressing $GABA_A$ receptors. We thank Dr. J. Boulter for the neuronal nicotinic cDNAs and Drs. N. Hamilton and R. Logan for the isomers of etomidate. We are grateful to Mr. I. Mair for the preparation of illustrative material.

References

Akaike, N., Hattori, K., Inomata, N.& Oomura, Y. (1985) *J. Physiol.* **360**, 367-386.

Andrews, P.R. & Mark, L.C. (1982) *Anesthesiology* **57**, 314-320.

Barker, J.L. & Ransom, B.R. (1978) *J. Physiol.* **280**, 355-372.

Barker, J.J., Harrison, N.L., Lange, G.D. & Owen, D.G. (1987) *J. Physiol.* **386**, 485-501.

Belelli, D., Balcarek, J.M., Hope, A.G., Peters, J.A., Lambert, J.J. & Blackburn, T.P. (1995) *Mol. Pharmacol.* **48**, 1054-1062.

Belelli, D., Callachan, H., Hill-Venning, C., Peters, J.A. & Lambert, J.J. (1996) *Br. J. Pharmacol.*, **118**, 563-576.

Birnir, B., Tierney, M.L., Dalziel, J.E., Cox, G.B. & Gage, P.W. (1997) *J. Memb. Biol.* **155**, 157-166.

Chen, R., Belelli, D., Lambert, J.J., Peters, J.A., Reyes, A. & Lan, N.C. (1994) *Proc. Natl. Acad. Sci., U.S.A.*, **91**, 6069-6073.

Child, K.J., Gibson, W., Harnby, G. & Hart, J.W. (1972). *Postgrad. Med. J.*, **48** (Suppl.), 37-43.

Christenson, H.D. & Lee, I.S. (1973) *Toxicol. Appl. Pharmacol.* **26**, 495-503.

Cottrell, G.A., Lambert, J.J. & Peters, J.A. (1987) *Br. J. Pharmacol.* **90**, 491-500.

Doenicke, A., Löffler, B. Kugler, J.. Suttmann, H. & Grote, B. (1982) *Br. J. Anaesth.* 54, 393-400.

Doenicke, A. & Ostwald, P. (1997) *In Textbook of Intravenous Anesthesia* (Ed. White, P.F.) pp 93-109. Williams & Wilkins, Baltimore.

Franks, N.P. & Lieb, W.R. (1987) *Trends Pharmacol. Sci.* **8**, 169-174.

Franks, N.P & Lieb, W.R. (1994) *Nature* **367**, 607-614.

Hadingham, K.L., Garrett, E.M., Wafford, K.A., Bain, C., Heavens, R.P., Sirinathsinghji, D.J.S. & Whiting, P.J. (1996) *Mol. Pharmacol.* **49**, 253-259.

Hadingham, K.L., Wingrove, P.B., Wafford, K.A., Bain, C., Kemp, J.A., Palmer, K.J., Wilson, A.W., Wilcox, A.S., Sikela, J.M., Ragan, C.I. & Whiting, P.J. (1993) *Mol. Pharmacol.*, **44**, 1211-1218.

Harris, R.A., Mihic, S.J., Dildy-Mayfield, J.E. & Machu, T. (1995) *FASEB J.* **9**, 1454-1462.

Harrison, N.L. & Simmonds, M.A. (1984) *Brain Res.* **323**, 287-292.

Harrison, N.L., Vicini, S. & Barker, J.L. (1987) *J. Neurosci.* **7**, 604-609.

Heykants, J.J.P., Meuldermans, W.E.G., Michelis, L.J.M., Lewi, P.J. and Janssen, P.A. (1975). *Arch. Int. Pharmacodyn.* **216**, 113-129.

Hill-Venning, C., Belelli, D., Peters, J.A. & Lambert, J.J. (1997) *Br. J. Pharmacol.* **120**, 749-756

Hill-Venning, C, Lambert, J.J., Peters, J.A. & Hales, T.G. (1991) In *Neurosteroids and Brain Function* (eds. Costa, E. & Paul, S.M.) pp 77-85. Thieme, New York.

Huang, L.-Y.M. & Barker, J.L. (1980) *Science*, **207**, 195-197.

Kapur, J. & MacDonald, R.L. (1996) *Mol. Pharmacol.* **50**, 458-466.

Koblin, D.D., Chortkoff, B.S., Laster, M.J., Eger, E.I. II, Halset, M.J. & Ionescu, P. (1994) *Anesth. Analg.* **79**, 1043-1048.

Lambert, J.J., Belelli, D., Hill-Venning, C. & Peters, J.A. (1995) *Trends Pharmacol. Sci.* **16**, 295-303.

Lambert, J.J., Hill-Venning, C., Peters, J.A., Surgess, N.C. & Hales, T.G. (1991) In *Transmitter Amino Acid Receptors: Structures, Transduction and Models for Drug Development* (Eds. Barnard, E.A. & Costa, E.). pp 219-236. Thieme, New York.

Lambert, J.J., Peters, J.A., Sturgess, N.C. & Hales, T.G. (1990) In *Steroids and Neuronal Activity* (Eds Chadwick, D, & Widdows, K.) pp 56-82. Wiley, Chichester.

Little, H.J. (1996) *Pharmacol. Ther.*, **69**, 37-58.

Liu, J., Laster, M.J., Taheri, S., Eger, E.I. II, Chortkoff, B. & Halsey, M.J. (1994) *Anesth. Analg.* **79**, 1049-1055.

Majewska, M.D., Harrison, N.L., Schwartz, R.D., Barker, J.L. & Paul, S.M. (1986) *Science* **232**, 1004-1007.

Marszalec, W. & Narahashi, T. (1993) *Brain Res.*, 608, 7-15.

Meuldermans, W.E. and Heykants, J.J.P. (1976) *Arch. Int. Pharmacodyn.*, **221**, 150-162.

Olsen, R.W. Fischer, J.B. & Dunwiddie, T.V. (1986) In *Molecular and Cellular Mechanisms of Anaesthetics.* (eds. Roth, S.H. & Miller, K.W.) pp. 165-177. New York, Plenum Press.

Orser, B.A, Burtlik, M., Wang, L.-Y. & MacDonald, J.F. (1995). *Br. J. Pharmacol.* **116**, 1761-1768.

Ortells, M.O. & Lunt, G.C. (1995) Trends Neurosci. **18**, 121-127.

Peters, J.A., Kirkness, E.F., Callachan, H. Lambert, J.J. & Turner, A.J. (1988) *Br. J. Pharmacol.* **94**, 1257-1269.

Raines, D.E. & Miller, K.W. (1994) *Anesth. Analg.* **79**, 1031-1033.

Rushman, G.B., Davies, N.G.H. & Atkinson, R.S. (1996) *A Short History of Anaesthesia: The First Hundred and Fifty Years.* Butterworth-Heinemann, Oxford.

Sear, J.W., Phillips, K.C., Andrews, C.H.J. & Prys-Roberts, C. (1983) *Anaesthesia* **38**, 931-936.

Sieghart, W. (1995) *Pharmacol. Rev.*, **47**, 182-234.

Servin, F. (1988) *Anesthesiology*, **69**, 887-891.

Shafer, S.L. (1993) *J. Clin. Anesth.* **5** (Suppl. 1), 14S-21S.

Shyr, M.H., Tsai, T.H. Tan, P.P. Chen, C.F. & Chan, S.H. (1995) *Neurosci. Lett.* 184, 212-215.

Smith, G.B. & Olsen, R.W. (1995) *Trends Pharmacol. Sci.* **16**, 162-168.

Tanelian, D.L., Kosek, P., Mody, I. & MacIver, B. (1993) *Anesthesiology*, **78**, 757-776.

Ticku, R.K. and Rastogi, S.K. (1986) In *Molecular and Cellular Mechanisms of Anaesthetics* (Eds. Roth, S.H. & Miller, K.W.) pp. 179-188. New York, Plenum Press.

Wafford, K.A., Bain, C.J., Quirk, K., McKernan, R.M., Wingrove, P.B., Whiting, P.J. and Kemp, J.A. (1994) *Neuron*, **12**, 775-782.

Wingrove, P.B., Wafford, K.A., Bain, C. & Whiting, P.J. (1994) *Proc. Natl. Acad. Sci., U.S.A.*, **91**, 4569-4573.

Wisden, W., Laurie, D.J., Monyer, H. & Seeburg, P.H. (1992) *J. Neurosci.*, **12**, 1040-1062.

Wittmer, L.L., Hu, Y., Kalkbrenner, M., Evers, A.S., Zorumski, C.F. & Covey, D.F. (1996) *Mol. Pharmacol.* **50**, 1581-1586.

Xu, M & Akabas, M.H. (1993) *J. Biol. Chem.* **268**, 21505-21508.

Anesthetic Effects on the Spinal Cord

Joan J Kendig

DEPARTMENT OF ANESTHESIA, STANFORD UNIVERSITY SCHOOL OF MEDICINE, STANFORD, USA

1 Introduction

1.1 Definitions of the state of anesthesia

The state of anesthesia has been difficult to define, in part because it is a phenomenon with qualitative endpoints rather than one with a traditional pharmacologic dose-response curve. A number of such endpoints are considered to be part of an anesthetic state. In rough order of increasing anesthetic concentrations required for each, they are amnesia, analgesia, loss of consciousness defined in humans as cessation of response to verbal command, loss of spontaneous movement or 'sleep time', loss of righting reflex, absence of purposeful movement with response to a noxious stimulus and absence of autonomic response to a noxious stimulus. Concentrations for the last two endpoints vary according to the strength of the stimulus. For inhalation agents, which as inspired gases can easily be given continuously to approach equilibrium, the most widely used anesthetic endpoint has been the absence of withdrawal movement in response to a defined noxious stimulus. Absence of movement is the endpoint used to define MAC, the *M*inimal *A*lveolar *A*nesthetic *C*oncentration that just prevents movement in response to a standard painful stimulus in 50% of subjects (Quasha et al. 1980; White et al. 1974). MAC is the most commonly accepted standard for comparing anesthetic potencies among inhalation agents.

1.2 Evidence for the role of the spinal cord

The spinal cord plays a critical role in anesthesia according to this particular definition. Recent studies have provided strong evidence that MAC is due to anesthetic effects on the spinal cord. In rats, spinal cord transection does not affect MAC as measured by withdrawal ofa limb caudal to the transection (Rampil, 1994; Rampil et al. 1993). In goats, preferential delivery of an inhalation agent to supraspinal structures alone results in a MAC determination that is higher than when the agent is delivered to the whole animal

(Antognini and Schwartz, 1993) or preferentially to the spinal cord (Borges and Antognini, 1994). The latter finding is of interest, inasmuch as it suggests supraspinal effects may play a role in determining anesthetic potency, but that supraspinal actions are in a direction that opposes spinal effects.

In the history of neurophysiology, many of the early studies on anesthetic effects were carried out in spinal cord. Examples are the work of Somjen comparing ether with thiopental on the various forms of synaptic transmission recognized at that time (Somjen, 1967; Somjen, 1963; Somjen and Gill, 1963). One of the important studies suggesting that some anesthetics may hyperpolarize neurons was done in spinal cord (Nicoll, 1975). More recently a number of studies documenting inhalation anesthetic effects on properties of single neurons in the dorsal horn have been carried out with results suggesting complex alterations in receptive field properties (Collins et al. 19951 Collins and Ren, 1987; Namiki er al. 1980). Evidence has also been adduced that inhalation anesthetic agents may directly or indirectly alter motor neuron excitability at concentrations related to MAC O(ing and Rampil, 1994).

1.3 Inhalation agents and the Meyer-Overton correlation

For a long time structural variability among anesthetic agents within the class of gaseous and volatile agents led to exclusive reliance on the correlation between lipid solubility and anesthetic potency as a valid, and indeed the only known modifying characteristic of these agents. This relationship, independently described by Meyer (Meyer, 1899) and by Overton (Overton, 1901) nearly 100 years ago, presents lipid solubility as the defining characteristic of a general anesthetic agent. The corollary, particularly after the lipid nature of the cell membrane was understood, was that anesthetic agents exerted their effects by some non-specific action on membrane lipids. For many years this was the central hypothesis of anesthetic action. However, there are many studies showing that agents exert effects specific to certain membrane proteins (Franks and Lieb, 1994; Franks and Lieb, 1993; Richards, 1973; Richards et al. 1975; Richards and Smaje, 1976; Richards and White, 1975; Tanelian et al. 1993) and that inhalation agents differ among themselves (MacIver and Kendig, 1991; MacIver and Roth. 1988). It is now much more likely that direct actions on proteins are involved.

A more direct line of evidence suggesting that the long-standing Meyer-Overton correlation must be modified is provided by the discovery that a large number of compounds do not obey the correlation.

For inhalation agents in clinical use, the product of MAC × oil/gas partition coefficient is roughly constant, with a value approximately 2. For a series of clinical anesthetic agents, the product of MAC determined in rats and the olive oil/gas partition coefficient is 1.82 ± 0.56 (mean ± SD) (Taheri et al. 1991). However some compounds, highly halogenated structures, have lipid solubilities that predict they should be anesthetics but have no anesthetic properties (Koblin et al. 1994). Short straight-chain alcohols, on the other hand, violate the Meyer-Overton correlation in the opposite direction: in adult rats they are more potent than their lipid solubilities predict, with an olive oil/gas × MAC product of 0.1-0.2 (Fang, Ibnescu, Chortkoff et al. unpublished data). This constant is an order of magnitude less than the constant derived from the Meyer-Overton correlation.

1.4 Studies in spinal cord

1.4.1 Design of the study. We have used an *in vitro* spinal cord preparation to compare the actions of agents which are in violation of the Meyer-Overton correlation and those which obey it. The specific compounds are ethyl alcohol, the clinical anesthetic agent isoflurane, and two experimental halogenated cyclobutanes. The cyclobutane 1-chloro-1,2,2-trifluorocyclobutane ($(CH_2)_2CClF(CF_2)$) is an anesthetic at a partial pressure of 0.0143 atm and has an olive oil/gas partition coefficient of 248: it thus falls within the prediction of the Meyer-Overton correlation. A related compound, 1,2-dichlorohexafluorocyclobutane ($(CF_2)_2(CClF)_2$) has an oil/gas partition coefficient of 43.5 but does not immobilize rats at concentrations up to 0.11 atm, although the Meyer-Overton correlation predicts anesthesia at 0.046 atm. For the sake of brevity, these compounds will be referred to as 1A (the anesthetic) and 2N (the non-anesthetic) cyclobutanes respectively. The results with these agents are set in context with the profiles of several intravenous agents on the same preparation.

1.4.2 Properties of the spinal cord. The isolated spinal cord from neonatal rats was first developed by Otsuka and colleagues (Konishi and Otsuka, 1974) and shown to exhibit a slow ventral root potential (sVRP) response to noxious stimulation, mediated in part by substance P (Akagi et al. 1985; Yanagisawa et al. 1985). An early part of the sVRP is sensitive to NMDA antagonists (Evans et al. 1987; Woodlev and Kendig, 1991), whereas the fast monosynaptic reflex (MSR) is not, being mediated entirely by glutamate receptors of the non-NMDA types (Long et al. 1990). More recently we have shown that the excitatory postsynaptic potential(EPSP) which underlies the monosynaptic reflex is mediated by glutamate receptors of the AMPA class, with no contribution by kainate receptors (Tauck and Kendig, unpublished data). There is also a relatively small but distinct NMDA receptor-mediated component. In addition to ventral root potentials, dorsal root stimulation elicits a dorsal root potential (DRP) in adjacent dorsal roots. The DRP is evoked by glutamate released from primary afferent nerve terminals, acting on GABA-ergic interneurons; GABA released by the interneurons acts on $GABA_A$ receptors on primary afferent nerve terminals to depolarize them and inhibit transmitter release. A DRP can also be evoked directly by applying a GABA agonist such as muscimol, bypassing the glutamatergic receptors on the interneurons and evoking a pure $GABA_A$ response. Figure 1 shows a diagram of the circuitry in the spinal cord, the receptors which mediate several different evoked responses, and representative examples of the responses themselves.

The anatomical, temporal and pharmacological separation of the responses make it possible to determine the receptor-specific effects of anesthetic agents

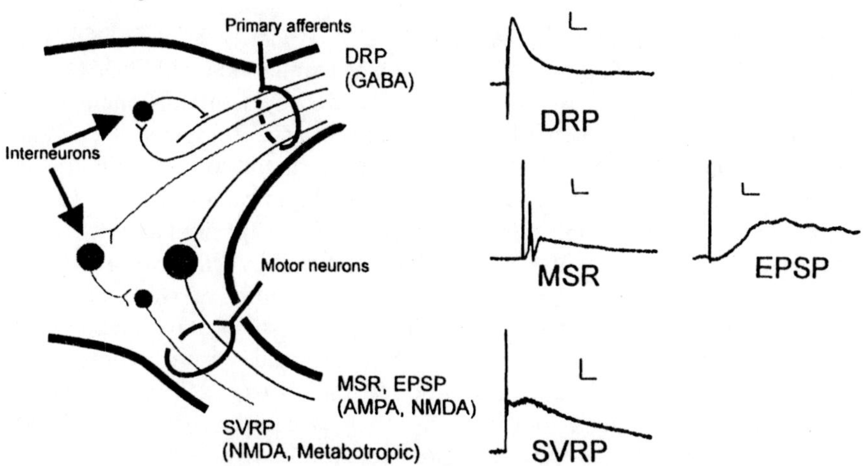

Figure 1 *Pathways in spinal cord, their associated neurotransmitters, and examples of the corresponding evoked responses. The dorsal root potential (DRP) is evoked when a stimulus to the dorsal root releases glutamate from primary afferent nerve terminals to glutamate receptors on GABA-ergic interneurons. The latter in turn release GABA to $GABA_A$ receptors on the primary afferent nerve terminals, with a resulting depolarizing conductance increase that inhibits further evoked glutamate release, a form of presynaptic inhibition. The monosynaptic reflex (MSR) is the compound action potential in motor neurons elicited by a glutamate AMPA receptor-mediated excitatory postsynaptic potential (EPSP) with a small NMDA receptor-mediated component. The slow ventral root potential (SVRP) is a very long response with an early phase sensitive to NMDA receptor antagonists and a late phase sensitive to a variety of metabotropic receptor antagonists. Calibration marks for DRP are 0.5 mV, 20 ms; for MSR 1 mV, 10 ms; for EPSP 0.5 mV, 10 ms; and for SVRP 0.1 mV, 1 s.*

on these pathways in relative isolation. We therefore examined the effects of the anesthetic agents on the MSR (AMPA), EPSP (AMPA and NMDA), early sVRP (NMDA), late sVRP (metabotrobic, possibly of more than one type, including NK1), DRP evoked by dorsal root stimulation (GABA, but with intervening glutamate activated interneurons) and DRP evoked by muscimol ($GABA_A$). The results permitted an evaluation of receptor-specific anesthetic actions, and in particular an evaluation of the relative importance of actions at the $GABA_A$ receptor. The original work on which these results are based was reported primarily in three papers and an abstract (Kendig and Gibbs, 1994a; Kendig et al. 1994; Savola et al. 1991; Wong et al. 1997)

2 Methods

Sprague-Dawley rat pups, 1-7 days old, were anesthetized with enflurane. decapitated, and spinal cords removed from mid-thorax through sacral regions in a protocol approved by the Institutional Animal Care and Use Committee. The cords were superfused with artificial cerebrospinal fluid (ACSF) of the following composition in mM: NaCl 123, KCl 5, $CaCl_2$ 2, NaH_2PO_4 1.2, $MgSO_4$ 1.3, $NaHCO_3$ 26, glucose 30. ACSF was equilibrated with 95% O_2-5% CO_2 to a pH of 7.3-7.4; temperature was measured by a thermistor near the cord and maintained at 27-28°C, which is physiological for pups of this age when not under the mother. Volatile agents were dissolved in ACSF to the desired concentration and delivered from glass syringes via an infusion pump. In some experiments with isoflurane the anesthetic was vaporized and delivered in the gas phase to a spinal cord thinly covered by circulating ACSF. Concentrations were confirmed by gas chromatography of compounds in equilibrium with the gas phase, or in the case of some experiments with isoflurane by infrared measurement. Suction electrodes were arranged to stimulate a lumbar dorsal root, typically L4, and to record from the corresponding ipsilateral ventral root or from an adjacent dorsal root. In some experiments muscimol, was applied by pressure ejection from a pipette with tip positioned near the insertion of the dorsal root.

After a recovery period during which stability of the preparation's responses was verified, control recordings were made and the preparation exposed for 30 min to a single concentration of one agent, followed by washing with drug-free ACSF. Responses were digitized, averaged in groups of 5, and recorded for later analysis. Responses to muscimol application were not averaged. Responses in the presence of anesthetic agent were expressed relative to control.

3 Results

3.1 Glutamate Receptor - Mediated Neurotransmission

3.1.1 The Glutamate AMPA Receptor. The monosynaptic reflex is made up of a compound action potential in motor neurons; triggering the action potential is an excitatory postsynaptic potential with a predominant component mediated by glutamate receptors of the AMPA class. Agents which act on the monosynaptic reflex therefore may be exerting their effects on impulse initiation and/or AMPA-receptor mediated synaptic transmission. Isoflurane, ethanol and the anesthetic cyclobutane 1A each depressed the MSR at concentrations within their anesthetic range in rats; the nonanesthetic cyclobutane 2N either had no effect or at some concentrations actually enhanced this response. Examples are shown in Figure 2.

The population EPSP which underlies the monosynaptic reflex can normally not be evoked without the compound action potential. Recently we have devised a way of eliciting this response in isolation from the action potentials, and have examined the effects of ethanol on the EPSP. Figure 3 shows examples of ethanol effects on the entire EPSP (3A), on its AMPA receptor-mediated components isolated by treatment with the NMDA receptor antagonist CPP (3B) and the dose-response curve for the AMPA receptor-mediated component (3C). It is very probable that the depression shown is sufficient to account for depression of the monosynaptic reflex by ethanol.

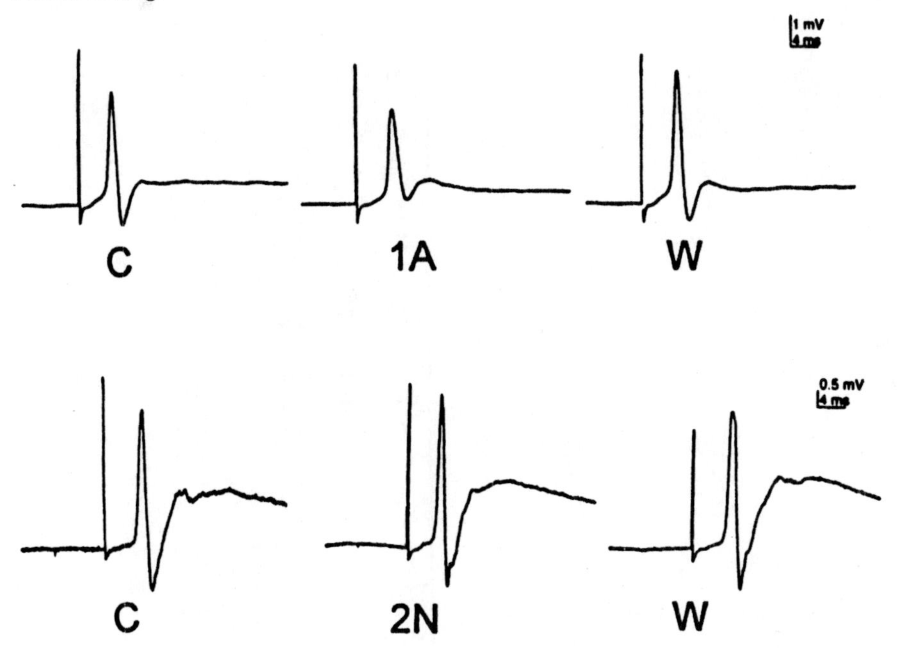

Figure 2 *Contrasting effects of two fluorinated cyclobutanes on the monosynaptic reflex. 1A is an anesthetic within the predictions of the Meyer-Overton correlation; 2N is not. C, control; W, wash. 1A reversibly depressed the monosynaptic reflex at its anesthetic concentration (MAC). 2N increased monosynaptic reflex amplitude and reduced latency at a concentration equivalent to 0.625 of its predicted anesthetic concentration according to the Meyer-Overton correlation. Taken by permission from Kendig et al., Eur. J. Pharmacol. 264: 427-436, 1994.*

3.1.2 The glutamate NMDA receptor. Although the population EPSP is predominantly an AMPA-receptor mediated response, there are two NMDA receptor-mediated components, one relatively short-latency and probably monosynaptic, which remains in the presence of an AMPA antagonist, the second longer latency, probably polysynaptic, abolished by either AMPA or NMDA antagonists and leading into the very long latency sVRP. Ethanol potently depresses the NMDA receptor-mediated response (Figure 3C, D).

We have examined the second, polysynaptic response by isolating the first 500 ms of the slow VRP, which is maximally sensitive to NMDA receptor antagonists. This too is sensitive to ethanol as well as to isoflurane and the anesthetic cyclobutane.

3.1.3 The inhibitory $GABA_A$ receptor. $GABA_A$ responses can be evoked by focal brief applications of a GABA agonist such as muscimol to a region near the insertion of the dorsal root. When such responses are evoked, inhalation agents such as isoflurane clearly enhance them (Figure 4). The same is true for

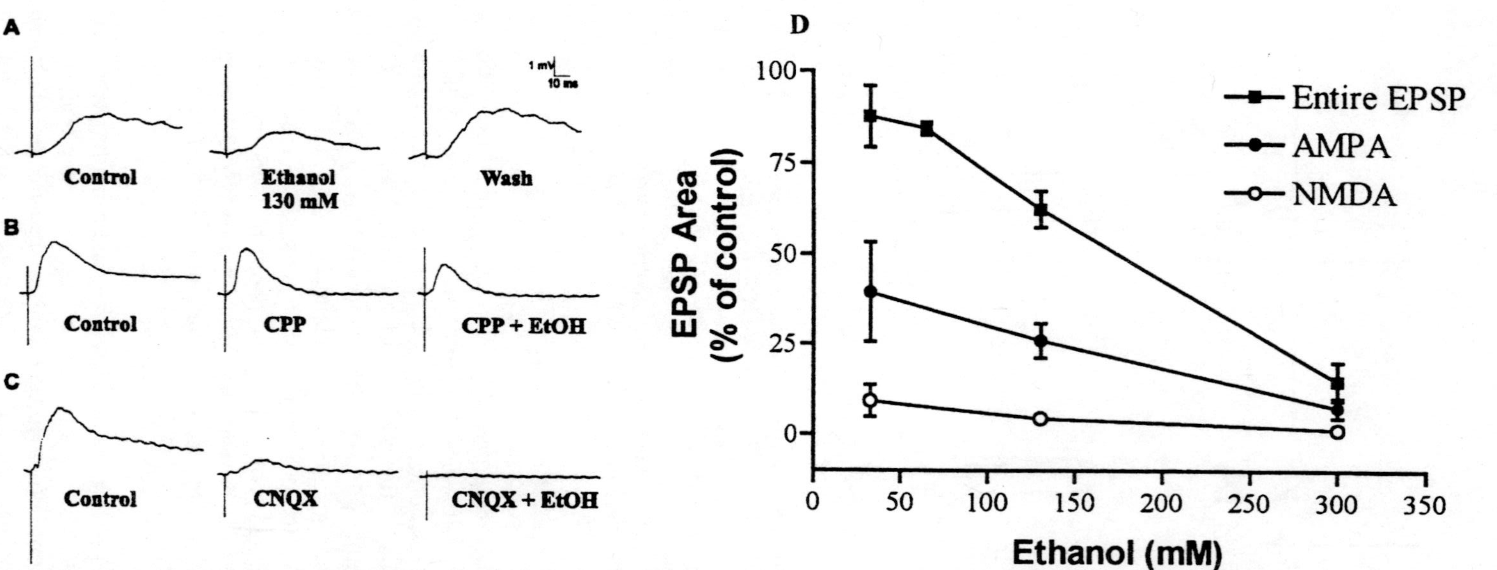

Figure 3 *Effects of ethanol on the population excitatory postsynaptic potential (EPSP) that underlies the monosynaptic reflex and on its glutamate AMPA and NMDA receptor-mediated components. A, the intact EPSP is reversibly depressed by 130 mM ethanol. B, the NMDA receptor antagonist CPP abolishes the relatively slow components of the EPSP; the remaining AMPA receptor-mediated responses is depressed by ethanol (CPP + EtOH) to an extent sufficient to account for depression of the monosynaptic reflex. C, the same experiment with the AMPA antagonist CNQX; the remaining NMDA receptor-mediated response is depressed by ethanol. D, dose-response curves for the entire EPSP and for the AMPA and NMDA components. The anesthetic concentration of ethanol in adult rats is 97 mM and in neonates 263 mM.*

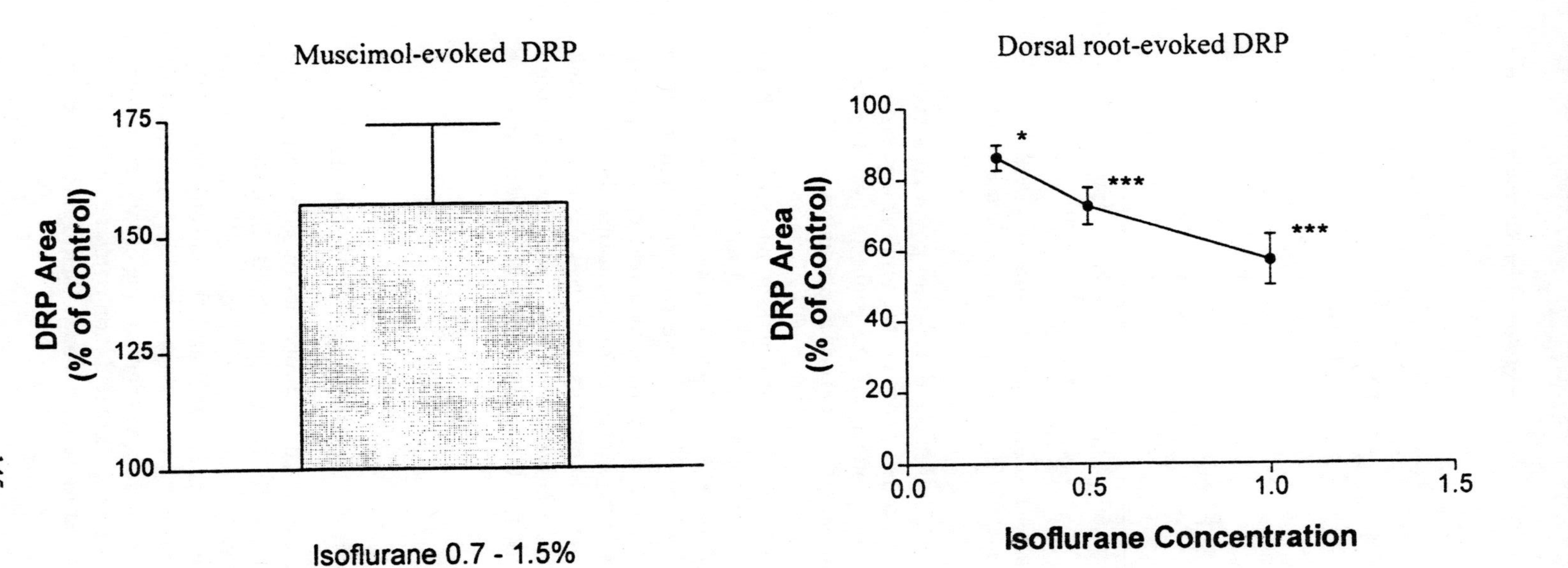

Figure 4 *Contrasting effects of isoflurane on a $GABA_A$ response evoked by direct application of the agonist muscimol or via the normal interneuronal circuitry. Isflurane increases the area of the muscimol-evoked DRP at concentrations which bracket the anesthetic partial pressure. When the response is evoked by dorsal root stimulation and includes the GABAergic interneurons, however, the effect of isoflurane is depression. Anesthetic partial pressure of isoflurane in rats is 1.4%.*

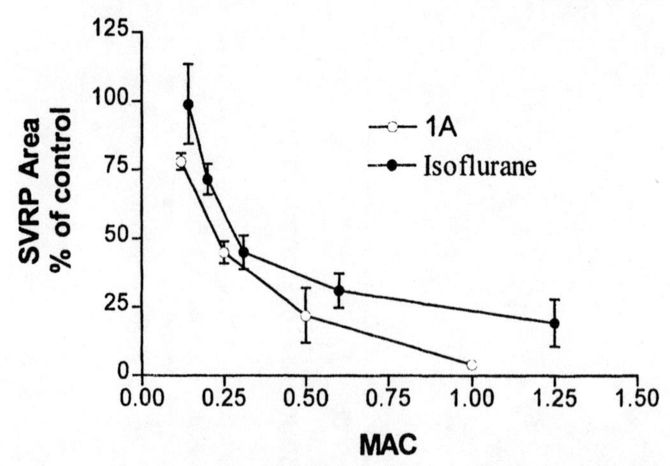

Figure 5 *Similar effects of isoflurane and the experimental anesthetic cyclobutane 1A on the slow ventral root potential (SVRP) at comparable anesthetic potencies.*

the anesthetic cyclobutane. However when the dorsal root potential is evoked by dorsal root stimulation via its normal pathway, the result is depression of the response (Figure 4), in the case of any of the 3 anesthetics. The non-anesthetic cyclobutane 2N is without effect or exerts slight depression.

3.1.4 Metabotropic receptors of the slow ventral root potential. All the anesthetic agents tested potently depress the slow ventral root potential. The dose-response curves for ethanol and the anesthetic cyclobutane are shown in Figure 5. When the response is separated into the early component sensitive to NMDA antagonists and the later component mediated by a variety of metabotropic receptors, there is some selectivity for the former but it is not pronounced. The non-anesthetic cyclobutane also depresses this response, but its effects can be attributed to the relatively low oxygen pressure in solutions containing high concentrations of this compound; the sVRP is very sensitive to changes in oxygen tension.

4 Discussion

4.1 Comparison to intravenous anesthetic and analgesic agents

In a number of studies we have examined the properties of sedative/hypnotic and analgesic agents on the same pathways in spinal cord (Brockmeyer and Kendig, 1995; Feng and Kendig, 1996; Feng and Kendig, 1995a; Feng and Kendig, 1995b; Feng and Kendig, 1996; Jewett et al. 1992; Savola et al. 1991). The comparative profiles of their actions are shown in Figure 6. Two main points should be noted. First, inhalation agents are unique in their capacity to diminish the glutamate AMPA-receptor mediated monosynaptic reflex. If the results with ethanol can be extended to isoflurane and other agents, this effect

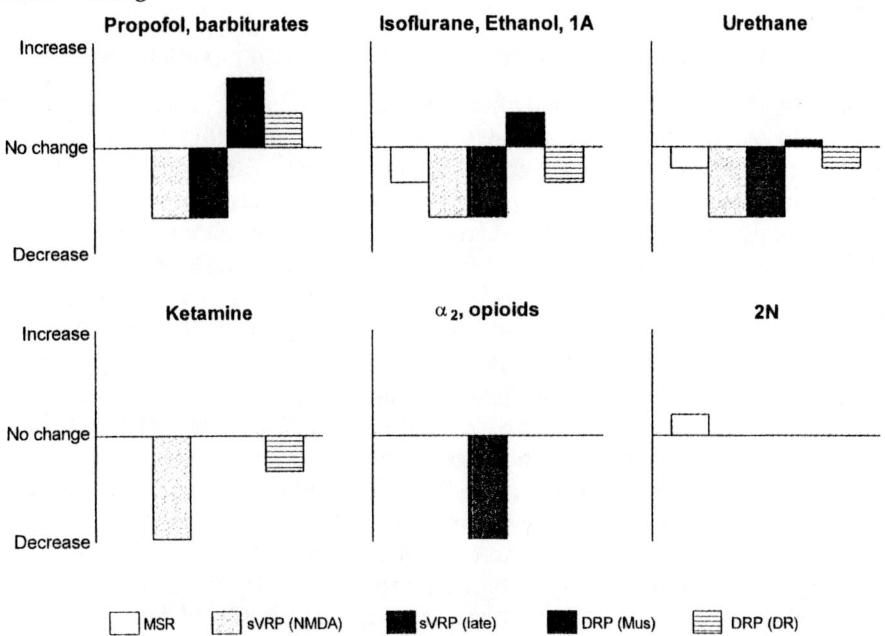

Figure 6 *Profiles of inhalation and intravenous agent effects on receptor-specific responses in spinal cord. Responses include the monosynaptic reflex (MSR), the early NMDA receptor-mediated slow ventral root potential (sVRP(NMDA)), the late metabotropic receptor-mediated slow ventral root potential (sVRP(late)), the dorsal root potential directly evoked by the agonist muscimol (DRP (Mus)) and the dorsal root potential evoked through the normal circuitry by stimulating a dorsal root (DRP (DR)). Depression of the monosynaptic reflex at anesthetic concentrations is a property of inhalation agents that distinguishes them from intravenous agents. This property is shared by ethanol and the experimental anesthetic cyclobutane 1A; the non-anesthetic cyclobutane acts in the opposite direction. Enhancement of the directly evoked GABA response is seen with both inhalation agents and propofol and barbiturates; however the dorsal root potential evoked by the normal pathway is depressed by inhalation agents.*

is due at least in part to inhibition of excitatory synaptic transmission. None of the sedative/hypnotics tested exert effects on this response at their effective anesthetic concentrations. Nor do the analgesic agents. In addition, this is the only response to offer a potentially interesting discrimination between the anesthetic and non-anesthetic cyclobutanes. Second, the relative potency of anesthetic agents in enhancing GABA-mediated responses varies. Propofol and barbiturates exert the largest effect, sufficient to enhance both the directly evoked response to a GABA agonist and the response evoked via the normal pathway containing glutamate-excited interneurons. The difference suggests a larger role for the GABA receptor in propofol or barbiturate anesthesia than is the case for inhalation agents or ethanol.

4.2 Depression of glutamate-mediated excitatory transmission

The methods used in the present study cannot discriminate between actions on the postsynaptic side and actions that inhibit transmitter release via a presynaptic effect. Arguing against a general role for interference with glutamate receptor functions as playing a role in anesthesia are two observations. First, halothane has been reported to reduce glutamate-mediated responses evoked by presynaptic transmitter release but not by direct agonist application (Perouansk et al. 1995; Richards and Smaje, 1976). This, and unselective effects on NMDA and nonNMDA receptor-mediated responses, argues against a role for glutamate receptors in the actions of this agent. On the other hand other inhalation agents including isoflurane do block glutamate currents at clinical concentrations (El-Beheiry and Pull, 1989; Richards and Smaje, 1976). Also arguing against a universal role for glutamate actions is the lack of correlation between effects at glutamate receptors expressed in oocytes and anesthetic potency for long-chain alkanols (Harris et al. 1995a). On the other hand glutamate actions for ethanol and for many of the inhalation agents are well known and may contribute to the effects of these agents if not all inhalation agents. NMDA-evoked responses are inhibited by ethanol at concentrations associated with intoxication but below those which cause general anesthesia in several preparations (Dildy-Mayfield et al. 1996; Lovinger et al. 1990; Lovinger et al. 1989; Morrisen and Swartzwelder, 1993; Peoples and Weight, 1995). Some studies suggest that glutamate non-NMDA receptors are also affected at modest ethanol concentrations (Dildy-Mayfield et al. 1996; Monisett and Swartzwelder, 1993), but other studies report only modest inhibition at high concentrations (Weight, 1992).

4.3 The importance of GABA receptors

Ethanol enhances activity at $GABA_A$ receptors at relatively modest concentrations (Harris et al. 1995b; Klein et al. 1995; Leidenheimer and Harris, 1992), and the ability of alkanols to enhance $GABA_A$ function has been correlated with their general anesthetic potencies (Dildy-Mayfield et al. 1996; Mihic et al. 1994). An important role for GABA enhancement in anesthesia is proposed for inhalation general anesthetics and some intravenous agents with known sites on the GABA receptor such as propofol, barbiturates, and benzodiazepines (Franks and Lieb, 1994; Tanelian et al. 1993). However in the spinal cord when the DRP is evoked via the native circuitry rather than by direct application of a $GABA_A$ agonist, the effect of ethanol (Wong et al. 1997) as well as that of inhalation agents (Kendig and Gibbs, 1994b) is depression. This is in contrast to propofol, barbiturates (Jewett et al. 1992) and benzodiazepines (Slarey ct al. 1994) which enhance both directly evoked GABA responses and responses evoked by dorsal root stimulation. Similar depressant effects on inhibitory transmission to motor neurons in spinal cord (Takenoshita and Yoshiya, 1994) and to hippocampal neurons (Perouansky et al. 1996) have also been reported. The depressant effects on the circuitry in the present study can be due to inhibition of glutamate receptors on the interneurons and/or to a non-specific inhibition of inhibitory as well as excitatory neurotransmitter release. Recent studies suggest that halothane blocks excitement of GABA-ergic interneurons (Perouansky et al. 1996). The observation that general anesthetic agents inhibit rather than enhance GABA, inhibition when the response is evoked by normal neurotransmission leads to questions about the commonly assumed role of enhancement of GABA inhibition in general anesthesia as a generality applicable to all agents. In particular, the opposing effects on enhancement of GABA receptor function and depressant effects on the neurons which release GABA suggest that anesthetic effects, even locally, are not unidirectional. The balance between enhancement and depression of total inhibition will be dependent not only on the properties of the anesthetics but on the circuitry.

5 Summary and Conclusions

Immobility in response to a noxious stimulus is one definition of the anesthetic state induced by volatile and gaseous anesthetic agents. There is strong evidence that immobility is produced by actions at the spinal level. In the spinal cord, an action unique to inhalation agents is ability to depress a response mediated by glutamate AMPA receptors. NMDA receptor-mediated responses are also sensitive. The contribution of enhancement of GABA-ergic inhibition is difficult to define because of opposing actions on the GABA receptors themselves and on GABA-ergic neurons.

Acknowledgments

Supported by National Institutes of Health Grants NS 13 108 and GM47818 to JJK.

References

Akagi H, Konishi S, Otsuka M, Yanagisawa M (1985) The role of substance P as a neurotransmitter in the reflexes of slow time courses in the neonatal rat spinal cord. Br J Pharmacol 84:663-673.

Antognini JF, Schwartz K (1993) Exaggerated anesthetic requirements in the preferentially anesthetized brain. Anesthesiology 79:1244-1249.

Borges M, Antognini JF (1994) Does the brain influence somatic responses to noxious stimuli during isoflurane anesthesia? Anesthesiology 81:1511-1515.

Brockmeyer D, Kendig JJ (1995) Selective effects of ketamine on amino acid-mediated pathways in neonatal rat spinal cord. British Journal of Anaesthesia 74:79-84.

Collins JG, Kendig JJ, Mason P (1995) Anesthetic actions within the spinal cord: contributions to the state of general anesthesia. TINS 18:549-553.

Collins JG, Ren K (1987) WDR response profiles of spinal dorsal horn neurons may be unmasked by barbiturate anesthesia. Pain 28:369-378.

Dildy-Mayfield JE, Mihic SJ, Liu Y, Deitrich RA, Harris RA (1996) Actions of long chain alcohols on GABAA and glutamate receptors: relation to in vivo effects. Br J Pharmacol 118:378-384.

El-Beheiry H, Puil E (1989) Postsynaptic depression induced by isoflurane and Althesin in neocortical neurons. Experimental Brain Research 75:361-368.

Evans RH, Evans SJ, Pook PC, Sunter DC (1987) A comparison of excitatory amino acid antagonists acting at primary afferent C fibres and motoneurones of the isolated spinal cord of the rat. Br J Pharmacol 91:531-537.

Feng J, Kendig JJ (1995a) Selective effects of alfentanil on nociceptive-related neurotransmission in neonatal rat spinal cord. British Journal of Anaesthesia 74:691-696.

Feng J, Kendig JJ (1995b) *N*-Methyl-D-aspartate receptors are implicated in hyperresponsiveness following naloxone reversal of alfentanil in isolated rat spinal cord. Neurosci Lett 189:128-130.

Feng J, Kendig JJ (1996) Synergistic interactions between midazolam and alfentanil in isolated neonatal rat spinal cord. Br J Anaesth 77:1-6.

Feng JQ, Kendig JJ (1996) The NMDA receptor antagonist MK-801 differentially modulates m and K opioid actions in spinal cord *in vitro*. Pain 66:343-349.

Franks NP, Lieb WR (1993) Selective action of volatile general anaesthetics at molecular and cellular levels. Br J Anaesth 71:65-76.

Franks NP, Lieb WR (1994) Molecular and cellular mechanisms of general anaesthesia. Nature 367:607-614.

Harris RA, Mihic SJ, Dildy-Mayfield JE, Machu TK (1995a) Actions of anesthetics on ligand-gated ion channels: role of receptor subunit composition. FASEB J 9:1454-1462.

Harris RA, Proctor WR, McQuilkin SJ, Klein RL, Mascia MP, Whatley V, Whiting PJ, Dunwiddie TV (1995b) Ethanol increases GABAA responses in cells stably transfected with receptor subunits. Alcohol Clin Exp Res 19:226-232.

Jewett BA, Tarasiuk A, Gibbs L, Kendig JJ (1992) Propofol and barbiturate depression of spinal nociceptive neurotransmission. Anesthesiology 77:1148-1154.

Kendig JJ, Kodde A, Gibbs LM, Ionescu P, Eger EI, II (1994) Correlates of anesthetic properties in isolated spinal cord: cyclobutanes. Eur J Pharmacol 264:427-436.

Kendig JJ, Gibbs LM (1994a) The $GABA_A$ receptor in anesthesia: Isoflurane. Anesthesiology 81:A1477

Kendig JJ, Gibbs LM (1994b) The GABAA Receptor in Anesthesia Isoflurane. Anesthesiology 81, No. 3A:A1477

King BS, Rampil IJ (1994) Anesthetic depression of spinal motor neurons may contribute to lack of movement in response to noxious stimuli. Anesthesiology 81:1484-1492.

Klein RL, Mascia MP, Whiting PJ, Harris RA (1995) GABAA receptor function and binding in stably transfected cells: chronic ethanol treatment. Alcohol Clin Exp Res 19:1338-1344.

Koblin DD, Chortkoff BS, Laster MJ, Eger EI, II, Halsey MJ, Ionescu P (1994) Polyhalogenated and perfluorinated compounds that disobey the Meyer-Overton hypothesis. Anesth Analg 79:6:1043-1048.

Konishi S, Otsuka M (1974) Electrophysiology of mammalian spinal cord *in vitro*. Nature 252:733-735.

Leidenheimer NJ, Harris RA (1992) Acute effects of ethanol on $GABA_A$ receptor function: Molecular and physiological determinants. In: GABAergic Synaptic Transmission (Biggio G, Concas ed), pp 269-279. New York: Raven Press.

Long SK, Smith DA, Siarey RJ, Evans RH (1990) Effect of 6-cyano-2,3-dihydroxy-7-nitro-quinoxaline (CNQX) on dorsal root-, NMDA-, kainate- and quisqualate-mediated depolarization of rat motorneurones in vitro. Br J Pharmacol 100:850-854.

Lovinger DM, White G, Weight FF (1989) Ethanol inhibits NMDA-activated ion current in hippocampal neurons. Science 243:1721-1724.

Lovinger DM, White G, Weight FF (1990) Ethanol inhibition of neuronal glutamate receptor function. Ann Med 22:247-252.

MacIver MB, Kendig JJ (1991) Anesthetic effects on resting membrane potential are voltage-dependent and agent-specific. Anesthesiology 74:83-88.

MacIver MB, Roth SH (1988) Inhalation anaesthetics exhibit pathway-specific and differential actions on hippocampal synaptic responses *in vitro*. British Journal of Anaesthesia 60:680-691.

Meyer HH (1899) Zur theorie der alkoholnarkose. I. Mit welch Eigenshaft der Anasthetika bedingt ihre narkotische Wirkung? Arch Exp Path Pharmak 42:109

Mihic SJ, Whiting PJ, Harris RA (1994) Anaesthetic concentrations of alcohols potentiate $GABA_A$ receptor-mediated currents: lack of subunit specificity. Eur J Pharmacol 268:209-214.

Morrisett RA, Swartzwelder HS (1993) Attenuation of hippocampal long-term potentiation by ethanol - a patch-clamp analysis of glutamatergic and gabaergic mechanisms. J Neurosci 13:2264-2272.

Namiki A, Collins JG, Kitahata LM, Kikuchi H, Homma E, Thalhammer JG (1980) Effects of Halothane on spinal neuronal responses to graded noxious heat stimulation in the cat. Anesthesiology 53:475-480.

Nicoll RA (1975) Pentobarbital: action on frog motoneurons. Brain Res 96:119-123.

Overton E (1901) Studien uber die narkose zugleich ein beitrag zur allgemeinen pharmacologie. Jena: G.Fischer.

Peoples RW, Weight FF (1995) Cutoff in potency implicates alcohol inhibition of *N*-methyl-D-asparate receptors in alcohol intoxication. Proc Natl Acad Sci USA 92:2825-2829.

Perouansky M, Baranov D, Salman M, Yaari Y (1995) Effects of halothane on glutamate receptor-mediated excitatory postsynaptic currents. A patch-clamp study in adult mouse hippocampal slices. Anesthesiology 83:109-119.

Perouansky M, Kirson ED, Yaari Y (1996) Halothane blocks synaptic excitation of inhibitory interneurons. Anesthesiology 85:1431-1438.

Quasha AL, Eger EI, Tinker JH (1980) Determination and application of MAC. Anesthesiology 53:315-334.

Rampil IJ, Mason P, Singh H (1993) Anesthetic potency (MAC) is independent of telencephalic structures in the rat. Anesthesiology 78:707-712.

Rampil IJ (1994) Anesthetic potency is not altered after hypothermic spinal cord transection in rats. Anesthesiology 80:606-610.

Richards CD (1973) On the mechanism of halothane anaesthesia. Journal of Physiology 233:439-456.

Richards CD, Russell WJ, Smaje JC (1975) The action of ether and methoxyflurane on synaptic transmission in isolated preparation of the mammalian cortex. Journal of Physiology 248:121-142.

Richards CD, Smaje JC (1976) Anaesthetics depress the sensitivity of cortical neurones to L-glutamate. Br J Pharmacol 58:347-357.

Richards CD, White AE (1975) The actions of volatile anaesthetics on synaptic transmission in the dentate gyrus. Journal of Physiology 252:241-257.

Savola MKT, Woodley SJ, Maze M, Kendig JJ (1991) Isoflurane and an a_2-adrenoceptor agonist suppress nociceptive neurotransmission in neonatal rat spinal cord. Anesthesiology 75:489-498.

Siarey RJ, Long SK, Tulp MTH, Evans RH (1994) The effects of central myorelaxants on synaptically-evoked primary afferent depolarization in the immature rat spinal cord in vitro. Br J Pharmacol 111:497-502.

Somjen G (1967) Effects of anesthetics on spinal cord of mammals. Anesthesiology 28:135-143.

Somjen GG (1963) Effects of ether and thiopental on spinal presynaptic terminals. J Pharmacol Exp Ther 140:396-402.

Somjen GG, Gill M (1963) The mechanism of the blockade of synaptic transmission in the mammalian spinal cord by diethyl ether and by thiopental. J Pharmacol Exp Ther 140:19-30.

Taheri S, Halsey MJ, Liu J, Eger EI, II, Koblin DD, Laster MJ (1991) What solvent best represents the site of action of inhaled anesthetics in humans, rats, and dogs? Anesth Analg 72:627-634.

Takenoshita M, Yoshiya I (1994) Inhibitory Synaptic Transmission (Monosynaptic Ipsc) is Depressed by Halothane. Anesthesiology 81, No. 3A:A890

Tanelian DL, Kosek P, Mody I, MacIver MB (1993) The role of the $GABA_A$ receptor/chloride channel complex in anesthesia. Anesthesiology 78:757-776.

Weight FF (1992) Cellular and molecular physiology of alcohol actions in the nervous system. Int Rev Neurobiol 33:289-348.

White PF, Johnston AR, Eger EI (1974) Determination of anesthetic requirements in rats. Anesthesiology 40:52-57.

Wong SME, Fong E, Tauck DL, Kendig JJ (1997) Ethanol as a general anesthetic: actions in the spinal cord. Eur J Pharmacol in press:

Woodley SJ, Kendig JJ (1991) Substance P and NMDA receptors mediate a slow nociceptive ventral root potential in neonatal rat spinal cord. Brain Res 559:17-21.

Yanagisawa M, Murakoshi T, Tamai S, Otsuka M (1985) Tail-pinch method *in vitro* and the effects of some antinociceptive compounds. Eur J Pharmacol 106:231-239.

Historical Session

Humphry Davy, Thomas Beddoes and the Introduction of Nitrous Oxide Anaesthesia

E. Brian Smith

UNIVERSITY OF WALES, CARDIFF, UK

At the end of the eighteenth century investigations into the breathing of exotic gases might well have led to the discovery of general anaesthesia some fifty years before the events at the Massachusetts General Hospital which marked its first clinical use. Professional historians are quite properly reluctant to indulge in speculation as to what might have been. However in this case it is the privilege of the amateur to explore these investigations that did not as might have been expected bear fruit. This paper is based not on primary sources but on readily available biographies and general medical histories, and is written from the view point of a research worker interested in anaesthesia and the breathing of exotic gases.

Thomas Beddoes and the Pneumatic Institution[1]

The driving force behind these early investigations into the physiological effects of breathing exotic gases was Thomas Beddoes (1760-1808). He was by any standards a remarkable character. Humphry Davy describes him as *'one of the most original men I ever saw - uncommonly short and fat, with little elegance of manners...'*[2]

He was educated at London and Edinburgh before obtaining an MD from Oxford. There he lectured in chemistry but rapidly became disillusioned. His lectures, he wrote, at first attracted the largest audience ever seen in any department but within a few years he experienced falling attendances which led to a reduced income. As well as his scientific interest he was, like the man in whose honour we are meeting, a confirmed radical. His sympathies with the French Revolution and his criticism of British Imperial designs were not compatible with a comfortable existence in Oxford. In 1793 he moved to Bristol with a new venture in mind. Beddoes had come to believe that the breathing of exotic gases might provide a cure for consumption. He was aware of Priestley's researches and had written on respiration and was anxious to found an Institute at which research in *'pneumatic'* sciences could be pursued. In 1794 he set about raising funds to establish his Pneumatic Institution which would have the capacity to synthesise gases and to deliver them to suitable patients. A superintendent was to be appointed. He was fortunate in his patrons. Erasmus Darwin and his father-in-law Richard Edgeworth had offered support. James Watt, whose daughter had died of consumption, offered not only help with seeking funds but was later to play a direct role in designing and constructing apparatus for the Institution. The search for funds was not helped by the radical views Beddoes held which could hardly endear him to rich potential patrons. In the context of the Priestley Conference it is interesting to note that members of the Lunar Society were active in his cause. I do not know if Joseph Priestley himself was a subscriber but as will be described later he was treated at the Institution. One suspects that it was not in Beddoes' nature to take the easy path and his fund raising was fraught with difficulty. However in 1798 he opened the Institute with minimal funds. He purchased a house at Dowry Square Clifton and hired a young, unknown Humphry Davy to act as his superintendent.

Humphry Davy had already exhibited an interest in breathing recently discovered gases. Joseph Priestley had in 1772 discovered nitrous oxide or nitrous phosoxyd or gaseous oxyd of azote as it was first known. Its properties attracted considerable attention and Dr Samuel Lathan Mitchill a Professor of Chemistry at Columbia had proposed that it was *'the oxide of septron - the principle of contagion'* which when breathed or brought in contact with the skin had the power of carrying disease. Davy read of this speculation in a book by Beddoes and Watt *'Consideration of Factious Airs'* 1794-96 which he may have found in the library of his benefactor Dr Davies Gilbert (or Giddy). This led to Davy's first experiment with the gas in his attic bedroom in the spring of 1798 when he was 20 years old and an apprentice to the surgeon-apothecary John Bingham Borlane of Penzance.

'The fallacy of this theory was soon demonstrated by a few coarse experiments made on small quantities of the gas procured from zinc and diluted nitrous acid. Wounds were exposed to its action. The bodies of animals were immersed in it without injury and I breathed mingled in small quantities with common air without remarkable effects. An inability to procure it in sufficient quantities prevented me at this time from pursuing the experiments to any greater extent. I communicated an account of them to Dr Beddoes.'[3]

Dr Beddoes was clearly interested by the communication and other correspondence on the *'subject of heat and light'*.

Davy was also supported by Davies Gilbert who was a one time prodigy of Beddoes and who was to become President of the Royal Society. In October 1798 Davy joined Beddoes at the Pneumatic Institution. He wrote to his mother

'Dr Beddoes has nothing characteristic externally of genius or science. Extremely silent and in a few words a very bad companion. His behaviour to me however has been particular handsome. He has paid me the highest compliments on my discoveries and has in fact become a convert to my theory which I little expected. He had given up to me the whole of the business of the Pneumatic Hospital and has sent to the editor of the monthly magazine a letter to be

published in November in which I have the honour to be mentioned in the Highest terms.'[4]

It is sometimes said that 'Sir Humphry Davy's greatest discovery was Michael Faraday'; it can be said with even more conviction that Thomas Beddoes' discovery was undoubtedly Humphry Davy. Davy undertook an extensive investigation into the effects of nitrous oxide and other gases on both man and animal. He was to make the institution a fashionable resort. Its *'patients'* were the leading lights of society and the arts. The results of his investigation were eventually published as *Researches Chemical and Philosophical; chiefly concerning Nitrous Oxide or Dephlogisticated Nitrous Air, and Its Respiration*[5] in the summer of 1800. His note-books of the time said '*the researches have been made since April 1799 the period when I first breathed nitrous oxide. Ten months of incessant labour were employed in making them, three months in detailing them*' (clearly Davy discounted the breathing of dilute nitrous oxide during the previous year.) His first experiments disclosed that breathing the gas was *'attended by a highly pleasurable thrilling, particularly in the chest and the extremities'* though

'whenever its operation (breathing) was carried out to its highest extent... impressions ceased to be perceived... and voluntary power was altogether destroyed so that the mouthpiece generally dropped from my unclosed lips'[6]

Others were induced to try the gas, eighteen of whom described their sensations in contributions to Davy's publication. Coleridge wrote

'The first time I inspired the nitrous oxide, I felt a highly pleasurable sensation of warmth over my whole frame, resembling that which I remember once to have experienced after returning from a walk in the snow into a warm room. The only motion which I felt inclined to make, was that of laughing at those who were looking at me'[7]

Southey wrote to his brother

'Oh, Tom! Such gas has Davy discovered, the gaseous oxide! Oh, Tom! I have had some; it made me laugh and tingle in every toe and finger tip. Davy has actually invented a new pleasure, for which language has no name. Oh Tom! I am going for more this evening; it makes one strong and so happy, so gloriously happy! Oh excellent air bag! Tom, I am sure the air in heaven must be this wonder-working air of delight.'[8]

One of Beddoes' patients said he *'felt like the sound of a harp'* on breathing the gas. Davy took moonlight walks and tried to emulate his poetic friends by composing verse under its influence

'Not in the ideal dreams of wild desire
Have I beheld a rapture-wakening form;
My bosom burns with no unhallow'd fire,
Yet is my cheek with rosy blushes warm'[9]

Not all were so entranced. Joseph Priestley had *'unpleasant fullness of the head and throbbing of the arteries'*[10] which prevented him from continuing with the gas and Josiah Wedgewood, Thomas Wedgewood, R Boulton and G Watt all failed to experience the pleasurable effects reported by others. The unadulterated pleasure with which Davy and most of his friends enjoyed the gas seems out of keeping with present day scientific attitudes but may have served to promote the much later utilitarian use of the gas.

Davy's Animal Experiments

Davy's attitude to his experiments with animals was, in contrast to his observations with humans determinedly scientific. He carried out fourteen experiments in which warm blooded animals - rabbits, mice, cats and a dog breath nitrous oxide. He was led to three conclusions:

'1st. That nitrous oxide is destructive when respired for a certain time to the warm-blooded animals, apparently previously exciting them to a great extent.

2ndly. That when its operation is stopped before complete exhaustion is brought on, the healthy living action is capable of being gradually reproduced, by enabling the animal to respire atmospheric air.

3dly. That exhaustion and death is produced in the small animals by nitrous oxide sooner than in the larger ones, and in young animals of the same species, in a shorter time than in old ones, as indeed Dr Beddoes had conjectured a priori would be the case.'[11]

He became convinced that *'nitrous oxide acted on animals by producing some positive change in the blood.'*

This view led him to perform three further experiments in which he compares exposure to nitrous oxide to that with hydrogen and with immersion in water using carefully matched pairs of animals. He observed that the time to death in nitrous oxide was over twice that observed in hydrogen or water. These experiments led him to the further conclusions

1 *'Warm-blooded animals die in nitrous oxide infinitely sooner than in common air or oxygen; but not nearly in so short a time as in gases incapable of effecting positive changes in the venous blood, or in non-respirable gases.'*

2 *'Peculiar changes are effected in the organs of animals by the respiration of nitrous oxide. In animals destroyed by it, the arterial blood is purple red, the lungs are covered with purple spots, both the hollow and compact muscles are* apparently *very inirritable, and the brain is dark-coloured.'*[12]

The modern view is not that nitrous oxide is capable of supporting life but that it may produce some protection from the anoxic conditions perhaps by conferring protection against convulsions and lowering the metabolic rate. The colour changes reported by Davy are not in keeping with modern observation. It is possible (though unlikely) that Davy's nitrous oxide was contaminated with nitric oxide which binds strongly to haemoglobin producing a bright red complex. Davy performed a number of experiments with cold-blooded animals: lizards, fish, earthworms and insects. These, more resistant to anoxia, should have enabled him to observe the effects of nitrous oxide more easily than in mammals. However, though some hyper-excitability was observed, none of the animals lost consciousness in less than fifteen minutes. He concluded that *'amphibious animals die in nitrous oxide in a much shorter time than in hydrogen or pure water.'*

Further experiments with fish confirmed his view that

'they are destroyed not by the privation of atmospheric air but from some positive change effected in their blood by the gas.'[13]

Again these observations do not accord with any modern observations or views. We must conclude, regretfully, that they most probably arose from impurities in his nitrous oxide.

Davy's Experiments on Himself

We have already referred to some of Davy's experiments breathing exotic gases. He breathed hydrogen, carbon dioxide, water gas (a mixture of carbon monoxide and hydrogen) and nitric oxide - the last two experiences he was lucky to survive. Two of his observations deserve mention. He diluted carbon dioxide with air and found that *'stimulated the epiglottis in nearly the same manner as pure carbonic acid.'*[14] Second, by careful analysis of expired gas, he concluded *'the exhausted capacity of my lungs was equal to about forty-one cubic inches.'*[15]

In addition he breathed nitrous oxide a large number of times.

'Between May and July [1799] I habitually breathed the gas, occasionally three or four times a day for a week together; at other times four or five times a week only.'

These experiments were apparently pursued for the pleasure they gave as much a the scientific insight obtained.

'Generally when I breathed from six to seven quarts, muscular motions were produced to a certain extent; sometimes I manifested my pleasure by stamping or laughing only; at other times, by dancing round the room and vociferating.

After the respiration of small doses, the exhilaration generally lasted for five or six minutes only. In one or two experiments when ten quarts had been breathed for near four minutes, an exhilaration and a sense of slight intoxication lasted for two or three hours.'[16]

However it was following this series of exposures, when he continued to breath the gas only occasionally that he stumbled on the analgesic properties of nitrous oxide.

'In one instance, when I had head-ache from indigestion, it was immediately removed by the effects of a large dose of gas; though it afterwards returned, but with much less violence. In a second instance, a slighter degree of head-ache was wholly removed by two doses of gas. In cutting one of the unlucky teeth called dentes sapientiae, I experienced an extensive inflammation of the gum, accompanied with great pain. On the day when the inflammation was most troublesome, I breathed three large doses of nitrous oxide. The pain always diminished after the first four or five inspirations.'[17]

He summarised his experiences...

'As nitrous oxide in its extensive operation appears capable of destroying

physical pain, it may probably be used with advantage during surgical operations in which no great effusion of blood takes place.'[18]

Had this observation been followed up it is tempting to believe that surgical anaesthesia could have been established some fifty years earlier than was the case. However, the reasons for the neglect of Davy's suggestions were far from simple. Davy's career developed in a manner that took him from his work on the respiration of gases. In 1801 Davy left Clifton to take up the position of Director of the Laboratory and Assistant Lecturer in Chemistry at the Royal Institution and though the famous Gilray cartoon shows that *'pneumaticks'* and *'elastic fluids'* still held his interest the era of his research on this subject was at an end. His career as a scientist was to take him to great heights. But had Davy remained at Clifton it is by no means certain that progress to surgical anaesthesia would have been made. Nitrous oxide is a far from ideal anaesthetic.

Nitrous Oxide Anaesthesia[19]

Though surgical and dental anaesthesia with nitrous oxide was not firmly established for some 65 years after Davy's experiments it was often employed for amusement in the intervening years. In 1818 Faraday noted the effects of breathing ether. He reported

'In trying the effects of the ethereal vapour on persons who are peculiarly affected by nitrous oxide, the similarity of sensation produced was very unexpectedly found to have taken place. One person who always feels a depression of spirits on inhaling the gas [nitrous oxide], had sensation of a similar kind produced by inhaling the vapour [ether].'[20]

Ether rapidly found its place and *ether frolics* became as popular as *laughing gas parties*. During these years some exploration of surgical anaesthesia did occur. Henry Hickman, a General Practitioner from Shropshire, induced 'suspended animation' in animals by the inhalation of carbon dioxide in order to carry out operations. He was also said to have used nitrous oxide. In 1828 he wrote to King Charles X of France claiming he had

'discovered the means of suspending sensation in animals who are forced to submit to serious surgical operations'[21]

The letter does not mention nitrous oxide. Hickman's observations do not seem to have made any impact and it was not until 20 years later that the exploitation of surgical anaesthesia began to gather momentum. The role of public entertainment with nitrous oxide played an important part in keeping alive interest in the subject. Amongst the showmen and entertainers who demonstrated 'laughing gas' was an ex professor of chemistry, Gardener Coulton, who travelled the USA exhibiting the gas. One such demonstration in Connecticut in 1844 was seen by a local dentist, Horace Wells, who participated in breathing the gas. He observed that one subject who gashed his leg claimed to feel no pain. The next day he had one of his own teeth extracted under the gas administered by Coulton without feeling pain. In the weeks that followed Wells made his own gas and used it on about a dozen occasions. His progress was checked, however, when he sought to demonstrate the technique to the staff and students a Harvard University. The exhibition was a failure and he stated

'it compelled me to relinquish my professional business entirely'[22]

However, he appeared to have continued his research and is reported to have employed the gas subsequently for the amputation of a leg and the removal of a fatty tumour. Wells has an important claim to be the first person to employ anaesthesia, but he became addicted to nitrous oxide and later chloroform and committed suicide in 1848, severing an artery while breathing chloroform.

The Coulton exhibition had another and perhaps even stronger link to the establishment of anaesthesia. Willliam Thomas Green Morton, a fellow dentist and one time partner of Wells, had assisted at the unhappy Harvard demonstration when he was a student. In July 1844 Morton used ether as a local anaesthetic when filling a tooth and noticing its effects were more general thought that it might be employed as a general anaesthetic. Following experiments on a number of animals and on himself and his assistants, he extracted an ulcerated tooth without causing pain. He then wrote, explaining his discovery to Dr JC Warren who had arranged Wells' abortive presentation at the Harvard Medical School, and requesting permission to use ether for an operation. On 16 October 1846 Morton administered ether to a patient with a tumour of the jaw. The report of Warren, the surgeon, confirmed that the patient had felt no pain on its removal. The next day a further operation took place and the practice of surgical anaesthesia was essentially established. In December, ether was used to amputate a leg at Bristol and also in an operation at University College Hospital London. Chloroform was introduced by Simpson in 1847 and the success of the volatile agents led to nitrous oxide being abandoned.

Its 'rediscovery' was due to Coulton who in 1863 returned to dental practice and extracted 2000 teeth and went on to encourage the use of nitrous oxide in dentistry.[23] But the essential problem of nitrous oxide anaesthesia would not go away. Its anaesthetic potency is so low that it requires one atmosphere of partial pressure to produce anaesthesia. Delivered in this way the patient is subject to anoxia if exposed for more than 2 minutes or so. This problem was addressed by Andrews, who in 1868, suggested the use of nitrous oxide/oxygen mixtures. Paul Bert (1878) tackled the problem in an imaginative if impractical way. He designed a pressure chamber in which nitrous oxide/oxygen mixtures could be delivered at 2 atmospheres pressure. The apparatus proved successful but not surprisingly did not catch on.

After much research it did become possible to obtain surgical analgesia with the presence of sufficient oxygen to prevent anoxia, and in the early years of the 20th Century and up to the Second World War nitrous oxide became the preferred dental analgesic. Surprisingly, nitrous oxide is used as an anaesthetic now more than ever, but not alone. It is used in conjunction with other more potent, volatile agents with the nitrous oxide providing some 50 - 70% of the anaesthetic dose. This provides a modern surgeon with adequate depth of anaesthesia, together with the potential for rapid recovery which is associated with nitrous oxide. It is true that some concerns have been expressed as to its safety, long assumed to be excellent.

However, the long story has reached a secure conclusion 200 years after those enthusiastic investigations of Humphry Davy. Nitrous oxide is now established as a central part of the anaesthetist's armoury.

References

1. T H Levere, Dr Thomas Beddoes and the Establishment of his Pneumatic Institution, Notes and records of the Royal Society 32 (1977)
2. ed. J Davy, Collected Works of Sir Humphry Davy, Smith, Elder and Cornhill, London (1839), Vol 1 p 46
3. Ibid Vol 3 p 269
4. Ibid Vol 1 p 46
5. H. Davy, Researches Chemical and Philosophical; chiefly concerning Nitrous Oxide or Dephlogisticated Nitrous Air, and Its Respiration, 1800.
6. Works Vol 3 p 272-273
7. Ibid Vol 3 p 306
8. Quoted in J Kendall, Humphry Davy; Pilot of Penzance, London (1954) p 46
9. Ibid p 46-47
10. Works Vol 3 p 317
11. Ibid Vol 3 p 201-202
12. Ibid Vol 3 p 212
13. Ibid Vol 3 p 217
14. Ibid Vol 3 p 281
15. Ibid Vol 3 p 241
16. Ibid Vol 3 p 274
17. Ibid Vol 3 p 276
18. Ibid Vol 3 p 329
19. For comprehensive review of modern practice see: Nitrous Oxide, ed E I Eger, Edward, Arnold London (1985)
20. W B Saunders Quoted by H Langa, Relative Analgesia in Dental Practice, Saunders Philadelphia (1968) p 7
21. E A M Frost, A History of Nitrous Oxide, Chapter 1 of Ref 19, p 8
22. WH Archer, Life and Letters of Horace Wells, J. Amer. Coll Dent., Vol 11 p 81-210, (1944) Quoted by Frost See Ref 21
23. Ref 21 p 14
24. Ref 20 Preface

William Morton and the Early Work on Anaesthesia in the USA

N. G. Coley

HISTORY OF CHEMISTRY RESEARCH GROUP, FACULTY OF ARTS, THE OPEN UNIVERSITY, MILTON KEYNES, UK

Introduction

Of the seven pioneers of inhalation anaesthesia (Davy, Hickman, Long, Wells, Morton, Jackson and Simpson) four were American. Long, Wells and Morton were dentists; Jackson was a physician and chemist.[1] All were familiar with the so-called 'ether frolics' popular among students and young people in the early nineteenth century. Itinerant showmen who travelled around country fairs often ended their performances by inviting members of the audience to inhale nitrous oxide, 'laughing gas', for amusement. From these experiences it was common knowledge that one could be injured while 'under the influence' without feeling pain. To some this suggested the possibility that surgery might be made painless and nineteenth-century surgeons had a strong incentive to find ways of reducing the trauma of surgical operations. The time was right for innovative and courageous men to try out the possibilities, though the risks were considerable.

Crawford Long

Crawford Long (1815-78) had graduated from Franklin College, University of Georgia in 1835. He then studied medicine at Transylvania University, Lexington, Kentucky and later at the University of Pennsylvania. In 1839 he went to New York where he observed the traumatic effects of surgery without anaesthetics. Two years later he was back in Georgia where, at Jefferson, a small town of only a few hundred people, he first used ether as an anaesthetic in minor surgery. He had a patient, James Venables, who had a sebacious cyst in his neck which he had long wished to have removed but repeatedly shied away from the pain of the operation. Having observed that under the influence of ether quite severe injuries caused no pain Long persuaded Venables to have the cyst removed under ether and the operation was carried out successfully on 30 March 1842. A second operation on another patient on 6 June 1842 was more complicated and took longer so that the patient began to regain consciousness before it was complete. Long therefore suggested that the ether should be administered throughout the operation. These are the earliest known surgical operations using ether and there is evidence to confirm Long's claim. He had purchased his ether from Robert Goodman, a pharmacist in Athens, Georgia, and the order for it still exists. Goodman also recorded that Long had told him in November 1841 that he believed a surgical operation could be performed without pain using ether.

Long later said that though he had 'performed one or more surgical operations annually, on patients in a state of etherisation' since 1842, he did not publicise the fact because he wanted to make sure that 'anaesthesia was produced by the ether and was not the effect of the imagination or owing to any peculiar insusceptability to pain in the persons experimented on'. He remarked that there were eminent physicians who advocated mesmerism, but he [Long] thought that if the mesmeric state could be produced at all, it was only in 'those of strong imagination and weak minds'.[2] Since Long did not report his discovery until anaesthesia had become known through the work of Wells and Morton, his right to be considered the discoverer was weakened due to this extraordinary reticence.[3] There is no doubt that he was the first to practise ether anaesthesia, pre-dating Wells by two years and Morton by four, yet without their well-publicised efforts, and especially Morton's public demonstrations, Long's use of ether anaesthesia might have remained unknown outside his private surgery and would have died with him.

Horace Wells

Horace Wells (1815-48) was a leading dentist in Hartford, Connecticut. On December 10, 1844 he attended a public entertainment by G.Q.Colton, an itinerant chemical lecturer, during which members of the audience breathed nitrous oxide and in the capers which followed one of the volunteers severely 'barked his shins'. Later, this man declared that he had felt nothing and it was this incident which led Wells to think that nitrous oxide might be used to remove the pain of dental extractions. On the following day, 11 December 1844, he asked Colton to administer the gas to him whilst a former pupil, John M.Riggs, extracted one of his molars. The extraction was painless. Thus Wells had applied constructive reasoning to his casual observation and had bravely undertaken to test his idea personally. He wanted to make the process available to his patients, but pure nitrous oxide was not readily available, it had to be free from other oxides of nitrogen, especially nitric oxide, and Wells had to master the art of preparing the gas in a pure state himself.

By mid-January 1845 Wells had extracted teeth under nitrous oxide for about fifteen patients and nearly all the extractions were painless. If failures did occur there is no record of them.[4] His colleagues in medicine and dentristry were informed and some also tried the procedure with success. Within the month Wells sought an opportunity to make a public demonstration at the Massachusetts General Hospital in Boston. In preparation for this he had to make his own pure nitrous oxide, administer it and then carry out the extraction before an audience of sceptical medical men and students. Wells administered the nitrous oxide from a rubber bag and when he judged that the patient was sufficiently anaesthetised he began the operation. All went well at first, but near the end of the procedure the patient groaned and this was taken as failure by his sceptical audience.

Wells was labelled an imposter. Yet the patient said he had felt no pain, or almost none, and Wells blamed himself for removing the gas-bag too soon. It is not surprising that he should be worried about the length of time for which pure nitrous oxide was given. Alone the gas cannot support life, the patient was receiving no air or oxygen with the anaesthetic and there was no way of measuring how much gas was being inhaled. The depth of anaesthesia could only be judged on the basis of physical and physiological appearances. In the hostile atmosphere of the operating room Wells was anxious not to overplay his hand and, nervous individual that he was, he may well have erred on the side of leniency. However, it is now well known that patients under nitrous oxide anaesthesia often groan, moan or even scream involuntarily, but this is not from pain and does not mean that the anaesthesia is no longer working. Consequently, it seems evident that Wells had actually demonstrated inhalation anaesthesia successfully, but the patient's groan was taken by his sceptical audience to indicate what they expected and perhaps even wished to see.

The nervous strain Wells had placed himself under in his attempt to demonstrate his discovery of general anaesthesia produced an illness which forced him to retire from practice between April and September 1845, though he was still in contact with other dentists in Hartford who continued the successful use of nitrous oxide as an anaesthetic. Wells also experimented with ether, following his own observations of 'ether frolics', but he made no further attempt to demonstrate the effects publicly. However, he did reveal what he knew of anaesthesia to the two others who would later cause him to lose his claim as the discoverer of the phenomenon. One of these was his former pupil William T.G. Morton, a dentist in Boston. Morton had been present at Wells's demonstration and he must have been impressed by the mild reactions of the patient compared with the usual distress caused by dental extractions without anaesthesia. Wells and Morton had set up a partnership to sell a dental solder invented by Riggs for use in making dentures and in the course of their experiments they had consulted Charles T. Jackson, a physician and prominent chemist in Boston. Jackson, who readily endorsed the solder, was later to gain a reputation for his unscrupulous attempts to claim priority for the inventions and discoveries of others. He was to become one of the contestants in the controversy over the invention of inhalation anaesthesia, but in January 1845 neither Morton nor Jackson showed much enthusiasm for Wells' claims for the pain-relieving effects of nitrous oxide.

Charles T. Jackson

Charles T. Jackson (1805-80) had very good connections and graduated MD with honours at Harvard in 1829. He travelled to Europe where he spent three years studying medicine and geology and was present at about two hundred autopsies on cholera victims in Vienna. Besides his medical work he became a prominent geologist, completing numerous important surveys. He was the first to study the geology of several regions in America. A prodigious worker, he is said to have published more than four hundred papers. In geology no-one seemed better informed than Jackson; his reputation was world-wide. Sadly, with an over-reaching ambition Jackson employed unscrupulous methods of enhancing his reputation which led to his being despised and denounced. For example, he attempted to claim the credit for Morse's invention of the electric telegraph and Schonbein's invention of gun cotton. He also tried to get William Beaumont's subject, Alexis St Martin sent to Boston instead of St Louis where he was due to go with Beaumont, so that he could carry on Beaumont's experiments on digestion. Yet, Jackson was respected by Crawford Long, while both Wells and Morton brought their work on anaesthesia to his notice. As a bold experimenter Jackson tried the effects of ether on himself in February 1842, but he neither publicised the fact nor did he use ether in any surgical operations. Nevertheless, he later made a vigorous claim for priority over Morton and Wells.

William T.G. Morton

William T.G. Morton (1819-68), a dentist in Boston, was present at Wells's demonstration with nitrous oxide in January 1845 and later, in July of that year at a Conference in Hartford, Wells told him how to prepare pure nitrous oxide and how it could be administered. Wells also advised Morton to seek assistance from their mutual friend, Jackson. At the same time a woman who had had a tooth extracted by Wells using nitrous oxide, enthusiastically told Morton about her experience. Morton applied to Jackson on September 30 1846 for pure nitrous oxide to be used for anaesthetic tests but Jackson told him there was none available. On the other hand, ether could be purchased at a drugstore. Morton therefore began to use ether, his first patients inhaling it from the corner of a towel.[5] So, from the beginning ether was breathed mixed with air.

Morton carried out a number of painless extractions and then requested an opportunity to demonstrate ether anaesthesia publicly at the Massachussetts General hospital. J.C.Warren, an eminent surgeon agreed to try out ether in an operation, but no opportunity arose until it became necessary to excise a small tumour in the neck of a young printer named Gilbert Abbott, aged 20, who had had this tumour under on the left side of his neck since birth. It streched along the jaw-bone and could also be seen inside the mouth; it seemed to be linked with the sub maxilliary gland where a small hard lump could be felt, though the rest of the tumour was soft. Removal would be a considerable operation and Warren decided that this was a proper case for the use of ether. On October 16 1846, with Morton administering ether vapour from a glass flask, Warren excised the tumour without pain. On the day of the operation the patient was prepared and the audience were in their places, but Morton was late and Warren was about to perform the operation without anaesthetic when Morton arrived. He had been to obtain his apparatus which he had ordered from a surgical instrument maker. It consisted of a glass flask with two necks, filled with gauze soaked in ether and it ensured that the patient would breathe air saturated with ether vapour. After applying the apparatus for 4 or 5 minutes the patient appeared to be asleep and the operation was performed.

To Warren's surprise the patient did not shrink, nor cry out, but later he began to move his limbs and utter strange expressions which seemed to indicate pain.[6] However, after he regained consciousness he said that he had felt no pain but only a sensation like that of scraping the part with a blunt instrument. An eyewitness account of this operation was given by a student[7] and it was also described in a paper by Henry Jacob Bigelow, professor of materia medica, which was read to the Boston Society of Medical Improvement on 9 November - an abstract had already been read before the *American Academy of Arts and Sciences.* Bigelow's report, published on 18 November,[8] seemed to suggest that Wells's demonstration two years before had after all been successful. After Warren's operation ether anaesthesia was used in other cases; within a few days its success was established and it began to be used in every considerable surgical operation in Boston and its vicinity.

Methods of Administering Ether

In 1847 Morton wrote to *The Lancet* on his methods of administering ether. He says that at first he used a simple sponge, then he placed it under a conical glass tube (like an inverted glass funnel) and afterwards in a glass flask, but he remained dissatisfied with the results. Morton suspected that anaesthesia in Abbott's case had been only partially successful and he wanted to improve the method of administering the anaesthetic. On the 17th October, the day following Abbott's operation, he had a second opportunity. Late on the 16th he conferred with a Dr Gould, a distinguished physician of Boston about the best means of administering the ether and a new apparatus was devised with valves to control the amount of ether given. This was used in the second demonstration at Massachussetts General Hospital. Before his demonstrations surgeons and physicians, with few exceptions, were incredulous, but the success of the second operation confirmed his discovery and Morton published a pamphlet on the subject in the following year.[9]

Yet he claimed never to have been wholly satisfied with any apparatus and in making further experiments he was led to abandon all inhalers and return to using the sponge alone. He now used a sponge the size of an open hand, made concave to fit over the nose and mouth and saturated with ether. The patient was then told to breathe through the nose and open mouth. This he said gave more certain and satisfactory results.

Less than two weeks after the first successful demonstration a patent was sought in the joint names of Morton and Jackson. To protect his discovery Morton tried to disguise the smell of ether with aromatic essences, and he called the mixture 'Letheon'. When challenged he agreed that there was ether in it but denied that this was responsible for causing insensibility. It was only when the Hospital refused to continue its use because Morton would not divulge his secret, that he admitted that letheon was in fact highly rectified sulphuric ether. He tried to profit from his discovery by granting licences, but it soon became clear that it would be impossible to sustain his hold over the use of such a readily available substance. When Jackson saw that the likely profits would be far less valuable than the prestige of being designated the discoverer of anaesthesia, he assigned his patent rights to Morton and set out to esablish himself as the discoverer.

He now revealed that in the winter of 1841-42 he had made some observations of the effects of ether on himself and had as a result 'deduced' the phenomenon of anaesthesia. On 21 December 1846 he wrote to his friend Èlie de Beaumont at the Paris Académie, where he was already known, announcing himself as the sole discoverer of surgical anaesthesia. On 2 March 1847 he read a paper at the *American Academy of Arts and Sciences* which he had already published in the *Boston Daily Advertiser* the day before so that he could send copies of the newsaper abroad. This gave the appearance that the Academy had endorsed Jackson's claim and the impression remained in Europe long afterwards.[10] On 31 January 1849, after due deliberations, the Institut Française awarded Jackson the Cross of the Légion d'Honeur as the discoverer of etherization.

The Ether Controversy

As soon as he heard this Morton wrote to the French Institute, sending separately a package of papers supporting his own claim to the discovery. His letter arrived, but for some reason which is not clear the papers did not. However, it was the award to Jackson by the French which caused Morton to begin his fight for recognition. The French Academy of Medicine offered their Monthyon Prize jointly to Morton and Jackson, but Morton refused to accept it, insisting that the credit for the discovery belonged solely to him.

On December 6 1852 Morton made a petition for public recognition as the discoverer of ether anaesthesia to the United States Senate at Washington. It was signed by seven surgeons and five physicians of the Massachusetts General Hospital, headed by John C.Warren and including Oliver Wendell Holmes, who had suggested the terms 'anaesthesia' and 'anaesthetics' in a letter to Morton. Four consultants and three physicians of the Charitable Eye and Ear Hospital and 104 members of the Massachusetts Medical Society also signed Morton's petition. A Select Committee of the Senate was set up and the evidence was heard beginning on 21 January 1853.[11]

Morton began his case by stating that it was through him alone that ether anæsthesia was given to the world. Reports of the trustees of the Massachusetts General Hospital and of the House of Representatives had awarded the discovery to him, yet he had suffered the most malicious attacks. He said that it was his idea to use ether and he was wholly engrossed in testing it for some months prior to September 30 1846 when Jackson claimed to have given him the first hint of the possibility. Furthermore Jackson had ridiculed the idea of anaesthesia, even while it was being tested, and washed his hands of all responsibility. He remarked that even if Jackson's statements about trying out ether on himself were literally true, he was not entitled to make the deduction he claimed from his observations. This for two reasons: first, physiological science does not admit of exact reasoning like mathematical, mechanical or chemical science, and second, the same effects which he described are produced by other agents which do not withstand the test of surgery.

The Boston surgeons insisted that Jackson had no connection with any of the experiments either at the hospital or in private practice and that neither Dr Warren, nor anyone else connected with the hospital knew or suspected that Jackson had anything to do with the discovery until after the second operation. Warren specifically denied that Jackson had requested him to perform the experiment and stated that the surgeons relied solely on Morton's dental experiments, not even knowing for certain what they were administering by Morton's direction as he called it 'Letheon'. Moreover, Jackson had not even attended an operation until two months after ether was first used. Each of these points was supported by numerous statements, letters and other references.

To prove his case as fully as possible Morton also challenged the claims made on behalf of Horace Wells. He went to Connecticut where he had every witness within reach who was mentioned in support of Wells's claim called before a United States Commissioner, so as to have them examined in full. The results were laid before the Select Committee. Morton contrasts his own openness and wish to have his claim examined by impartial witnesses with the secrecy surrounding the claim submitted by representatives for Horace Wells whose witnesses had been examined under lock and key, their names withheld. Moreover, after nearly a month neither of his opponents had submitted their claims and notice was given to them to present their evidence to the Select Committee. Time still passed and so Morton himself presented printed copies of two minority reports in favour of Jackson and two pamphlets printed in favour of Wells one of which ran to 132 pages.

Morton also submitted a translation of the report of the French Committee which had made the award 'of all honour and originality' to Jackson. Morton pointed out that the French Committee agreed that he [Morton] had the 'preoccupation of mind' and 'the engrossing idea' of this discovery and that having the original idea he had 'completed' the discovery which, without his 'audacity' would probably have remained 'fruitless and without effect' in Dr Jackson's mind, yet the French report named Jackson as the discoverer rather than Morton because:

'Mr Jackson had observed that *some persons* on being exposed for a certain period of time to the action of ethereal vapours were momentarily deprived of all sensibility.'[12]

Morton pointed out that Jackson had tested the effects of ether on *himself* alone. He referred to his own letter of 15 March, 1849 in which he had stated that the authorities of the Massachusetts General Hospital had formally attributed to him the honour of the discovery in their report for 1848,[13] but as the documents backing his claim failed to arrive, his case was weakened. The French Committee had only his own statement. Jackson was supported by his friend M. Élie de Beaumont, whereas Morton was unknown. He suspected that the fact that his papers and collection of pamphlets, which he sent for circulation to journals and distribution among surgeons and others, were never received was due to the intervention of someone who had become prejudiced against him.

Repeated solicitations to the US Congress from the physicians and surgeons of the Massachusetts General Hospital as well as others, and from Morton himself, led to the appropriation of $100,000 to him for the discovery of practical anaesthesia, but this was voted down at three separate sessions of the legislature, largely due to Jackson's efforts, the Congressional supporters of Crawford Long and vigorous polemics from Truman Smith, Senator for Connecticut, on behalf of Wells's widow and infant son. Morton demanded an impartial examination of all the evidence, but it seemed unlikely that he would get his wish in view of a statement made by Truman Smith, who said in a debate on 28 August 1848:

'I pledge whatever reputation I may have that if the Senate will allow me, at the next session of Congress, an opportunity to be heard on the subject, I will make out a case for the family of Dr Horace Wells, deceased [...] I denounce this attempt to filch money from the Treasury as an outrage upon the rights of others [...] I believe that this Morton is a rank imposter- that there is no justice or truth in his pretended claim...'[14]

The arguments dragged on until the Civil War broke out in 1861 when the issue was shelved, never to be reopened. Nevertheless, ether was used widely by surgeons of both sides on the battlefields. The first surgical operation under ether in Britain was carried out by Robert Liston at University College Hospital on 2 December 1846[15] and the use of ether anaestheia quickly spread. Unfortunately in the hands of untrained surgeons there were some fatalities and doubts about its safety began to arise.[16]

Conclusion

The dilemma of the Senate Committee arose from the fact that it was not made clear what the proposed award was for. Wells certainly had a strong claim to the discovery of general anaesthesia and this has recently been revived,[17] but he used nitrous oxide, although he later experimented with ether. On the other hand Long had used ether four years before Morton, though he was more interested in receiving credit for his use of ether anaesthesia than in the award itself. In fact each of the contestants had a claim to part of the discovery, but none had an outright claim in the view of the Committee. Jackson's case seemed very shaky, especially in view of his reputation, yet it was strong enough to deny Morton the award which he regarded as his right. In the end it was never made. Morton, who had demonstrated the value of ether anaesthesia in surgery, felt let-down. He wore himself out with bitter disappointment, resentment and enmity towards Jackson and died of a stroke on July 15 1868, at the age of 48.

Francis Darwin (1848-1925), the Cambridge botanist once said:

'In science the credit goes to the man who convinces the world, not to the man to whom the idea first occurred'.

In this case, despite all the evidence and arguments, neither the man who first had the idea, nor the man who most clearly demonstrated its utility was able to convince the Select Committee of the United States Senate. Indeed, the argument still continues; new books and articles appear from time to time setting out again the claims of one or other of these four men. There are also monuments to each of them claiming the honour of discovery. In Bushnell Park, Hartford, Connecticut there is a statue of Horace Wells, 'discoverer of anaesthesia'; in the Smithsonian Institution in Washington, a bust of Morton declares him to be the 'discoverer of surgical anaesthesia'; on the public square at Jefferson, Georgia an obelisk is inscribed, 'Dr Crawford Long, the first discoverer of anaesthesia' and in Pilgrim Hall, Plymouth, Massachusetts an old rocking chair on display is labelled with a brass plate, 'Seated in this chair Dr Charles T.Jackson discovered etherization, February, 1842'. The ether controversy lives on, but after all, it is the value of the discovery itself which matters - by comparison the question of priority pales into insignificance.

References

1. Victor Robinson, *Victory over Pain; a history of anaesthesia*, Henry Schuman, New York, 1946, pp. 83-137; Barbara M. Duncum, *The Development of Inhalation Anaesthesia*, OUP, 1947, repr. Royal Soc. Med. Ltd., London and New York, 1994, pp. 99-129.
2. Frank Kells Boland, *The first anaesthetic: the story of Crawford Long*, Univ. of Georgia Press, Athens, GA, 1950.
3. Possible reasons for Long's failure to publicise his use of ether have been suggested by W. Stanley Sykes, *Essays on the first hundred years of anaesthesia*, 3 vols., Churchill Livingstone, London, Edinburgh, New York, 1982, vol. III, pp. 1-24.
4. A recent account of Wells's work is given by Richard J. Wolfe and Leonard F. Menczer, *I awaken to Glory: essays celebrating the sesquicentennial of the discovery of anaesthesia by Horace Wells, December 11, 1844–December 11, 1994*, Boston Med. Library in the Francis A. Countway Library of Med., Boston in association with the Historical Museum of Med. and Dentristry, Hartford, Conn., 1994.
5. For a detailed account of Morton's life and work see Betty MacQuitty, *The battle for oblivioun: the discovery of anaesthesia*, Harrop, London, 1969.
6. The scene was recorded and was later recreated in a well-known painting by Robert C. Hinckley; Richard J. Wolfe, *Robert C. Hinckley and the recreation of: 'The first operation under ether'*, Boston, 1993.
7. Washington Ayer, 'Account of an eye-witeness', *Semi-centennial of anaesthesia*, Boston 1897, pp. 89-90; repr. in F. Cole, MD., *Milestones in Anaesthesia: readings in the development of surgical anaesthesia 1665-1940*, Univ. of Nebraska Press, Lincoln, 1965, pp. 30-32.
8. H.J.Bigelow, *Boston Med. & Surg. J.*, 1846, 35: 32-42.

9. W.T.G.Morton, *Remarks on the proper mode of administering sulphuric ether by inhalation*, Dutton and Wentworth, Boston, 1847.
10. Robinson, op.cit., pp. 134-37.
11. W.T.G.Morton, *Statements supported by evidence of W.T.G.Morton, MD., on his claim to the discovery of the anaesthetic properties of ether, submitted to the select committee appointed by the Senate of the United States...*, Washington, 1853.
12. Ibid., p. 6.
13. W.T.G. Morton, *Comptes Endu.*, 23 April 1849, p. 556.
14. Morton, op.cit., 1853, p. 11.
15. L.J.Ludovici, *Cone of oblivion: a vendetta in science*, London, 1961.
16. James M.Ball, 'The ether tragedies', *Annals of Medical history*, (1925), 7: 264-66.
17. Richard J.Nolfe and Leonard F.Menczer, (eds), *I awaken to glory: essays celebrating the sesquicentennial of the discovery of anaesthesia by Horace Wells*, Boston, Mass., Medical Library in the Francis A.Countway Library of Medicine, Boston, in assoc. with the Historical Museum of Medicine and Dentistry, Hartford, 1994; Richard B.Gunderman, 'Dr Horace Wells and the conquest of surgical pain: a Promethan tale', *Perspectives in Biology and Medicine*, (1991-92), 35: 531-48; P.H.Jacobsohn, 'Dentistry's answer to the "humiliating spectacle"', *J. Amer. Dental Assoc.*, (1994), 125 (2): 1576-84.

Objections to Anaesthesia: the Case of James Young Simpson

Colin A. Russell

THE OPEN UNIVERSITY, UK

1 Simpson and the Introduction of Chloroform Anaesthesia

The subject of this paper is probably the most involatile anaesthetic material discussed at this conference: chloroform. It is historically important as the first to be used on any considerable scale, and dominated the field for half a century and more. Yet the circumstances of its introduction have been attended by a mystery so remarkable that only in the last 20 years has a solution begun to appear.

In a sense the story began with the man who was to become Sir James Young Simpson, President of the Royal College of Surgeons of Edinburgh, with a bust in Westminster Abbey and a statue on Princes Street, Edinburgh.

James Simpson was born 7 June 1811, at Bathgate, 15 miles west of Edinburgh. He was the seventh son of a baker. His middle name 'Young' was added later (to acknowledge a nickname derived from his youthful appearance, 'Young Simpson').

Simpson went up to Edinburgh University in 1825 (at the age of 14), and after two years at the classics moved over to medicine. He qualified MD in 1832. The Edinburgh medical school had the best of continental teaching systems and an international reputation, sending men to English universities, Dublin and USA.[1] It had the Royal College of Surgeons of Edinburgh (founded 1505), the Royal College of Physicians of Edinburgh (founded 1681),and the Royal Medical Society (founded 1737). Simpson was appointed to the City Lying-in Hospital in 1836, and became Professor of Midwifery in 1840, being elected by the Town Council with a majority of one. He was much in demand in the area and established a flourishing practice. He devised an 'air tractor', or suction device to be applied to the foetus' head to assist in difficult deliveries. What was new about his technique was the use of a brass air-pump to reduce the pressure. Though popular for some years it was long abandoned until returning in another form with the method of Ventouse (1954).

Sir James Young Simpson

The University of Edinburgh

Simpson was greatly concerned with limitation of pain during operations. He had flirted (unsuccessfully) with mesmerism, though found that many patients were unsusceptible to 'suggestion'. The news of ether anaesthesia reached Britain from the USA by late 1846 and was first used in London at University College by Robert Liston for a thigh amputation on 21 December, and a few days later by James Miller at the Edinburgh Infirmary. Simpson responded enthusiastically and applied ether on 17 January 1847 for a complicated delivery (the mother but not the child survived). In words attributed to Liston, 'This Yankee dodge beats mesmerism hollow'.

At this time many new volatile chemicals were becoming available. Simpson experimented with a range of others in a series of self-experiments, usually at home. He tried the new 1,2-dichloroethane in 1847,[2] and also its analogue 1,2-dibromoethane (supplied by the Professor of Chemistry, Lyon Playfair). This was first administered to rabbits; they were anaesthetised, recovered, – and then died! Had Playfair not insisted on a preliminary trial on animals Simpson would probably have shared the same fate.[3] Several years later he was still experimenting, and wrote to the Manchester chemist Edward Frankland for a supply of 'hydruret of amyl', *i.e.* pentane.[4]

Chloroform had been discovered in 1831 by Liebig and (independently) Souberain. Its formula was established by Dumas in 1835. Its use as an anaesthetic seems to have been suggested to Simpson by Dr David Waldie, a physician from Liverpool and also a chemist, who had used it in dilute ethanolic solution as an anti-spasmodic.[5] It was first tested by Simpson on himself and two medical colleagues at his home 52 Queen Street in the evening of 4 November 1847. All three experienced 'an unwonted hilarity', followed by inebriation and collapse. A niece present fell asleep crying improbably 'I'm an angel!'[6]

Simpson immediately ordered chloroform from a local manufacturer, Duncan and Flockhart. Its successful clinical use was first reported in a paper 6 days later. It was easier to administer than ether, caused no irritation or excitement, and more readily maintained deep anaesthesia. Also, less of it was needed (as ether is so volatile), a consideration when heavy bottles had to be carried up tenement staircases. And, as Simpson specially stressed, chloroform was non-inflammable. Duncan and Flockhart were soon retailing it at 3/- per ounce.[7]

The news spread like wildfire. By the next September chloroform was being used as far away as India, but it was in Edinburgh that its use first became routine practice. Within 28 months of Simpson's first announcement it had been employed in that city for between 80 000 and 100 000 successful operations.[8] Its final sanction may have come in 1853 with the birth of Prince Leopold to Queen Victoria.

However human fatalities did occur. The first case, in 1848, was Hannah Greener (15) of Newcastle-upon-Tyne, in a minor operation (removal of toe-nail). At the post mortem 'Sir John Fife blamed congestion of the lungs, which he was compelled to ascribe to the inhalation of chloroform'.[9] It seems to have been only the first of many cases. However Simpson was reluctant to admit

Simpson's house at 52 Queen Street, Edinburgh, where chloroform anaesthesia was first tried

The surgical hospital

dangers of chloroform. He knew nothing of possible damage to renal functions, for example. With the benefit of hindsight we can see the propriety of his own rules for administering chloroform:

1. absolute quietude and freedom from excitement;
2. avoidance of a 'stage of exhilaration';
3. waiting until the patient 'is thoroughly and indisputably soporised'.[10]

Nor were its subjects only humans. An early example from veterinary practice came with its successful application to a cantankerous horse at Rothesay.[11] Frankland, then working at Queenwood College and having become 'teetotally drunk' on chloroform in January 1848, was using it two months later to anaesthetise a cow in difficulties after giving birth to a calf.[12] *Punch* reported 'oysters opened and their beards taken off under the influence of chloroform', and that at Billingsgate eels were skinned under similar sedation![13]

2 Opposition to Anaesthesia

Considerable opposition began to appear, even within 1 month of Simpson's first use of chloroform. It tended to focus on the use in midwifery rather than

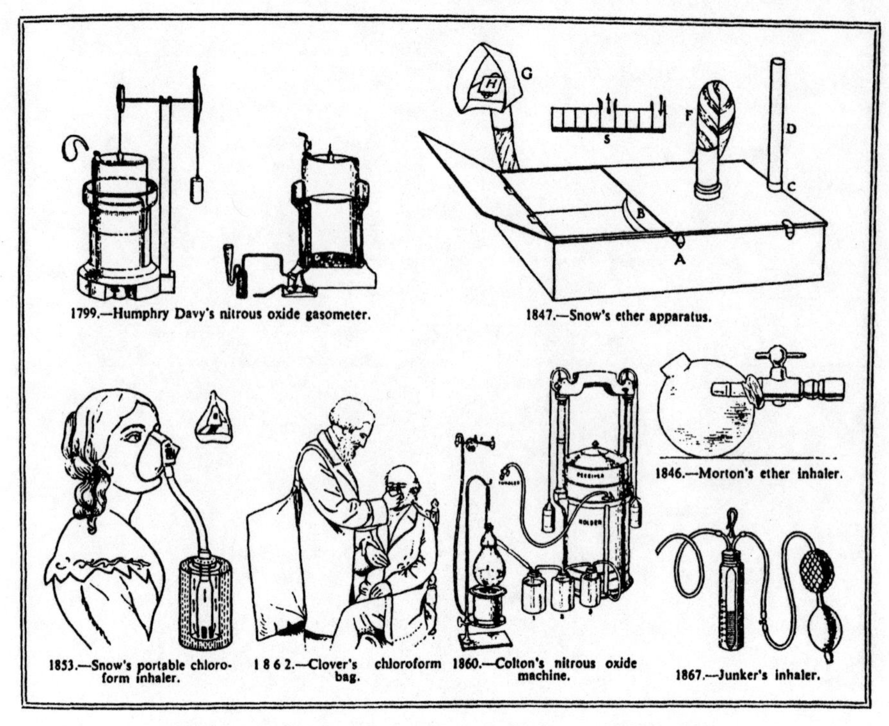

Apparatus for the administration of ether and chloroform

general surgery. One of the most familiar features of the traditional picture was the implacable opposition of the church. Many years after the events J. W. Draper put it like this:

When the great American discovery of anaesthetics was applied to obstetrical cases, it was discouraged, not so much for physiological reasons, as under the pretense that it was an impious attempt to escape from the curse denounced against all women in Genesis iii.16 [*i.e.*, 'in sorrow thou shalt bring forth children'].[14]

Almost 100 years later another author asserted:

In the opposition to the Darwinian theory we can find a parallel illustrating the ferocity with which anything challenging Victorian ethical or religious beliefs was fought.[15]

Yet as early as December 1847 Simpson had published a pamphlet Answer to the religious objections advanced against the employment of anaesthetic agents in midwifery and surgery.[16] His arguments were, broadly:

1. That the primeval curse was not only upon Eve but also upon Adam, in child-bearing and agriculture respectively; therefore relieving pain in the one case was comparable to eliminating weeds in the other;
2. That under the Christian dispensation the moral necessity of anguish in labour has terminated, for any such curse was abrogated by the death and sacrifice of Christ

John W. Draper

3. That there was even a divine precedent for anaesthesia as when 'the Lord God caused a deep sleep to fall upon Adam' prior to removal of a rib to form Eve!
4. That medicine is concerned not only to preserve life but also to alleviate human suffering, and that to do good in that sense was a positive Christian duty.

So whose were 'the religious objections' to which Simpson refers? Some of the first objections that appealed to the Bible were voiced by colleagues in the medical professions. Thus G. T. Gream, obstetrician at Queen Charlotte's Lying-in Hospital, quoted an eminent theologian 'perhaps unequalled in learning' (but unnamed) who regarded pain as 'a blessing of the Gospel'.[17] Robert Barnes (Lecturer in midwifery at Hunterian School of Medicine) wrote of 'that wise and necessary purpose' associated with labour pains.[18] Robert Lee (Lecturer in midwifery at St. George's Hospital): complained of chloroform anaesthesia as a futile attempt to abrogate an ordinance of the Almighty 'In sorrow thou shalt bring forth children'.[19]

These are all well-documented cases of 'religious' opposition, though it is noteworthy that all came from medical not clerical, men (and, be it noted, from men and not women). However the great bulk of subsequent commentaries on these events places the burden of responsibility on aggressive clerics rather than on conservative medics. It stresses the purely religious objections to anaesthesia in midwifery.

At this point in my argument a personal note creeps in. Aware of the established view of intense religious opposition to the anaesthetic use of chloroform, I asked one of my research students, Derek Farr, to identify precisely the location of such opposition, looking for evidence in material produced in 1847 and the next few years. He sought first of all records of sermons and of the Acts of the General Assemblies of the two main churches in Scotland. Finding nothing significant he examined the theological press and then contemporary newspapers and journals in Britain and America. Again he largely drew a blank. Finally, letters and papers of Simpson himself and of others close to the controversy were examined.

As a result of an exhaustive enquiry[20] he emerged with the astounding result that a purely religious criticism of Simpson's use of anaesthetics was virtually non-existent. Of the seven references in print that were discovered four were reviews of Simpson's own pamphlet (and favourable ones at that). The others were more general review articles expressing no religious qualms. The evidence, negative though it was, pointed dramatically to the possibility that we have all been led astray by inaccurate analyses and that the legend of persistent religious and clerical opposition to Simpson is just literally that. In reporting these results Dr Farr has challenged and demolished one of the most established myths of recent medical history.[21] But for his untimely death last summer he would have been here today, personally reporting on his research.

This revisionist view, that religious opposition was minimal, is at least partly confirmed by numerous cases of religious support for chloroform. A colleague of Simpson's was James Miller, Professor of Surgery, and the first in Edinburgh to operate using ether anaesthesia. He explained to the theologian Thomas Chalmers 'that *some* had been urging opposition to the use of anaesthesia in midwifery on the ground of it so far improperly enabling women to avoid part of the primeval curse'. Chalmers, prominent leader of the Secessionist movement and of the new Free Church, was at first unconvinced that he was serious. He then simply advised Miller and his friends to ignore 'small theologians'.[22] The Anglican minister Charles Kingsley wrote 'It is a real delight to my faith, as well as my pity, to know that the suffering of child-birth can be avoided'.[23] A recent historian could therefore say that in general the appeal to Genesis *etc.*, 'contrary to what might be supposed, was not particularly popular with the ministers of the church'.[24]

The only publication on the subject in England was a tract by the obstetrician Protheroe Smith (founder of London's first Hospital for Women). A staunch evangelical (like Chalmers and Simpson) Smith published a strong defence for anaesthesia based on Biblical principles, *Scriptural authority for the mitigation of pains in labour*.[25]

James Miller

3 Origins of a Myth

We therefore have a major problem on our hands. How could such a tissue of falsehood and absurdity have become the received 'wisdom' of western culture in the last 125 years? Farr presents a list of 31 books in the century beginning 1873 alleging substantial religious objections to obstetric anaesthesia. How could they all make the same mistake? Several reasons suggest themselves.

1. There is confusion between theological and technical objections. Admittedly a very few authors attempted to boost their scientific arguments by appeal to religious prejudice but that is very different from theological objection as such. It may be helpful to consider some of the non-religious but technical arguments put forward against chloroform anaesthesia in midwifery.

First was the view that *it was 'unnatural'*. Thus C. D. Meigs, professor of obstetrics at Pennsylvania, opposed the use of chloroform but did not cite scripture. He argued chiefly that labour pains were a natural feature of childbirth, that patient response was supremely important, and that anaesthesia was comparable to alcohol in its dangers.[26] When an Irish lady addressed Simpson with the question 'How unnatural it is for you doctors in Edinburgh to take away the pains of your patients when in labour,' he rejoined 'How unnatural it is for you to have swum over from Ireland to Scotland in a steam-boat!'.[27]

Then there was some anxiety that it could lead to *retarded delivery* (thus Gream). In the case of chloroform there was a real *danger of contamination* of substances like phosgene which could have lethal effects if inhaled. Finally, though this was more speculative, were concerns about *possible effects on unborn generations*. The chemist Sheridan Muspratt regarded the use of chloroform as 'most reprehensible' for that very reason.[28] A revisionist view has well been expressed thus:

The loudest and most persistent objections came not from the church, but from the members of Simpson's own profession.[29]

2. There is confusion between theological and ethical objections. Most remarkable was a fear that chloroform could induce *sexual arousal*. Gream quoted the case of a patient who 'drew an attendant to her to kiss as she was in the second state of narcotism'. There were other reports of *erotic dreams*, which Tyler Smith said reduced the patient 'to the level of the brute creation'.[30] Even before Darwin this phrase touched a raw nerve in Victorian society. The utterance of *obscene or disgusting language* gave further ammunition to the protesters, none more tellingly than the one who revealed his fear that 'delicate ladies will use language which it would be thought impossible they should ever have had an opportunity of hearing'![31]

These objections owed nothing to religion as such but everything to the prudery that was such a characteristic feature of Victorian society.

3. There is confusion between theological and institutional objections. It has been pointed out that the argument at times smacked of the known rivalry between obstetricians and surgeons. Thus Gream? It was supremely enshrined in the hostility between Simpson and Syme, and illustrated by an incident in which both men disagreed violently after being called to a case by the family doctor:

For a short time the most distinguished surgeon and the most distinguished obstetrician in Britain stood exchanging insults on the patient's doorstep.[32]

More obvious was the well-known rivalry between London and Edinburgh. This was rooted in deep antipathies going back a long way, particularly between the English and Scottish Royal Colleges. Also some migration of English students to obtain a Scottish MD with minimal examination and no residence requirements (especially St Andrews). Thus on the appointment of Simpson's Edinburgh colleague, James Syme, as Professor of Clinical Surgery at University College, London, in 1847 the *Lancet* declaimed:

When *Scotchmen* have any good offices to give away in Scotland, do *they* send to London for persons to occupy the vacant posts? No indeed! Journeys of such a character are only made from South to North, and not from North to South.[33]

James Syme

As if to confound his critics Syme moved back to Scotland after only 6 months in England!

Then again Scottish *practice* was different. Thus in London Snow used an inhaler, while in Edinburgh Simpson poured chloroform on to an absorbent material as folded handkerchief, placed over nose and mouth. He was careful to check respiration as well as pulse. And in Scotland chloroform was a lot cheaper owing to the lower excise duty on the ethanol from which it was made. This institutional rivalry goes far to explain Simpson's difficulties without resorting to mythical categories of religious opposition.

4. There is confusion between actual and anticipated objections. Simpson's own pamphlet looks much more like an attempt to head off suspected trouble in the future than a reaction to attacks in the past. It was issued *within one month of his first announcement.* No doubt there were rumblings of dissent among a few pious Scots but nothing remotely like an organised clerical opposition.

5. There has been much uncritical acceptance of unsubstantiated data. To account for over 30 major books perpetuating this legend one needs to examine the detailed documentation and discover the authorities from whom

HISTORY OF THE CONFLICT

BETWEEN

RELIGION AND SCIENCE

BY

JOHN WILLIAM DRAPER, M.D., LL.D.

LATE PROFESSOR IN THE UNIVERSITY OF NEW YORK

AND AUTHOR OF 'A TREATISE ON HUMAN PHYSIOLOGY'

NINETEENTH EDITION

LONDON

KEGAN PAUL, TRENCH, & CO., 1 PATERNOSTER SQUARE

1885

Title pages of the two most notorious books expounding the 'conflict' view of religion and science (this page and opposite)

the information was derived. It is a remarkable fact that nearly all statements until quite recently rely on those made by a previous generation of authors, most of whom reproduce the material from even earlier allegations, though without checking primary sources. To give a single example, one writer of the 1960s reported that Simpson was 'inundated with abusive letters from cranks, from clergymen and from colleagues',[34] but gives not a single reference in corroboration.

Much reliance is placed on statements of A. D. White. He wrote:

Simpson . . . was immediately met with a storm of opposition . . . From pulpit after pulpit Simpson's use of chloroform was denounced as impious and contrary to Holy Writ; . . . he seemed about to be overcome.[35]

Close inspection of his remarks reveals just one reference to the early biography by Simpson by Duns which, though quoting letters between

A HISTORY OF

THE WARFARE OF

SCIENCE WITH THEOLOGY

IN CHRISTENDOM

BY ANDREW D. WHITE

LL.D. (Yale and St. Andrews), L.H.D. (Columbia), Ph. Dr. (Jena), D.C.L. (Oxon.)
LATE PRESIDENT AND PROFESSOR OF HISTORY AT CORNELL UNIVERSITY

Simpson and various doctors, is silent on specifically clerical opposition (and indeed records the conversation between Miller and Chalmers mentioned above). White makes no distinction between opposition from theological sources and that on other grounds, and, indeed never even mentions Simpson's own pamphlet on the subject. His account is as one-sided and skewed as much of his other examples from 'history'.

6. *The myth of religious opposition fits well with the conflict thesis.* In fact White's book, like that of Draper, is today acknowledged as a polemic tract masquerading as history. They are the chief literary manifestations of the now notorious 'conflict thesis' that posited religion and science locked in a state of more or less permanent warfare. Recent historical research has demonstrated not only the untenability of such a thesis but also located its origins in the efforts of the Victorian scientific community to attract public and government support and, in general, to keep its end up by crowing over the supposed triumphs of Darwinism over the church.[36] Towards the end of the nineteenth century there was something perilously near to a conspiracy that failed in most respects but succeeded in establishing an inaccurate view of history that has survived until very recently in the scholarly world and is still the staple diet of ordinary people nourished by the mass media. It is a salutary reminder of how easy it is to repeat uncritically concepts that fit well into an existing framework but are unsupported by empirical data, whether in the laboratory or the library. The real challenges to James Young Simpson, and his responses to them, are as important for our understanding of the complex relationships between science and Christianity as they were for the rise of medical anaesthesia.

References

1. J. A. Shepherd, *Simpson and Syme of Edinburgh*, Livingstone, Edinburgh, 1969, pp.9-13.
2. J. Duns, *Memoir of Sir James Y. Simpson*, Edmonston & Douglas, Edinburgh, 1873, p.231.
3. Myrtle Simpson, *Simpson the obstetrician*, Gollancz, London, 1972, p.126.
4. J. Simpson to E. Frankland, 30 January 1856 [Raven Frankland Archives, Open University microfilm 01.02.0230].
5. Shepherd (*op. cit.*), , pp.91-3.
6 L. Gordon, *Sir James Young Simpson and chloroform (1811-1870)*, T.Fisher Unwin, London, 1897, pp.106-108.
7. Duns (*op. cit.*), p.215.
8. M. Simpson (*op. cit.*) pp.140 and 149.
9. A. J. Youngson, *The scientific revolution in Victorian medicine*, Croom Helm, London, 1979, pp.79-82.
10. *Ibid.*, p.83.
11. M. Simpson (*op. cit.*), pp.138-9.
12. E. Frankland, *Diary* for 23 January and 11/12 March 1848 [Joan Bucknall Archive, and Open University microfilm 02.02.1304-1484].
13. *Punch*, Jan.-June 1848, pp. 78 and 87 (twice).
14. J. W. Draper, *History of the conflict between religion and science*, H. S. King & Co., London, 1875, pp.318-9.
15. Shepherd (*op. cit.*), p.98.
16. J. Y. Simpson, *Answer to the religious objections advanced against the employment of anaesthetic agents in midwifery and surgery*, Sutherland and Knox, Edinburgh, 1847.
17. G. T. Gream, *Lancet*, 1848 (i), 228 ff.
18. R. Barnes, *Lancet*, 1850 (ii), 39.
19. R. Lee, *Lancet*, 1853 (ii), 611.
20. A. D. Farr, 'Medical developments and religious belief', Ph.D. thesis, The Open University, 1977.
21. A. D. Farr, 'Early opposition to obstetric anaesthesia', *Anaesthesia*, 1980, **35**, 896-907; 'Religious opposition to obstetric anaesthesia: a myth?', *Ann. Sci.*, 1983, **40**, 159-177.
22. Duns (*op. cit.*), p.260.
23. Letter dated 15 June 1852 from Charles Kingsley to unidentified recipient, in *Charles Kingsley, his letters and memories of his life*, King, London, 1877, vol. i, p.324.
24. Youngson (*op. cit.*), p.101.
25. P. Smith, *Scriptural authority for the mitigation of pains in labour*, Highley, London, 1848.
26. Youngson (*op. cit.*), pp.106-9.
27. A. M. Claye, *The evolution of obstetric analgesia*, Oxford University Press,, 1939, pp.17-18.
28. J. S. Muspratt, *Chemistry, theoretical, practical and analytical*, W. Mackenzie, Glasgow, n.d., vol. i., p.471.
29. M. Simpson (*op. cit.*), p.142.
30. J. Tyler Smith, *Lancet*, 1847 (i), 322.

31. *Idem, Lancet*, 1856 (ii), 424.
32. Youngson (*op. cit.*), p. 65.
33. Anon., *Lancet*, 1848 (i), 48.
34. Shepherd (*op. cit.*), p.99.
35. A. D. White, *A history of the warfare of science with theology in Christendom*, Macmillan, London, 1896; repr. Arco, London, 1956, vol. ii, pp.62-3.
36. C. A. Russell, 'The conflict metaphor and its social origins', *Science and Christian Belief*, 1989, 1, 3-26.

The Manufacture of Anaesthetic Nitrous Oxide N_2O – a Study in Technology Blending

W. A. Campbell

INORGANIC CHEMISTRY, UNIVERSITY OF NEWCASTLE,
NEWCASTLE UPON TYNE, UK

In 1762 Joseph Black heated ammonium nitrate and made nitrous oxide. Ten years later Joseph Priestley obtained the same gas by reducing nitric oxide NO with moist iron filings. The transformation of these laboratory experiments into the commercial production of the pure gas drew together techniques and equipment from many sources.

J.B. van Helmont coined the word gas in 1620 as a transliteration into Dutch of the Greek word chaos. The naming was apt, for when gases were prepared in closed systems to prevent their escape, chaotic explosions were likely to occur[1].

Before gases could be studied in the laboratory, special designs of apparatus were needed These took two forms, those in which the gas pushed aside a liquid boundary of water, mercury or saturated brine, and those in which the receiver itself expanded to accommodate the increasing volume of gas. The former arrangement led to the gas jar, beehive shelf, and pneumatic trough which by the end of the eighteenth century had displaced the alchemist's furnace as the most salient feature of the chemical laboratory. The latter system depended on bladders, usually ox or pig, and gas bags. Both methods of collecting gases achieved industrial importance.

The influence of ballooning

The craze for aeronautics in the 1780s saw the bladder of the laboratory grow into the balloon. The hot air balloons of the Montgolfiers had been made from paper, but the hydrogen balloon released in 1783 by the Charles brothers in Paris was made from silk painted with rubber solution. Subsequently, balloons were of varnished linen, or oiled silk coated with gilt paint. Argand, inventor of the eponymous lamp, advocated a varnish made from gum elastic dissolved in spirit of turpentine and boiled linseed oil. After varnishing, the silk segments of the balloon were to be sewn together, the joints pressed with a hot iron and further varnished[2].

These attempts to make gas-tight balloons helped to establish the oiled silk bags which were so widely used in the administration of gases to patients. The silk was stretched over a frame and painted several times with boiled linseed oil. The drying of the oil was accelerated by the incorporation of litharge or more frequently verdigris; this is the origin of Coleridge's reference to the 'little green bag' at the laughing gas frolics[3].

As balloons became larger and aeronauts more intrepid, the scale of production of the necessary hydrogen increased. Because the filling of a balloon was a social occasion, and because an accident would have been impossible to conceal, great attention was paid to making the apparatus gas-tight. Escapes did occur, and one Northern writer complained of 'stuff that would make a dog sick'. Nevertheless, under the glare of publicity the techniques of gas handling on a large scale were quickly refined.

From Woulfe's bottles through barrels painted with tar to lead cisterns inspired by Roebucks sulphuric acid chambers, the plant for generating hydrogen became larger and more stable. A balloon of thirty feet diameter called for one and a half tons of sulphuric acid to fill it with hydrogen, so useful devices such as flexible lead piping came into use, vulcanised rubber tubing being not yet available.

The Influence of Gas Lighting

In 1792, while minding Boulton and Watt's pumping engines in Cornwall, William Hurdock illuminated his cottage at Redruth by means of coal gas. His employers grasped the potential of what he had done, and in 1802 they celebrated the Peace of Amiens by staging a grand display of gas flares over their Soho factory near Birmingham[4]. As gas lighting spread, the problem of joining lengths of iron piping became more acute, and the technique of luting which had been practised in the laboratory since alchemical times was adapted to an industrial situation. Lutes were made essentially by beating china clay with linseed oil to the consistency of putty, though sometimes the worker would add bizarre ingredients of his own choice such as horse dung, blood, chopped hair, cheese or sour milk. By such means expertise on moving gases over distances was slowly gathered.

The difficulty and expense of laying pipes led for a time to the use of domestic gas holders. Gas bags of up to 100 cu ft capacity were filled at the gas works and transferred to the user's premises where they were discharged into his gas holder. The 'gasometers', large concentric cylinders of iron suspended in troughs of water, in which gas was stored at the works were direct descendants of Priestley's gas jars.

Apparatus at the Pneumatic Institution

Under the influence of the young Humphry Davy, nitrous oxide was made at Thomas Beddoes' Pneumatic Institution at Bristol by the action of heat on ammonium nitrate. James Watt had been attracted to the Institution on

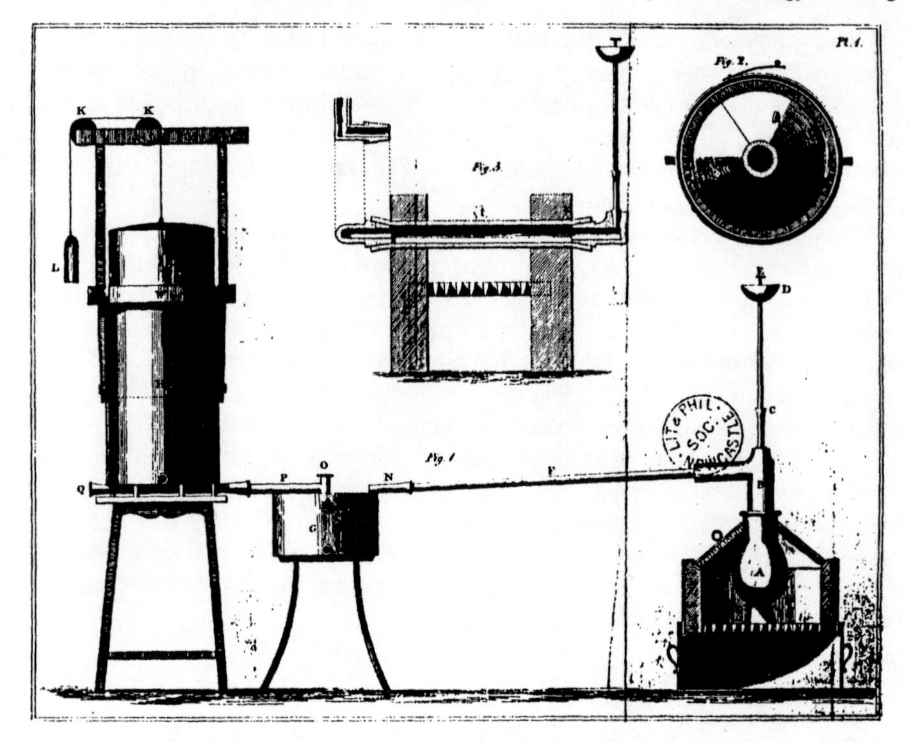

Figure 1 *James Watt's apparatus for preparing, storing and dispensing medicinal gases*

account of the illness of his son, and he soon set about designing apparatus for administering gases to patients. Watt's complete apparatus consisted of retorts and 'fire tubes' in which the starting material was decomposed, cooling and washing equipment, and a telescopic gas holder which he called a hydraulic bellows, and which resembled in miniature the gasometer at the gas works. The gas holder was filled with the washed gas, and by opening the outlet tap and shutting off the inlet, gas could be forced into a bladder or gas bag as required (Figure 1).

Joints were sealed with a cement of beeswax and rosin, easily softened by warming. Watt recommended oiled silk bags which had not been treated with verdigris, as he believed that this rotted the silk. The bags were made by sewing pieces of silk together, the joins being sealed with japanner's gold size - a transfer from balloon technology. To remove the smell of boiled oil which some patients found disagreeable, Watt stacked the silk pieces in alternate layers with charcoal for a minimum of five days.

Boulton and Watt made the complete sets of pneumatic apparatus at their Soho works (Figure 2). They also provided Beddoes with brass taps and nozzles for the internal fittings which conveyed the gas around the Institution. Taps and nozzles for the small bags used by patients were made of ivory or wood[5].

THE SIMPLIFIED AND PORTABLE

PNEUMATIC APPARATUS,

ARE MANUFACTURED BY

BOULTON & WATT,

OF SOHO, NEAR BIRMINGHAM,

AT THE FOLLOWING PRICES.

THE LARGE SIZE SIMPLIFIED APPARATUS.

The furnace, 18 inches diameter, lined with the beſt fire bricks, tongs and poker, two fire tubes, two end pieces, two rings, iron plug, water pipe and cup, iron conducting pipe and its tin end piece, with one large airholder and funnel £6 16 6

AUXILIARY ARTICLES FOR DITTO.

Two large-ſized airholders, one ſpare fire tube, caſt-iron pot for a ſand heat, two oiled ſilk bags and bellows to fill them with common air - 3 6 0

THE SECOND SIZE SIMPLIFIED APPARATUS.

The furnace, 13 inches diameter, and other articles as above, ſuitable - - - - - - - - £4 15 9

AUXILIARY ARTICLES AS ABOVE.

But the two airholders, ſecond ſize - , - - 2 15 0

THE PORTABLE APPARATUS.

One oxygene and one hydro-carbonate fire tube, with end pieces, water pipe and cup, conducting pipe, one ſecond-ſized airholder and funnel, and an oiled ſilk bag - - - - £2 12 6

AUXILIARIES.

One ſecond ſize airholder, two ſpare fire tubes and bellows to fill the bag with common air - 1 2 6

*** Packing Boxes and Carriage to be charged extra.

The Pneumatic Apparatus, with Hydraulic Bellows and Refrigeratories, continue to be made as uſual.

See Part II. for the Prices.

Figure 2 *Price list for pnuematic apparatus Boulton & Watt 1795.*

The Influence of the Limelight

The widespread use of limelight, obtained by playing an oxygen-coal gas flame onto quicklime, in theatres, opera houses and lecture rooms called for the commercial generation of oxygen and a means of carrying it to the place of use. Oxygen was usually purchased from a pharmacist who made it by heating 'oxygen mixture' (potassium chlorate with manganese dioxide). Joseph Wilson Swan of electric light bulb fame ran a flourishing trade in oxygen for theatres from his Newcastle pharmacy. The oxygen was carried to the theatre in a large leather gas bag of some fifty cubic feet capacity. In use, an oxygen bag and a coal gas bag were connected to a jet, the pressure being controlled by flat iron weights similar to those which an organ builder places on his bellows.

Explosions occurred from careless mixing of the gases and from ignorance; one theatre manager said 'I always light my oxygen first'.[6] This however is one way in which a knowledge of the properties of gases spread to a wider public. It is clear that the pharmacist who made oxygen could just as easily make nitrous oxide and convey it to a local surgery or hospital.

Preparation of Nitrous Oxide

1. By Priestley's reduction of nitric oxide:
   $2NO + Fe = N_2O + FeO$
2. From nitric oxide and sodium sulphite:
   $2NO + Na_2SO_3 = N_2O + Na_2SO_4$
3. From sodium nitrite and hydroxylamine hydrochloride:
   $NaNO_2 + [HONH_3]Cl = N_2O + NaCl + 2H_2O$
   A valuable laboratory method but too costly for commercial use
4. By heating ammonium nitrate:
   $NH_4NO_3 = N_2O + 2H_2O$

This last has been the preferred industrial method of preparing the gas, though it is not without risk. The decomposition begins gently enough, but unless it is carefully controlled the rate of reaction increases with temperature until an explosion occurs. The explosion at the BASF plant at Oppau on the Rhine took place on September 21 1921 when a caked stockpile of ammonium nitrate was being loosened. 561 lives were lost, many more injured and 7000 lost their homes[7]. To minimise such risks, mixtures of ammonium sulphate and sodium nitrate, either solid or in solution, were sometimes preferred.

For making a gas of anaesthetic quality, the ammonium sulphate must be free from organic impurities and from chloride. In the former case the gas is contaminated with carbon monoxide, and in the latter with chlorine.

After the rise of the synthetic dyestuffs industry, all nitrates were under severe competition; this was further compounded by the use of nitrates in explosive manufacture. Although the Chile nitre trade helped, the pressure was relieved only by the arc process of nitrogen oxidation due to Birkeland and Eyde, and to Ostwald's ammonia oxidation process.

Manufacture of Ammonium Nitrate

5. Direct combination of ammonia and nitric acid:

If the concentration of nitric acid is properly adjusted, the enthalpy of reaction drives off the water and molten ammonium nitrate is formed.

6. Modification of ammonia-soda process using sodium nitrate solution in place of brine:
   $NH_3 + CO_2 + H_2O + NaNO_3 = NH_4NO_3 + NaHCO_3$

This has the advantage of two saleable products instead of one.

7. A further modification of ammonia-soda technology:
   $2NH_3 + CO_2 + H_2O + Ca(NO_3)_2 = 2NH_4NO_3 + CaCO_3$
8. From nitrous fumes, air, steam and ammonia:
   $4NO_2 + 2H_2O + O_2 + 4NH_3 = 4NH_4NO_3$

Throughout the nineteenth century the first method remained in wide use, though Michael Faraday prepared his own ammonium nitrate from nitric acid and ammonium carbonate as Glauber had done in 1659.

The Liquefaction of the Gas

The efficient storage and transport of a gas is improved if the gas is liquefied. In 1823 Faraday liquefied nitrous oxide by a general method of his own design. Ammonium nitrate was heated in one leg of an inverted V-shaped sealed tube, while the other leg was cooled in a bath of ice and salt. Sufficient pressure was built up to liquefy the gas. Faraday was aware of the presence of another constituent and painstakingly showed that it was water.

In later work Faraday used a glass syringe with metal valves to withdraw successive portions of gas from a storage vessel and pump it into the liquefaction tube. The latter was made of thick green glass able to withstand high pressures. Using a better cooling bath of ether and carbon dioxide, he was able to freeze nitrous oxide to a clear crystalline solid at twenty-two atmospheres[8].

At such pressures Faraday had trouble with leaks. The syringe, first of glass and then of brass, was cemented on one side to the reservoir and on the other to the receiver with sealing wax. The liquefaction of gases was now sufficiently established for one of Faraday's contemporaries, a classical scholar name Thomas Northmore to order a complete kit of liquefaction apparatus, including brass condensing pumps, glass and metal receivers, taps and connecting tubes from a London instrument maker. A telling feature of the apparatus was a wooden screen to protect the operator from possible explosion[9]. Northmore's experiments were sufficiently interesting to be watched by Friedrich Accum, food analyst and gas lighting pioneer, but it is clear that anyone who wished to liquefy nitrous oxide could have obtained the necessary equipment at an early date.

Conclusion

The possibility of performing a surgical operation on an anaesthetised patient was demonstrated in 1844. By that date, the art of preparing, handling and employing gases had been enriched by the introduction of apparatus, skills and experience from a number of technical and social advances. Almost as soon as its value had been established, nitrous oxide could be obtained by any anaesthetist who needed it, when, where, and in what quantity he wished.

References

1. Van Helmont, J .B. , Ortus Medicinae, Amsterdam 1652, 59
2. Clow, A.& N., The Chemical Revolution, London 1952, 156
3. Cottle, J., Early Recollections of S.T. Coleridge, vol.2, London 1837, 37
4. Murdock, W., Phil. Trans. Roy. Soc., *98,* 1808, 124
5. Beddoes, T. & Watt, J., Considerations on the Medicinal Use and on the Production of Factitious Airs,2nd ed. , Bristol 1795, vol. 1, part 2, 4
6. Trans. Newcastle Chem. Soc., *3*, 1874-7, 194
7. BASF, In the Realm of Chemistry, Dusseldorf 1965, 93
8. Faraday, M., Papers on the Liquefaction of Gases, Alembic Club Reprints No.12, Edinburgh 1904, 16, 35; see also Faraday's Diary 1820-1862, ed. T. Martin, London 1936, vol.1, 98; vol.4, 152, 187, 214
9. Northmore, T., Nicholson's Journal, *12*, 1805, 368

The Priestley Lecture 1997
Nitric Oxide

S. Moncada

THE CRUCIFORM PROJECT UNIVERSITY COLLEGE LONDON, UK

Joseph Priestley is generally regarded to be the discoverer of nitric oxide (NO), although there are reports that it was synthesised by Mayow in 1669 and that Robert Boyle had observed in 1671 that it formed reddish brown fumes in contact with air. The earliest report of NO came in the early 1600s from Jan Baptist van Helmont, who described the gas but seemed to confuse it with carbon dioxide[1].

In 1772, Priestley examined the properties of NO, which he called 'nitrous air', and which he generated by adding nitric acid ('*spirit of nitre*') to various metals. He found that nitrous air would not support plant life, that it reduced putrefaction, that it was soluble in boiled water and that water impregnated with it had an acid taste. Priestley devised a method of measuring the purity or 'goodness' of common air from different sources. Air was mixed with nitrous air in a wide jar over water and the contraction in volume that occurred when the NO and oxygen combined to form the water-soluble nitrogen dioxide was measured and used as a 'standard of air'. He developed an apparatus with which changes in volume could be measured accurately: this later became known as a eudiometer or '*purity measurer*'. The development of this instrument represented one of the most important advances in the accurate measurement of gases and its use was later extended to many other chemical studies.

Priestley was particularly pleased with this test of the purity of air using nitrous air since it obviated the need to use mice for respiration studies. In 1774, he wrote that '*every person of feeling will rejoice with me in the discovery of nitrous air, which supersedes many experiments with the respiration of animals*'[2].

Nitric oxide later became recognized as an atmospheric pollutant associated with acid rain which results from the release of sulphur dioxide and nitrogen oxides from automobile exhausts and industrial operations; these gases combine with water vapour in clouds and the resulting precipitation is highly acidic. Photochemical smog, which occurs mostly in urban areas, results from the interactions of these nitrogen oxide emissions with hydrocarbon vapours in the presence of sunlight. Ozone is one product of this photochemical reaction and is a major irritant, causing plant damage, irritation of the eyes and respiratory distress.

In this context, the story of how NO was found to be a biologically generated and active substance is a fascinating tale at the interface between chemistry and biology. In 1985 I became interested in an unstable substance that had been found in vascular rings by Furchgott and Zawadzki some five years earlier[3]. This substance, known to be responsible for endothelium-dependent relaxation, was called 'endothelium-derived relaxing factor' (EDRF) and was found to be highly unstable. EDRF was later shown to act

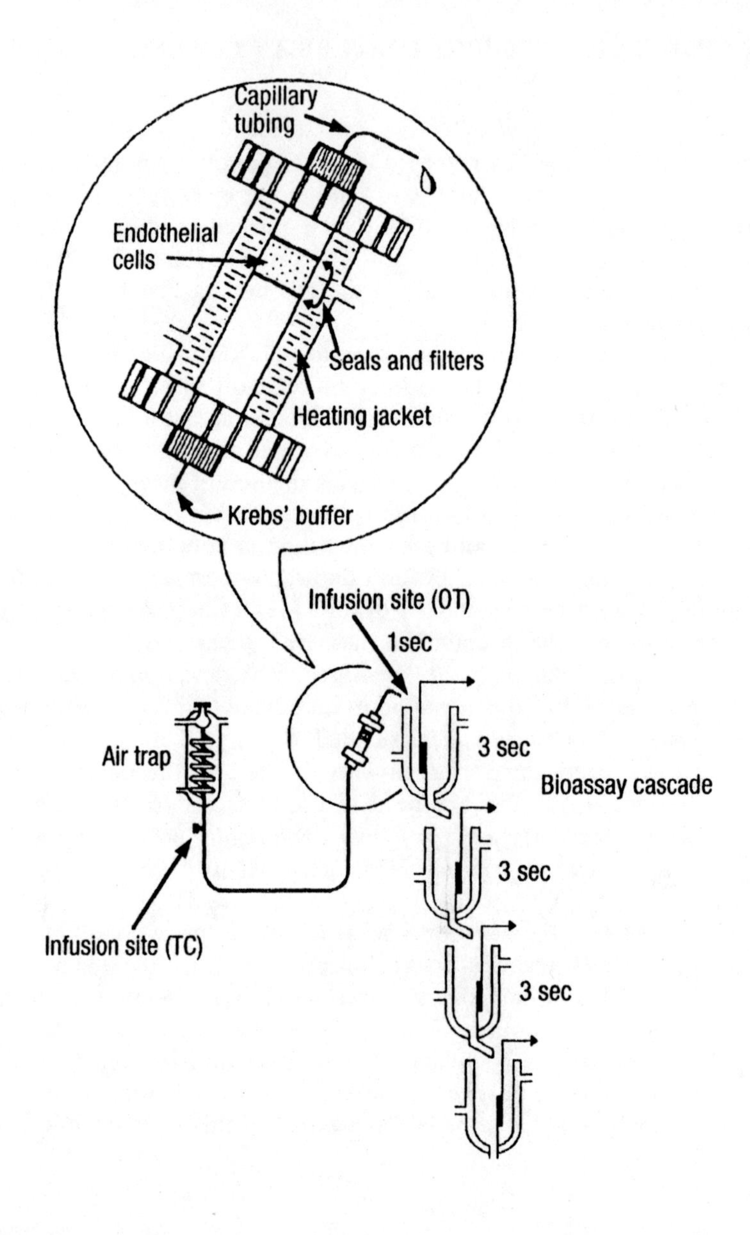

biologically by activating the enzyme soluble guanylate cyclase and to be inactivated by haemoglobin (see reference 4). By the mid-1980s EDRF had been hypothesized to be many different things including a novel metabolite of arachidonic acid, a product of cytochrome P-450 oxygenases, a carbonyl containing compound, and even ammonia[4,5].

A number of experiments were carried out at that time in the attempt to transfer biologically active EDRF from one piece of vascular tissue to another. Some of these were sufficiently successful for a half-life for EDRF to be calculated. Many other experiments were carried out attempting to identify the chemical structure of EDRF and to inhibit its activity (see 4,5).

The difficulties associated with the research on EDRF at that time were related to the indirect nature of the majority of the experiments, since endothelium-dependent relaxation was measured as an indication of EDRF release and furthermore, when direct measurements were attempted, very small amounts of EDRF were generated from vascular strips or rings.

We were particularly well placed to carry out studies on a substance as unstable as EDRF since, during the mid-1970s, we had gained a great deal of experience doing bioassay of unstable metabolites of arachidonic acid such as thromboxane A_2[6] and prostacyclin[7]. We thought that if extensive pharmacological studies were to be carried out and the chemical structure was to be identified then we needed a larger quantity of material and a reliable and, as much as possible, a quantitative bioassay system. We decided to culture endothelial cells on microcarrier beads and to perfuse them in a modified chromatography column, thus allowing the culture of millions of endothelial cells for the generation of EDRF (see Figure 1). We used the perfusate from these cells to superfuse a series of bioassay tissues, usually strips of rabbit aorta denuded of endothelium, which conveniently separated the EDRF generating system (the vascular endothelial cells in the column) from the

Figure 1 *Diagram showing a cascade for the bioassay of EDRF Porcine aortic endothelial cells were grown in culture on microcarrier beads. Between 1-3 mi of beads containing 1-9 × 107 endothelial cells were packed into a modified chromatographic column which was maintained at 37 °C and perfused with Krebs buffer at 5 mi/min. The perfusate was allowed to flow over a cascade of 4 strips of rabbit arterial tissue from which the endothelial cells had been removed. The arterial strips were mounted in heated (370 °C) glass chambers and superfused (5 mi/min) with Krebs buffer for 2-3 h before superfusion with effluent from the column. Changes in length of the tissues were detected by auxotonic levers attached to transducers, the display of which was recorded. The arterial strips were separated from each other by a delay of 3s and from the chromatographic column by a delay of is.*

EDRF was released from the column by 1 min infusions of bradykinin (10-50 nM) through the column (TC). Compounds were also infused directly over the tissues (OT). Using this apparatus, the release of EDRF from endothelial cells can be monitored, its half-life down the column of tissues can be determined, and its biological activity compared with that of known biological agents applied directly over the tissues. The effluent can also be collected and subjected to chemical analysis (see reference 8):

bioassay detector system in the superfusion cascade[8]. The introduction of this system opened the door to the many fundamental studies that followed.

Within a few months of our development of this method for the bioassay of EDRF, the technique helped us to discover that superoxide anions (O_2^-) inactivate EDRF[9]. In time, this proved to be a seminal observation for within a few months it allowed us to discover the mechanism of action of many EDRF antagonists (inactivators) as generators of O_2^- (ref. 10) and later on, inactivation of EDRF and NO became a crucial criterion for the identity of NO in many experiments. A few years later it opened the field of research related to the interaction between NO and O_2^- in biological systems[11], one of the most interesting areas of NO research as I am writing this review (see below).

Among the many hypotheses about the nature of EDRF was the suggestion that EDRF might be NO or a related molecule, since both EDRF and NO are unstable, sensitive to inactivation by O_2^- and haemoglobin and were known to relax vascular smooth muscle via stimulation of the enzyme, soluble guanylate cyclase. This struck me as an extremely interesting and attractive suggestion and thus we decided to investigate whether EDRF was in fact NO using two approaches; first, by studying the comparative pharmacology of authentic NO gas and EDRF, and second, by trying to develop a method to measure the release of NO by vascular endothelial cells.

Experience of bioassay techniques convinced me, after the initial experiments, that EDRF and NO are one and the same substance, since their behaviour was identical in a large number of comparative pharmacological experiments that we carried out in bioassay tissues (Figure 2) and later on in platelets (Figure 3). These encouraging results led us to try to develop a method for measuring the release of NO from vascular endothelial cells. We did not want to rely solely on the measurement of stable oxidation products of NO, such as nitrite or nitrate; instead we were interested in measuring NO directly. The method which was closest to our requirements was that of measuring NO by a chemiluminescence technique originally developed to analyse car exhaust and to monitor ambient air. We therefore contacted Dr C. L. Waiters at the University of Surrey, who had used a chemiluminescence analyser to detect the release of NO from nitrosamines. This machine did not have the sensitivity that would be required to detect NO released from biological tissues. Nevertheless, we decided to see whether the experiment could be carried out and found that stimulation of endothelial cells with bradykinin generated a reproducible, but barely detectable, NO signal. Later, with the help of the Department of Physical Sciences at Wellcome Research Laboratories, where we were working at that time, we were able to enhance the sensitivity of this method sufficiently to demonstrate that NO is indeed released from endothelial cells and that it accounts for the biological actions of EDRF. The results of these studies were summarized in a paper submitted to the journal *Nature* in January 1987[12]. Some results in support of this demonstration were published within a few months by other laboratories[13,14].

We continued our research by investigating how NO is formed in biological

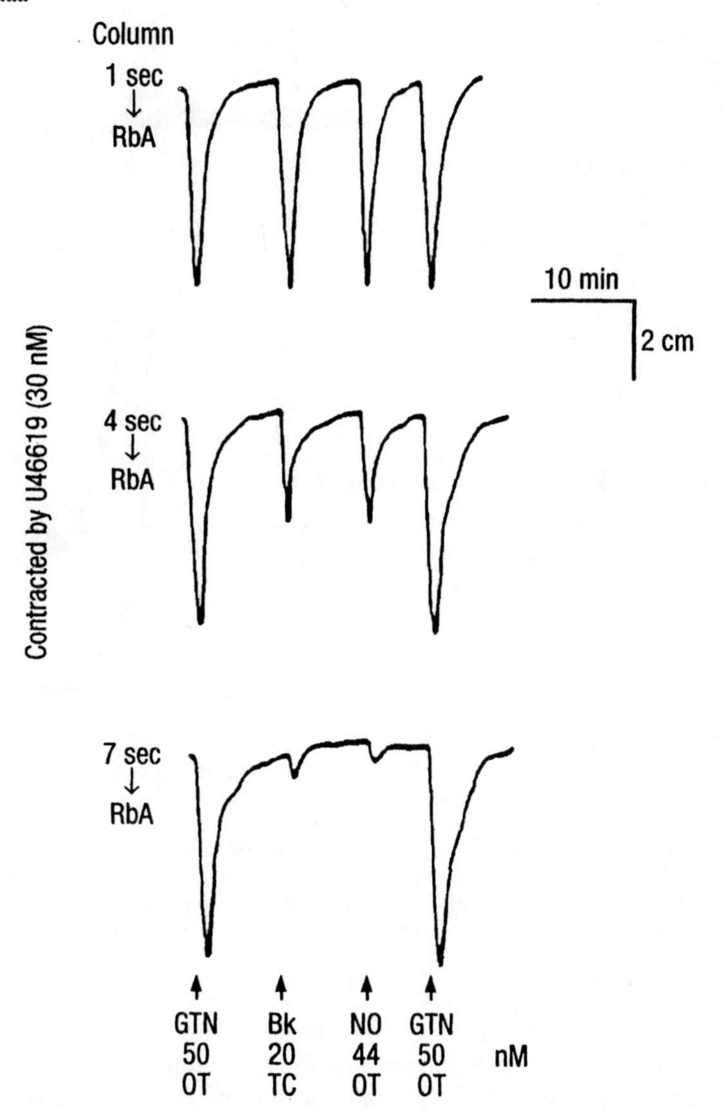

Figure 2 *Relaxation of rabbit aortae by EDRF and NO. The effluent of a column of endothelial cells on microcarriers perfused with Krebs' buffer was used to superfuse three spiral strips of rabbit aorta (RbA) which were separated from the cells by delays of 1, 4 and 7 s respectively. The tissues were contracted submaximally by a continuous infusion of 9,11-dideoxy-Sa, 11a-methane epoxy-prostaglandin FZrr (U46619; 30 nM). The sensitivity of the tissues was standardised with glyceryl trinitrate (GTN; 50 nM) administered over the tissues (OT). The tissues were relaxed to a similar extent by EDRF released from the cells by a 1 min infusion through the column (TC) of bradykinin (Bk 20 nM) and by NO (44 nM) administered as a 1 min infusion (OT).*

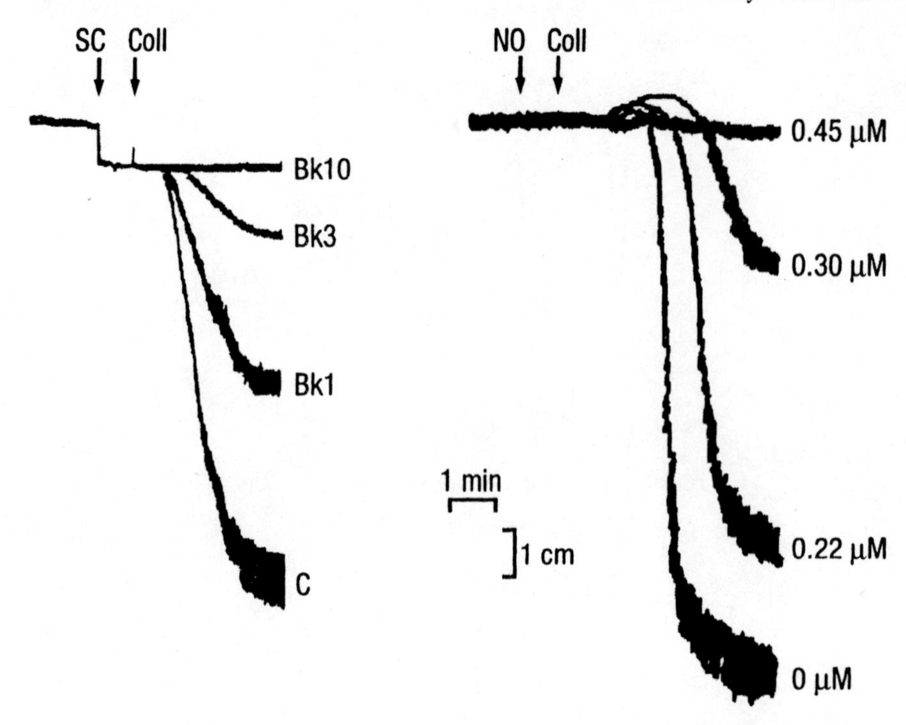

Figure 3 *Inhibitory actions of EDRF and NO on washed human platelets. Left hand panel: an aliquot (0.1 ml) of supernatant from 2 mi incubates containing 1.5 mi of unstimulated porcine aortic endothelial cells (C) does not inhibit platelet aggregation induced by collagen (Cell, 4 μg ml^{-1}). Aliquots of the same volume of cells stimulated (SC) with 1, 3 and 10 nM of bradykinin (Bk) to release EDRF inhibit aggregation in a concentration-dependent manner. Right hand panel: concentration-dependent inhibition of aggregation induced by 0.22, 0.30 and 0.45 μLM Of NO added to the platelets 1 min prior to the addition of collagen.*

systems. Initial studies involved feeding endothelial cells with different potential precursors, such as nitrite, nitrate, ammonia or amino acids, none of which was successful. Searching the literature we then came across the description of the formation of nitrite and nitrate from the amino acid L-arginine by activated macrophages[15,16]. These papers prompted us to try, again unsuccessfully, to feed our endothelial cells with L-arginine to release NO. The project was abandoned until a few weeks later when we concluded that probably the L-arginine already present in the culture medium was 'swamping the system' and decided to culture the cells in an L-arginine-free medium for 24h before the experiment. This solved the problem, since it led to experiments showing that we could enhance the production of NO by the cells by feeding L-arginine. These and other more sophisticated experiments using labelled L-arginine and mass spectrometry demonstrated that NO was synthesised from the guanidino nitrogen atom(s) of L-arginine[17].

These twin discoveries, the identification of NO[12] and of its biological precursor[17] were the foundations of a new biochemical pathway that we started to call the L-arginine: NO pathway[18]. Within a few months we identified an enzyme, the NO synthase (NOS), which converts L-arginine to L-citrulline and NO via the incorporation of molecular oxygen and the formation of the unstable N^G-hydroxy-L-arginine (see 19). Three isoforms of NO synthase have now been described - endothelial (eNOS), neuronal (nNOS) and an inducible NO synthase (iNOS); eNOS and nNOS are usually expressed constitutively in a variety of cells, while iNOS is synthesised by cells following exposure to certain inflammatory cytokines. The L-arginine:NO pathway is now known to be a ubiquitous biochemical pathway in both mammalian and non-mammalian tissues and the synthesis of NO has been shown to underlie a wide variety of biological functions (for reviews see 18-20).

A number of analogues of L-arginine have been found to be inhibitors of the synthesis of NO and have proved to be valuable tools in understanding the biological actions of NO. N^G-monomethyl-L-arginine (L-NMMA) a compound that we identified as a competitive inhibitor of NO synthesis, has proven to be a most invaluable pharmacological tool to investigate the roles of NO in biological systems. L-NMMA is a potent vasoconstrictor *in vivo*; it constricts blood vessels, produces a hypertensive response in animals and causes vasoconstriction of the forearm arterial circulation in humans. Its action is entirely endothelium-dependent, and its vasoconstrictor properties result from the inhibition of an endogenous vasodilator mechanism which is NO-dependent (see 20). Furthermore, it is orally active and it produces a sustained increase in blood pressure in laboratory animals when administered in the drinking water for prolonged periods[21]. These findings led us to conclude that there is a physiological, NO-dependent vasodilator tone that is essential for the regulation of blood flow and pressure[22]. Experiments confirming all these findings have now been carried out in humans and animals in many laboratories around the world and, more recently, additional evidence has been provided for a fundamental role of NO in the regulation of blood flow and pressure by the generation of eNOS knockout mice which are hypertensive[23].

The discovery of this vasodilator tone indicated the existence of an endogenous donor NO system, the actions of which are imitated by compounds such as glyceryl trinitrate and sodium nitroprusside (Figure 4). These NO donors exert their pharmacological actions after their metabolism, by enzymic or non-enzymic processes, into NO. The NO thus liberated reacts with the haem group in the soluble guanylate cyclase of the vascular smooth muscle cell, and the activated enzyme produces more cyclic GMP, leading to vascular relaxation[24,25].

With the growing knowledge of the biological actions of NO have come new uses for NO donors. Nitric oxide is a potent inhibitor of platelet aggregation and adhesion and this activity is shared by NO donors *in vivo*. Studies using S-nitroso-glutathione have shown that this compound potently inhibits platelet aggregation and adhesion at doses that cause only minimal vasodilatation[26]. Selective targeting to platelets without accompanying hypotension will allow

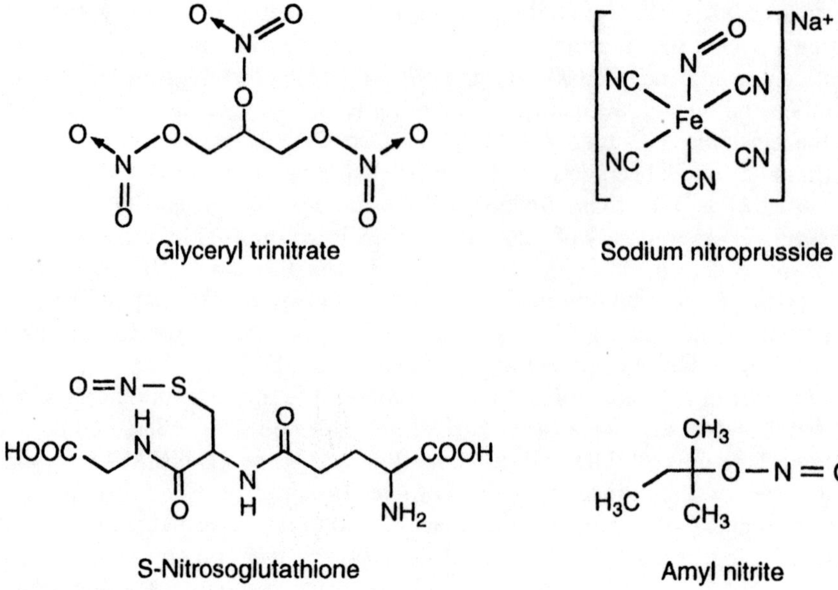

Glyceryl trinitrate

Sodium nitroprusside

S-Nitrosoglutathione

Amyl nitrite

Figure 4 *Structures of some nitric oxide donors.*

the use of this type of compound in thrombotic disorders, where the NO donor may be used alone or in combination with other antithrombotic agents.

Since NO plays a number of roles in the physiological functioning of the cardiovascular system, such as inhibition of platelet and white cell activation and inhibition of vascular smooth muscle cell proliferation, it is hardly surprising that reduced generation of NO has been linked with a number of clinical disorders of this system, including hypertension and atherosclerosis (see 27). In this context it will be extremely useful to investigate the precise nature and the consequences of the hypertension of eNOS knockout mice, as well as the response of the vessel wall and other organs to different types of injury including mechanical or cholesterol-induced vascular dysfunction.

An NO synthase is induced in activated macrophages and the NO produced by this iNOS accounts for the L-arginine-dependent cytostatic and cytotoxic actions of these cells against a number of microorganisms including *Cryptococcus neoformans*, *Toxoplasma gondii*, *Corynebacferium parvum*, *Leishmania major* and *Schisfosoma mansonii.* Thus the L-arginine:NO pathway acts as a primary defence mechanism against intracellular microorganisms as well as pathogens that are too large to be engulfed by the white cells[28,29].

The biochemical mechanisms of NO-induced cytostasis/cytotoxicity are not completely understood. There is an NO-dependent inhibition of key enzymes in the respiratory cycle and in the synthesis of DNA in the target cells, in some cases involving reaction with iron-sulphur centres in these enzymes[30,31]. Furthermore, NO inhibits key enzymes in the synthesis of DNA such as ribonucleotide reductase[32]. More recent studies suggest that NO may reversibly inhibit cytochrome c oxidase, the terminal electron transport protein in mitochondria. In large quantities, however, NO may also produce toxic effects due to the generation of peroxynitrite ($ONOO^-$), a product of the reaction between NO and O_2^- (ref. 33). At the mitochondrial level $ONOO^-$ inhibits complexes I-III irreversibly, thus causing cytotoxicity[34]. It may also cause damage by nitrating tyrosine and other residues on proteins. However, $ONOO^-$ may be inactivated by combining with a range of biomolecules such as thiols, antioxidants and sugars[35]. Thus the net effect of $ONOO^-$ appears to be critically dependent on the local concentration of thiols and other molecules that can act as scavengers.

The expression of iNOS is induced in endothelial and vascular smooth muscle cells, as well as other cells and tissues in a number of pathological situations such as septic shock. Nitric oxide generated by this enzyme accounts for the profound vasodilatation, resistance to vasoconstrictors and vascular leak syndrome associated with this condition. In patients with septic shock, low doses of L-NMMA added to standard therapy have been shown to restore blood pressure. Selective inhibitors of iNOS are being developed that are likely to provide improved therapy since, unlike L-NMMA, they will allow NO to continue to be produced by the constitutive NO synthases[36]. Overproduction of NO by iNOS contributes to a number of acute and chronic inflammatory conditions including ulcerative colitis, rheumatoid arthritis and osteoarthritis (see 27). Interestingly, knockout mice for iNOS have a reduced inflammatory response and are resistant to LPS-induced mortality while having a reduced capacity to fight infection[37].

We also found that NO is present in the exhaled air of animals and humans[38]. At that time we speculated that this may come from the vasculature of the lung, where it may be playing a physiological function, or from 'activated cells' and thus play a defence role; since then it has been found that NO is also generated in the nasopharynx and might be physiologically inhaled to dilate the pulmonary circulation[39]. The amount of NO exhaled has been shown to be increased in patients with untreated asthma, suggesting that iNOS is induced in this condition (see 40).

Although NO reacts with oxygen to form the toxic gas nitrogen dioxide, this gas phase reaction is of the third order, second order in NO and first order in O_2. This explains why in ambient air NO is more stable at lower concentrations. This fact, together with the faster rate of reaction of NO with haemoglobin than with oxygen, means that low concentrations of NO gas can be safely administered by inhalation. When inhaled at 5 - 80 ppm NO has been shown to reverse persistent pulmonary hypertension of the newborn, pulmonary hypertension induced by hypoxia or after surgery, and chronic pulmonary hypertension. The beneficial effects of NO last throughout the inhalation period and in some cases persist after termination of treatment (reviewed in 41). Inhalation of NO by patients with severe adult respiratory distress syndrome (ARDS) alleviates the pulmonary hypertension and hypoxaemia associated with this condition. This is achieved by NO being distributed selectively to the ventilated pulmonary areas, thus increasing blood flow preferentially to the well-ventilated alveoli and improving the ventilation-perfusion ratio. Studies have shown that even very low concentrations of inhaled NO, comparable to those measured in the atmosphere (~100 ppb) or generated in the nasal cavity, are beneficial in ARDS (see 41).

In 1988 it was demonstrated that rat cerebellar cells stimulated with N-methyl-D-aspartate (which activates certain receptors for the neurotransmitter glutamate) release an EDRF-like material and have elevated levels of cyclic GMP[42]. At around this time we were looking at the possibility that the L-arginine:NO pathway might be present in the brain, since we had learned from the literature that L-arginine can activate the soluble guanylate cyclase in brain cells and tissue (see 20). We found that addition of L-arginine to cytosol from rat brain synaptosomes does indeed result in the formation of NO; this process was inhibited by haemoglobin and L-NMMA, showing that the brain possesses the NO synthase[43]. Now it is known that nNOS is widely distributed in the central nervous system where it is stimulated by the action of glutamate on an NMDA receptor. The NO thus produced acts by mediating a variety of functions which include synaptic plasticity, regulation of cerebral circulation and cerebrospinal fluid production, induction and regulation of the circadian rhythm, the induction of hyperalgesia, and the development of tolerance to and withdrawal from morphine. Nitric oxide also appears to play a role in the development of the nervous system, either via a trophic action on developing neurons or by eliciting programmed cell death (apoptosis) (see 44,45).

In the peripheral nervous system, NO is now known to be the mediator released by a widespread system of nerves, previously recognised as nonadrenergic and non-cholinergic, and now known as nitrergic: These nerves mediate some forms of neurogenic vasodilatation and regulate certain gastrointestinal, respiratory and genitourinary functions (see 46-48). These physiological actions of NO are mediated by activation of the soluble guanylate cyclase and consequent increase in concentration of cyclic GMP in target cells. These nitrergic nerves are proving to be as important as adrenergic, cholinergic and peptidergic nerves, and their dysfunction may lead to a variety of disorders, including hypertrophic pyloric stenosis in infants, achalasia, hyperactivity of the urinary bladder and male impotence (see 41).

Nitric oxide has been found in similar locations in primitive and in higher species. For example, nNOS is present in abundance in the pyloric stomach of the starfish (*Marthasterias glacialis*) where it may play a role in muscle relaxation, as it does in the mammalian gastrointestinal tract[49].

When we embarked on the project to investigate the identity of EDRF in 1986, we little imagined that we were about to link an old and illustrious field of chemical work with so many fundamental aspects of biology. Now, 11 years later, it is clear that an immense area of biological research was opened up by these early experiments. We now know that the simple gaseous molecule NO has a wide variety of biological roles and that the L-arginine:NO pathway is extensively distributed throughout the animal kingdom. It is remarkable, therefore, that until so recently its production by the body and its physiological importance were not even imagined. Indeed, as Francis Bacon said in 1620:

'The human mind is often so awkward and ill-regulated in the career of invention that it is at first diffident, and then despises itself. For it appears at first incredible that any such discovery should be made, and when it has been made, it appears incredible that it should so long have escaped men's research'.

References

1. J.W. Mellor. *Mellor's Modern Inorganic Chemistry*. Fifth Edition, 1961,Longmans, Green & Co., p. 459.
2. J.R. Partington. '*A History of Chemistry*', McMillan & Co., London,1962, Vol. 3, pp. 253.
3. R.F. Furchgott and J.V. Zawadzki, *Nature*, 1980, **288**, 373.
4. R.F. Furchgott, *Ann. Rev. Pharmacol. Toxicol.*, 1984, **24**, 175.
5. S. Moncada, R.M.J. Palmer and E.A. Higgs. '*Thrombosis and Haemostasis*', ed. M. Verstraete, J. Vermylen, H.R. Lijnen and J.Arnout, Leuven University Press, 1987, pp. 597.
6. P. Needleman, S. Moncada, S. Bunting, J.R. Vane, M. Hamberg and B. Samuelsson, *Nature*, 1976, **261**, 558.
7. S. Moncada, R.J. Gryglewski, S. Bunting and J.R. Vane, *Nature*, 1976, **263**, 663.
8. R.J. Gryglewski, S. Moncada and R.M.J. Palmer, *Br.J. Pharmacol.*,1986, **87**, 685.
9. R.J. Gryglewski, R.M.J. Palmer and S. Moncada, *Nature*, 1986, **320**, 454.
10. S. Moncada, R.M.J. Palmer and R.J. Gryglewski, *Proc. Natl. Acad. Sci. USA*, 1986, **83**, 9164.
11. T.B. McCall, N.K. Boughton-Smith, R.M.J. Palmer, B.J.R. Whittle and S. Moncada, *Bicchem. J.*, 1989, **261**, 293.
12. R.M.J. Palmer, A.G. Ferrige and S. Moncada, *Nature*, 1987, **327**, 524.
13. M.R. Khan and R.F. Furchgott, '*Pharmacology*' ed. M.J. Rand and C. Raper, Elsevier, Amsterdam, 1987, pp. 341.
14. L.J. Ignarro, G.M. Buga, K.S. Wood, R.E. Byrns and G. Chaudhuri, *Proc. Natl. Acad. Sci. USA*, 1987. **84**, 9265.
15. J.B. Hibbs, R.R. Taintor and Z. Vavrin, *Science*, 1987, **235**, 473.
16. R. lyengar, D. Stuehr and M.A. Marletta, *Proc. Natl. Acad. Sci USA*, 1987, **84**, 6369.
17. R.M.J. Palmer, D.S. Ashton and S.Moncada, *Nature*, 1988, **333**, 664.
18. S. Moncada, R.M.J. Palmer and E.A. Higgs, *Biochem. Pharmacol.*, 1989, **38**, 1709.
19. S. Moncada, *Acta Physiol. Scand.*, 1992, **145**, 201.
20. S. Moncada, R.M.J. Palmer and E.A. Higgs, *Pharm. Revs.*, 1991, **43**,109.
21. S.M. Gardiner, T. Bennett, P.A. Kemp, R.M.J. Palmer and S. Moncada In: *The Biology of Nitric Oxide. 1. Physiological & Clinical Aspects*, Eds. S. Moncada, M.A. Marletta, J.B. Hibbs Jr. & E.A. Higgs, Portland Press, London 1992, pp. 127.
22. D.D. Pees, R.M.J. Palmer and S. Moncada, *Proc. Natl. Acad. Sci. USA*, 1989, **86**, 3375.
23. P.L. Huang, Z. Huang, H. Mashimo, K. D. Bloch, M.A. Moskowitz, J.A.Bevan and M.A. Fishman, *Nature,* 1995, **377**, 239.
24. S. Moncada, R.M.J. Palmer and E.A. Higgs, *Hypertension*, 1988, **12**,365.
25. L.J. Ignarro, *Blood Vessels*, 1991,**28**, 67.
26. M.W. Radomski, D.D. Rees, A. Dutra and S. Moncada, *Br.J.Pharmacol.*, 1992, **107**, 745.

27. S. Moncada and E.A. Higgs, *N. Engl. Med.*, 1993, **329**, 2002.
28. C.F. Nathan and J.B. Hibbs, Jr., *Curr Opin. Immunol.*, 1991, **3**, 65.
29. K. Nussler and T.R. Billiar, J. *LeukosBiol.*, 1993, **54**, 171.
30. J.B. Hibbs, Jr., R.R. Taintor, Z. Vavrin, D.L. Granger, J.-C., Drapier, I.J. Amber and J.R. Laneaster Jr., '*Nitric Oxide from L-arginine: A Bioregulatory System*', ed. S. Moncada and E.A. Higgs, Elsevier, Amsterdam, 1990, pp. 189.
31. T. Nguyen, D. Brunson, C.L. Crespi, B.W. Penman, J.S. Wishnok, and S.R. Tannenbaum, *Proc. Natl. Acad. Sci. USA*, 1992, **89**, 3030.
32. M. Lepoivre, B. Chenais, A. Yapo, G. Lemaire, L. Thelander and J-P.Tenu, J. *Biol. Chem.*, 1990, **265**, 14143.
33. J.S. Beekman, T.W. Beekman, J. Chen, P.A. Marshall and B.A. Freeman, *Proc. natl. Acad. Sci. USA*, 1990, **87**, 1620.
34. Lizasoain, M.A. Moro, R.G. Knowles and S. Moncada, *Biochem. J*, 1996, **314**, 877.
35. M.A. Moro, V.M. Darley-Usmar, D.A. Goodwin, N.G. Read, R. Zamora Pino, M. Feeliseh, M.W. Radomski and S. Moncada, *Proc. Natl. Acad.Sci USA*, 1994, **91**, 6702.
36. P. Valiance and S. Moncada, *New Horizons*, 1993, i, 77.
37. X-Q. Wei, I.G. Charles, A. Smith, J. Ure, G-J. Feng, F-P. Huang, D.Xu, W.A. Sands, S.A. Baylis, S. Moncada and F.Y. Liew, *Nature*, 1995, **375**, 408. 18
38. L.E. Gustafsson, A.M. Leone, M.G. Persson, N.P. Wiklund and S. Moncada, *Biochem. Biophys. Res. Commun.*, 1991,**181**,852.
39. J.O.N. Lundberg, E. Weitzberg, S.L. Nordvall, R. Kuylenstierna, J.M. Lundberg and K. Alving, *fur. Resp. J.*, 1994,**7**, 1501.
40. S.A. Kharitonov, D. Yates, R.A. Robbins, R. Logan-Sinclair, E.A. Shinebourne and P.J. Barnes, *Lancet*, 1994, **343**, 133.
41. S. Moncada and E.A. Higgs, *FASEB J.*, 1995, **9**, 1319.
42. J. Garthwaite, S.L. Charles and R. Chess-Williams, *Nature*, 1988, **336**,
43. R.G. Knowles, M. Palacios, R.M.J. Palmer and S. Moncada, *Proc. NaN. Acad. Sci. USA,* 1989, **86**, 5159.
44. J. Garthwaite, *Trends Neurosci.*, 1991, **14**, 60.
45. S.H. Snyder and D.S. Bredt, *Sci Am.*, 1992, **266**, 68.
46. J.S. Gillespie, X. Liu and W. Martin. '*Nitric Oxide from L-arginine: A bioregulatory system*', ed. S. Moncada and E.A. Higgs, Elsevier, Amsterdam, 1990, pp. 147.
47. M.J. Rand, Clin. Exp. *Pharmacol. Physiol*, 1992, **19**, 147.
48. N. Toda. '*Nitric Oxide in the Nervous System*', ed. S. Vincent,Academic Press, Orlando, 1995, pp. 207.
49. Martinet, V. Riveros-Moreno, J.M. Polak, S. Moncada and P.Sesma, *Cell Tissue Res*, 1994, **275**, 599.
50. F. Bacon, *Novum Organum*, 1620, Book 1.

Anaesthetics and Other Medical Gases

Other Gases Used Medically

M. E. Garrett

BOC GASES, THE PRIESTLEY CENTRE, SURREY RESEARCH PARK, UK

The use of gas in medicine is very strongly associated with anaesthesia. This is unsurprising when the impact that painless surgery made on the medical profession is considered. Gases however, have many other properties and it is unsurprising that several have proved to be useful in other aspects of medicine and that further uses are also being discovered.

This paper will examine other uses for gases in medicine and examine them from the aspect of treatment, diagnostics and other uses. Although not exhaustive possible future uses of gases will also be considered.

Gases Used in Treatment

Oxygen

The gas which has the widest use in treatment must surely be life supporting oxygen, which is used extensively in patient care. The discovery of oxygen is attributed to Priestley who is reputed to have remarked that, the only creatures that had breathed this new gas were himself and a mouse and he wondered if someday it would have a much wider application. Oxygen has now been used medically for many years in a very similar way, although the apparatus for dispensing it has changed significantly from user unfriendly, heavy and rather formidable cylinders in the early part of the century to the life-line type emergency set (see Figure 1) which is user friendly and requires only minimal training.

Another change is that if oxygen is good, more might be better and chambers in which patients can breathe oxygen at two or three times atmospheric pressure have been tried on a wide variety of diseases. Although initial claims were somewhat extravagant it is becoming accepted as an effective treatment for wounds, carbon monoxide poisoning and ulcers.

Heliox

Oxygen is not always used alone and when mixed with helium it is given the generic name 'Heliox'. This gas mixture is very much easier to breath than air

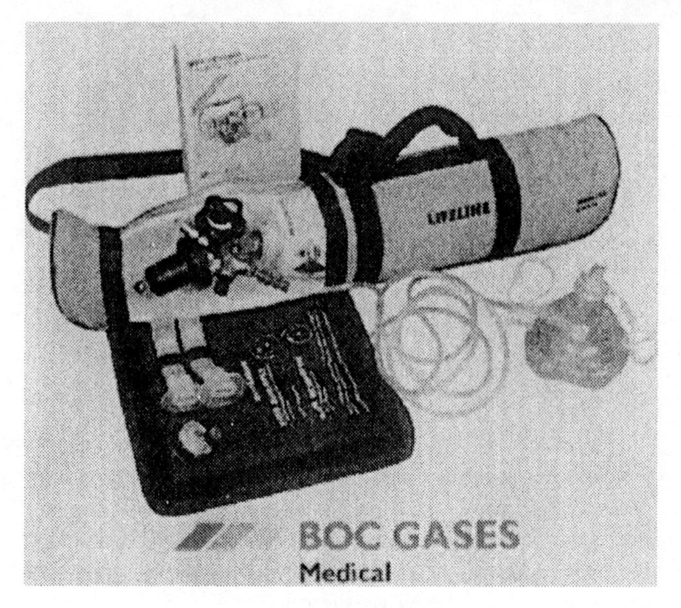

Figure 1

because of its low mass and is used where there is obstruction in the trachea and even in the treatment of Status Asthmaticus[1].

Nitric oxide

The new wondergas used in treatment is nitric oxide and this has been described fully in other papers associated with this conference.

Historical

Air

The gas that must have been prescribed more often than any other in history both by the medical profession and parents is 'fresh air'. Although becoming a much scarcer commodity these days there is still some of this to be found in the wide world and it is interesting that recently that hospitals have been requiring compressed gas cylinders filled with synthetic air, namely a mixture of oxygen and nitrogen.

Ammonia

Ammonia was often associated with the use of fresh air but mainly in the form of smelling salts. Treatment usually used to consist of a strong dose of smelling salts to revive a fainting person followed by a breath of fresh air. It seems to have been remarkably efficient and its decline in use has been attributed to the decline in whale-bone corsets.

Argon

Argon has also been used as a treatment gas mixed with oxygen in the ratio 80:20[2]; again it was thought to be easier for a patient to breath than conventional air, although its use seems to have been discontinued.

Radon

Another gas used in the early part of the century was radon, either dissolved in water for baths to treat rheumatism or even breathed as a treatment for the lungs. Today its presence seems to be inadvertently limited to cottages in Cornwall and its medical use discontinued.

Sulphur dioxide

This was another gas used as a fumigant in controlling the spread of typhus and similar illnesses. Typically the illness effected people living in close proximity, such as jails and warships and, at least in the case of ships, sulphur dioxide was generated by burning gun powder wetted with vinegar on the mess decks.

Oxygen

Although the discovery of oxygen has been attributed to Priestley with Scheele and Lavoisier also being strong contenders, it is intriguing to think that the gas may have been prepared and used many years earlier. There is a report of a submarine travelling from Westminster to Greenwich, a journey of more than an hour in the reign of King James I with the air in the vessel being replenished from 'jars'. This can be traced back to Drebbel a dutch inventor and a contemporary woodcut in his Treaties on the Elements of Nature shows a patient breathing gas that has been bubbled through water and prepared by heating a retort (See Figure 2)[3]. The chemical would be nitre or potassium nitrate as it is known today, which was manufactured as a key component of gun powder from urine soaked manure beds, and the gas produced would be oxygen.

Gases Used in Diagnosis

This is an area which shows an increasing use of gas as an aid to modern sophisticated diagnostic techniques. Examples include ^{133}xenon in CAT scans (Computerised Axial Tomography), carbon dioxide and ^{129}xenon in magnetic resonance imaging and ^{13}nitrogen, ^{11}carbon, ^{18}fluorine and ^{15}oxygen in PET (Positron Emission Tomography). The action of CO_2 in MRI has been to

Figure 2

increase oxygenation of the chaotic vasculature in tumours, enabling the technique to identify cancerous growths even though they are too small to be palpated (See Figures 3 and 4)[4].

Other Uses

Cryosurgery

This covers a range of areas in which gases play an important part for example cryosurgery makes use of liquid nitrogen, solid CO_2 or nitrous oxide as the cooling agent and in some cases cold can be achieved by the Joule-Thomson cooling effect obtained by expanding compressed air. Liquid nitrogen again appears as the medium for tissue preservation, and for semen and embryo

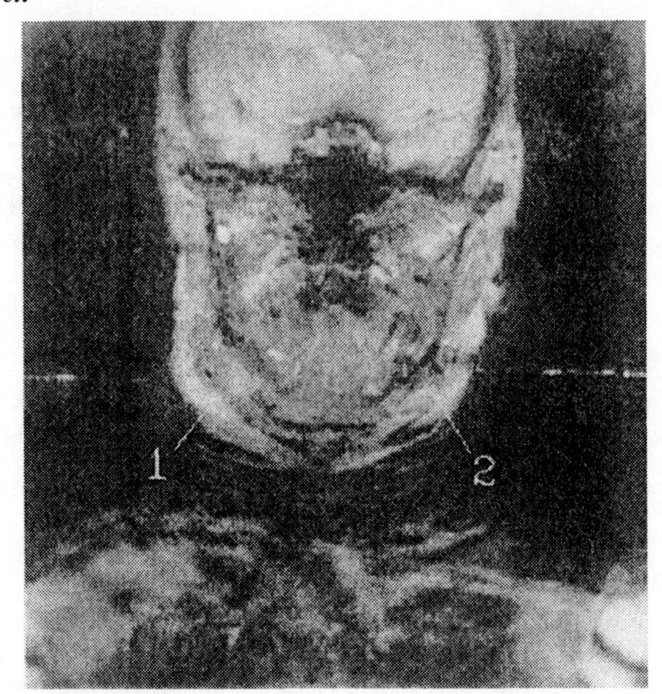

Figure 3 *MRI Scan of head*

Figure 4 *False colour enhancement showing tumor, and heightened response and second tumor with CO_2 addition.*

storage. The use of liquid nitrogen to preserve corpses in the hope of resurrection at some future date is not strictly a medical use nor likely to be very effective; however, it does provide hope and generate money.

Local anaesthesia

An early technique of minor surgery was to numb the affected area by spraying it with an evaporating liquid gas such as ethyl chloride. Strangely this became incorporated into the language and analgesic injections are often said to 'freeze' the affected part.

MRI cooling

The wide use of MRI diagnostics using nuclear magnetic resonance generated by radio frequency pertivations on large superconducting coils was only made possible by the use of liquid helium to keep the coils at superconducting temperatures. This largely remains the case today, although refrigerators are now sometimes incorporated to avoid the need of topping up with liquid helium.

Synthetic air

Synthetic air was mentioned earlier and can be provided by mixing oxygen and nitrogen gas, vaporised from liquid storage tanks as a substitute for hospital compressed air. There are many advantages, the gases are sterile and clean and oil free, reducing the risks of contamination within the hospital.

Sterilisation

Sterilisation of disposables and other materials is often carried out by the use of gases as an alternative to irradiation and autoclaving. The gases used are ethylene oxide, chlorine and ozone, usually singularly. Another piece of equipment which depends on gases for its function is the laser such as the krypton type used in eye surgery (see Figure 5) and the more powerful sulphur hexofluoride and perfluropropane lasers.

Future Uses

Variable air

Reference was made above to the use of synthetic air made by mixing together oxygen and nitrogen in the same proportions as in ambient air. Of course the key factor is not the percentage of oxygen but rather the partial pressure of oxygen in the gas: there is still 21% oxygen at the top of Mount Everest but not enough to breath.

Consequently at sea level these effects can be reproduced by varying the oxygen concentration, for example reducing the concentration to 15% oxygen is roughly equivalent to 8,000 feet elevation and this technique has been used to increase red blood cell count in athletes without the need for mountain top training. Similarly, increasing the oxygen concentration by only a few percent has been reported as having a beneficial effect on wound healing and may one day be the norm in surgical wards.

Dissolved oxygen

The use of hyperbaric oxygen in pressure chambers has already been mentioned, but because of its relatively high solubility in water, if used dissolved in

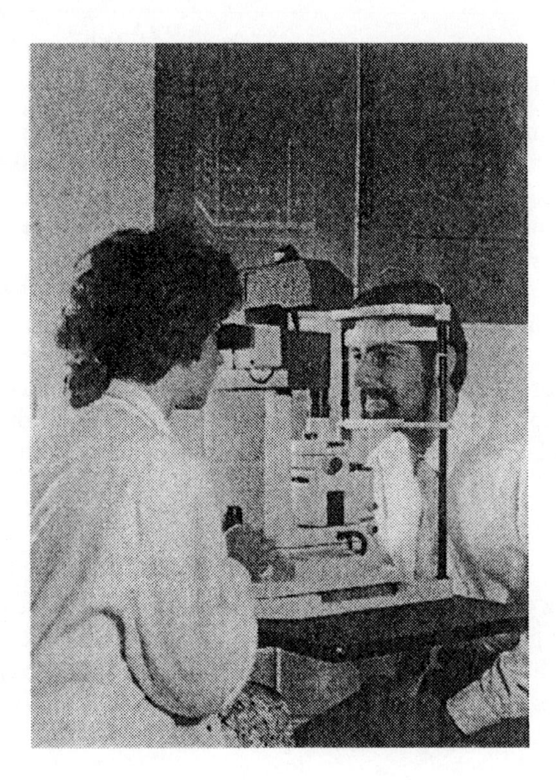

Figure 5

a hyperbaric chamber filled with air, then the equivalent of 5 bar could be achieved, and is likely to be more efficacious for ulcer treatment.

Ozone

The allotrope of oxygen, ozone, is known to be a highly effective bactericide and viricide and therefore could be used as a gaseous disinfectant. It is more likely however, to be used dissolved in water as the gas is aggressive to the lungs and in water the ozone has a half life of about 20 minutes avoiding any residual effects.

Hydrogen

Asthma is an illness currently affecting more and more people and even with the best treatments available today some deaths still occur. Patients in extremis, even when incubated, can die from asphyxiation and although helium has been mentioned as having a low drag because of its low mass, it is still greater than that of hydrogen. A hydrogen oxygen mixture therefore, could prove life-saving in these situations and although at first sight this would seem to be a hazardous mixture it would not be of any greater difficulty than cyclopropane anaesthesia a few years ago.

Supercritical gases

Another phenomenon associated with gases is supercriticality. This is the state of a gas when the liquid is heated above its critical temperature and beyond this point the gas can no longer be liquefied by pressure. When this is observed the meniscus can be seen to disappear and the density throughout the vessel becomes uniform.

Gases such as CO_2 under these conditions still retain much of their capacity to dissolve other materials and the liquid/supercritical transition is likely to prove an important step in the manufacture of pharmaceuticals. It is also possible therefore that the gas might prove a good delivery medium for drugs, particularly because the particles formed by the evaporation of the gas will be extremely small and uniform making them particularly suitable for inhaled medicines.

Liquid gases

Liquid CO_2 is a good solvent in its own right and will still have the property of generating fine particles and may present an alternative to mechanical nebulisation. It also has been used to dissolve a range of other chemicals such as teetree oil making it an effective bactericide as well as a deodorant.

Conclusion

Other mixtures of gases are possible in which the properties of the individual gases such as density or physiological effects can be combined synergistically. A few of these mixtures have been tried but the mounting evidence that some of the simplest molecules play major roles in physiology should lead to more interesting discoveries in the near future.

The gases described here interface more or less directly with the patient, industrial gases however, are a key contributor to the manufacture of pharmaceuticals, metals alloys, fibres, plastics, glass etc, and even in the preservation and transportation of produce; without them very few of the things taken for granted in modern medicine would be as we know them today.

References

1. Helium-Oxygen Mixtures in Intubated Patients with Status Asthmaticus and Respiratory Acidosis. Eric H Gluck et al. Chest /98/3. September, 1993. Pages 693-697.
2. Tissue Gas and Blood Analyses of Human Subjects Breathing 80% Argon and 20% Oxygen. D.J. Horrigan et al. Aviation, Space, and Environmental Medicine. April, 1979. Pages 357-362.
3. Treaties on the Elements of Nature, Cornelis Drebbel 1688 Dutch Edition.
4. The Effect of Carbogen Inhalation on Tumour Oxygenation and Blood Flow, Monitored by Gradient-Recalled Echo Magnetic Resonance Imaging. John R Griffiths, N Jane Taylor et al. Paper submitted 28.11.95 to The Lancet.

Non-Hypnotic Effects of General Anaesthesia

Charles S. Reilly

UNIVERSITY OF SHEFFIELD, UK

The actions of anaesthetic agents on the central nervous system appear to be non-specific. That is, they are not directed at a single specific receptor in the way that an opioid analgesic can be made specific for a mu-opioid receptor. It is also clear that anaesthetic agents act on systems other than the central nervous system.

If we take as a starting point the properties that would be ascribed to an ideal anaesthetic agent we could divide these properties in to pharmaceutical, pharmacokinetic and pharmacodynamic[1]. The pharmaceutical properties would relate to the formulation, presentation and stability of the drug. The desirable pharmacokinetic properties would be a rapid onset and offset and lack of accumulation. In describing the pharmacodynamics of an ideal agent the requirements would relate to its effects on the central nervous system and any other systems. If we look at the properties proposed for an ideal agent we would see that the list includes: no cardiovascular depression, no respiratory depression, not toxic to the liver or kidneys, no histamine release and an analgesic effect. The implication of such a list is that the current agents available do not meet these criteria and demonstrate some of these unwanted effects on other systems. In this review I will look at the current available inhalational and intravenous agents with respect to their action on the central nervous system and other systems looking at both their adverse effects and other potential therapeutic effects. There is a danger that this could become a list of adverse effects. I have therefore concentrated on effects on the central nervous system and cardiorespiratory systems. For a further description of adverse effects of anaesthetic agents readers may wish to consult a specific text dealing with this[2].

Adverse effects can be described using the broad classification proposed by Rawlins and Thompson[3]. This classification describes adverse reactions as either type A or type B. Type A reactions are predictable reactions which can be regarded as an extension of the normal pharmacological action of the drug and these effects are usually dose related. Type B reactions can be regarded as unpredictable actions which may or may not be directly related to the pharmacological action of the drug and are often not dose related. Type A reactions are relatively common and while they may be serious they generally carry a low mortality (e.g. fall in blood pressure on induction of anaesthesia). Type B reactions are less common but may carry a high mortality (e.g. an anaphylactic reaction to a drug). As many of the adverse effects are dose related attention is often paid in clinical practice using methods and rates of delivery of drugs which minimise the possibility of adverse reactions by using the smallest effective dose. However, in a review covering 25 years of anaesthesia in Australia, drug overdosage was ranked as the fourth highest in order of frequency of anaesthetic errors which contributed to mortality[4].

A beneficial action of a drug can be described as being a therapeutic action and it is possible that anaesthetic agents may have therapeutic uses other than induction and maintenance of anaesthesia. Where such actions have been identified they will be included in this review.

1 Central Nervous System

Anaesthesia produces a general suppression of the central nervous system activity which results in sleep. This cerebral metabolic rate is depressed by both intravenous and inhalational anaesthesia. With the inhalational agents isoflurane has the most marked effect and produces a reduction in oxygen consumption of around 50% at one MAC (MAC is the minimum alveolar concentration)[5]. A similar spectrum of effect is seen with the intravenous agents with propofol at a dose which produces an iso-electric EEG reducing cerebral metabolic rate by 35% and thiopentone at a similar end point reducing cerebral metabolic rate by 50%[6,7].

1.1 Electro-encephalographic Effects

Induction of anaesthesia with all of the intravenous agents other than ketamine is associated with a decrease in EEG activity which can be demonstrated by a decrease in the median frequency and a burst suppression pattern[8]. The initial pattern with inhalational agents differs in having an initial increase in frequency in voltage which is followed as depth of anaesthesia increases by a move to initially moderate voltage, low frequency activity and eventually to burst suppression. The majority of intravenous agents including ketamine have anti-convulsant properties.

It has been noted with a number of intravenous anaesthetic agents that patients demonstrate excitatory movement on induction of anaesthesia. The incidence of this is lowest with benzodiazepines and thiopentone where the incidence is less than 10% and highest with etomidate where some series have reported incidences of 70%. These effects have been found in some patients to be associated with epilleptiform changes on the EEG. It has been suggested that these proconvulsant effects of anaesthetic agents may be dose related in that lower doses are required to suppress the inhibitory neuronal activity than are required to suppress excitatory neurones. It is interesting to note, however, that methohexitone which has two asymmetric carbon atoms could have four

Table 1 *The incidence of postoperative nausea and vomiting following anaesthesia with different intravenous agents*

	Nausea (%)	*Vomiting (%)*
Thiopentone	6	11
Methohexitone	12	14
Diazepam	3	5
Ketamine	23	18
Etomidate	12	27
Propofol	5	0

McCollum and Dundee, 1986

isomers (α-RS and β-RS). A mixture of all four isomers while acting as an induction agent was associated with excessive excitatory activity and convulsions. This turned out to be due to the β isomer and therefore the α isomer is the one that is used clinically. Of the inhalational agents only enflurane appears to be associated with excitatory activity and seizure complexes have been reported on the EEG during enflurane anaesthesia[9]. This appears to be exacerbated by increasing depth of anaesthesia and also by hyperventilation. With enflurane and with the intravenous agents that are associated with epilleptiform changes these effects are seen more frequently if not exclusively in patients with a history of epilepsy. Therefore, their use is contraindicated in such patients.

1.2 Emetic and Anti-emetic Effects

Nausea and vomiting following surgery and anaesthesia is known to be a multi-factorial problem but some anaesthetic agents appear to carry a higher incidence of post operative nausea and vomiting than others[10] (see Table 1). In contrast the use of propofol seems to be associated with a lower incidence of post operative nausea and vomiting (PONV). This has led to investigation of anti-emetic properties of propofol in the post operative period and in other situations where emesis is a problem and it has the potential for a specific therapeutic use.

There is reasonable evidence from clinical studies that propofol tends to decrease the incidence of PONV. This effect is probably most marked in studies of day case surgery patients where the usual incidence of around 20% has been considerably reduced in a number of studies which have compared a propofol based anaesthetic technique with one based on the use of the inhalational agents. The effect is also reasonably well demonstrated in patients having general surgery particularly in surgery where post operative pain is not a major consideration. Good examples of this are the study by Doze *et al.*[11] who found an incidence of around 20% in a group of patients following general surgery using a propofol-nitrous oxide technique compared with 40% in patients who received a thiopentone-isoflurane-nitrous oxide technique.

Similarly, a study by Best *et al.*[12] found an incidence of nausea and vomiting of only 5% compared with 35% in patients who received methohexitone. The effect is most clearly seen in the immediate post operative period as with time the amount of propofol in the body decreases fairly quickly. An important confounding factor in the patients who have had major surgery is the use of opioids, particularly morphine, for post operative analgesia which may well over-ride any anti-emetic effect produced by propofol.

As a result of these observations an anti-emetic effect was proposed for propofol. Propofol has been used in the post operative period in low dosage to treat PONV and in one study was moderately successful in doing this[13]. However, a much more dramatic effect was seen in a study that looked at the effect of a low dose infusion of propofol during and after chemotherapy[14]. A group of patients who had had emesis which was refractory to treatment with ondansetron and dexamethasone during a previous course of chemotherapy received propofol during their next administration of chemotherapy. A low dose of propofol (1mg/kg/hr) was given as an infusion starting 4 hours before the chemotherapy and continued over the next 24 hours. The effect of this was to reduce the number of emetic episodes from >5 per day during the first treatment to none during the second treatment. This effect has been subsequently demonstrated within other groups receiving chemotherapy.

This subject has been reviewed recently[14] but the mechanism of this anti-emetic effect is not clear and requires further investigation. It does however seem to be a specific therapeutic effect of the anaesthetic agent propofol.

1.3 Analgesia

With the exception of ketamine none of the intravenous agents are associated with an analgesic effect. With regards to the inhalational agents none of the volatile agents are associated with analgesia but nitrous oxide has an analgesic effect. A number of potential mechanisms for ketamine analgesia have been postulated and its site of action has been proposed to be both at a supra-spinal and at a spinal level. The receptors proposed to be involved have included both opioid receptors and NMDA receptors. The analgesic effect of ketamine is not reversed by the administration of naloxone and it is now thought unlikely that the analgesic effects are mediated through opioid receptors[15]. Ketamine does have an effect on the NMDA receptors where it is thought to antagonise the excitatory neuro-transmitter glutamate[16]. This interaction is non-competitive which suggests that the drug produces indirect changes at the receptor site.

2 Cardiovascular System

As expected from agents which have a depressant effect on the central nervous system anaesthetic agents in general have a depressant effect on the cardiovascular system. Overall effect is a fall in cardiac output, decreased arterial pressure and a normal or raised heart rate. These effects are brought about by three mechanisms; central depression of the vasomotor centre, decreased myocardial contractility and alteration of vascular smooth muscle tone. These mechanisms are all altered by both inhalational and intravenous agents but the extent of involvement of each mechanism varies between the two groups.

All the inhalational anaesthetic agents cause a direct dose-dependent depression of myocardial contractility affecting both isotonic and isometric contraction. Enflurane followed by halothane appear to be the most potent agents at producing a decrease in the peak force and rate of rise of developed force during myocardial contraction[17].

At a cellular level the inhalational agents act on the myocardial cells by influencing the availability of calcium in a number of ways. The intracellular free calcium concentration is decreased and the influx of calcium through calcium channels in the sarcolemma is decreased[18]. The agents also decrease the accumulation of calcium within the sarcoplasmic reticulum and enhance the calcium release from the sarcoplasmic reticulum. This means that there is less calcium available for subsequent reactions[19]. The agents also appear to reduce the binding of calcium to the regulatory protein troponin and tropomyosin. Inhalational agents also cause a decrease in peripheral vascular resistance. This effect is less marked than is seen with the intravenous agents. This effect is greatest with isoflurane and desflurane and least with enflurane.

The normal response to a fall in blood pressure would be an increase in heart rate mediated through the baroreceptors (the baroreceptor reflex). This reflex is blunted by the use of inhalational agents. However, of the commonly used agents, isoflurane depresses the reflex the least and this in part explains why cardiac output is maintained during isoflurane anaesthesia.

The intravenous agents produce a fall in arterial pressure which is similar in magnitude to that produced by the inhalational agents. However, although all three mechanisms are involved the balance of influence differs across the three elements. The principal difference is that the effect on myocardial contractility is less than the effect on peripheral vascular resistance. There is a variation between agents on the extent of impairment of myocardial contractility and also for the vasodilatory effect.

Etomidate produces the least cardiovascular disturbance of the intravenous agents (Table 2) and induction results in a fall in arterial pressure of only 5-10% with little or no increase in heart rate. This is principally due to a fall in cardiac output of around 6% and a decrease in peripheral vascular resistance of around 8%[20]. The most marked effects are seen with thiopentone and propofol. Thiopentone produces a fall in arterial pressure of around 20% and this is due to both peripheral venodilation and a direct myocardial depressant effect. Induction of anaesthesia with propofol produces a fall of some 20-30% in arterial pressure[21]. This is due mainly to a fall in peripheral vascular resistance and there is less direct myocardial depression than is seen with thiopentone[10].

Table 2 *The effect of intravenous anaesthetic agents on systolic arterial pressure (SAP) and heart rate (HR) (% change from baseline)*

	Dose	*SAP*	*HR*
Thiopentone	4.0	−6	6
Thiopentone	5.0	−10	9
Methohexitone	1.5	−1	9
Etomidate	0.3	−5	3
Propofol	2.0	−15	5
Propofol	2.5	−17	5

3 Respiratory System

All the inhalational agents produce a direct depression of respiration in a dose related fashion. This can be demonstrated by measuring end tidal carbon dioxide concentration which will increase as the alveolar concentration of the inhalational agents increase[22]. Tidal volume is decreased but there is some increase in respiratory rate however overall minute ventilation is decreased.

All the inhaled anaesthetics including nitrous oxide depress the ventilatory response to carbon dioxide. Apnoea may occur with all agents with concentrations above 2 MAC. There is an even more profound effect on the ventilatory response to hypoxia. This response is mediated by peripheral chemoreceptors and the presence of even 0.1 MAC of isoflurane can reduce this response by 50% and it is entirely abolished at 1.0 MAC. The homeostatic protective mechanism of hypoxic pulmonary vasoconstriction is also depressed by all the volatile agents in a dose related manner[23].

The respiratory defence mechanism which clears mucus from the respiratory tract is known as the mucociliary escalator. The effectiveness of this mechanism has been shown to be decreased in a dose related manner by halothane[24] and enflurane.

An effect on the respiratory system which has been put to therapeutic use is the bronchodilatory effect of the inhalational agents, particularly halothane. This effect has been put to clinical use in patients with severe asthma requiring assisted ventilation.

Intravenous induction of anaesthesia is associated with respiratory depression which occurs in a dose related manner and with all agents. This is a result of central depression. There is a marked decrease in tidal volume which often proceeds to apnoea. While it occurs with all agents there is variation in the frequency with which this occurs with the different agents (see Table 3). The effect is smallest following etomidate[25] and greatest after thiopentone and propofol[26]. The ventilatory response to carbon dioxide is also markedly impaired but it has been noted that for a given pCO_2 the respiratory rate was greater following etomidate than following methohexitone[27].

Table 3 *The incidence of apnoea following induction of anaesthesia with different intravenous agents*

	Apnoea (%)
Thiopentone	50–80
Methohexitone	40–60
Midazolam	10–20
Ketamine	<1
Etomidate	30
Propofol	50–80

4 Summary

As a group the anaesthetic agents, both intravenous and inhalational, have a number of other effects. The majority of these could be classified as adverse events which are on the whole Type A adverse effects that is, they are predictable dose related effects. The non-specific nature of these effects which are produced mainly through central nervous system depression would tend to support a non-specific mechanism of action. However, some specific agent receptor interactions have been demonstrated.

There are relatively few additional therapeutic effects which can be attributed to the anaesthetic agents. Of these the anti-emetic effect of propofol and the bronchodilatory of the inhalational agents particularly halothane are the most useful.

The effects of the different classes of the agents tend to be similar but there are significant differences in the magnitude of some of the effects produced. It is possible that closer examination of these differences could give us insight in to their differing mechanisms of action.

References

1. J.W. Dundee. *New intravenous anaesthetics.* Br J Anaesth 1979, **51**, 641.
2. M.C. Berthoud and C.S. Reilly. *Adverse Effects of General Anaesthetics.* Drug Safety 1992, **7**, 434
3. M.D. Rawlins and J.W. Thomson. *Pathogenesis of adverse drug reactions. In Davies (Ed.) Drug Treatment.* Oxford University Press 1978, pp 9
4. R. Holland. *Anaesthetic mortality in New South Wales.* Br. J. Anaesth., 1987, **59,** 834
5. L.A. Newberg, J.H. Milde, J.D. Michenfelder. *Systemic and cerebral effects of isoflurane induced hypotension in dogs.* Anesth. 1984, **60,** 541
6. J.D. Michenfelder. The interdependency of cerebral functional and metabolic effects following massive doses of thiopental in the dog. Anesth., 1974, **41,** 231
7. H Stephan, H Sonntag, HD Schenk, D Kettler. *Effects of propofol on cardiovascular dynamics, myocardial blood flow and myocardial metabolism in patients with coronary artery disease.* Br J Anaes 1986, **58**, 969.

8. M.M. Ghonheim, T. Yamada. *Etomidate: a clinical and electroencephalographic comparison with thiopental.* Anesth. Anal., 1977, **56,** 479
9. J.L. Neigh, J.K. Garman, J.R. Harp. The electroencephalographic pattern during anaesthesia with enflurane. Anaesth. 1971, **35,** 482
10. J.S.C. McCollum, J.W. Dundee. *Comparison of the induction characteristics of four intravenous anaesthetic agents.* Anaesth., 1986, **41,** 995
11. VA Doze, A Shafer, PF White. *Propofol-nitrous oxide versus thiopental-isoflurane-nitrous oxide for general anaesthesia.* Anesth 1988, **69**, 63.
12. N Best, F Traugott. *Comparative evaluation of propofol or methohexitone as sole agent for microlaryngeal surgery.* Anaesth Int Care 1991, **19**, 50.
13. A Borgeat, OHG Wilder-Smith, M Saiah, K Rifat. *Subhypnotic doses of propofol possess direct anti-emetic properties.* Anesth Analg 1992, **74**, 539.
14. A Borgeat, OHG Wilder-Smith, PM Suter. *The non-hypnotic therapeutic applications of propofol.* Anesth 1994, **80**, 642.
15. P. Klepstad, A. Maurset, E.R. Moberg, I. Oye. *Evidence of a role of NMDA receptors in pain perception.* Eur. J. Pharmacol. 1990, **187,** 513
16. A. Maurset, L.A. Skoglund, O. Hustveit, I. Oye. *Comparison of ketamine and pethidine in experimental and postoperative pain.* Pain 1989, **36,** 37
17. P.R. Houseman, I. Murat. *Comparative effects of halothane, enflurane and isoflurane in eqvipotent anaesthetic doses on isolated ventricular myocardium of the ferret. I. Contractility.* Anesth., 1988, **69**, 451
18. Z.J. Bosnjak and J.P. Kampine. *Effects of halothane, enflurane and isoflurane on the S.A. node.* Anesth., 1983, **58**, 314
19. JY Su, JG Bell. *Intracellular mechanism of action of isoflurane and halothane on striated muscle of the rat.* Anesth Analg 1986, **65**, 457.
20. A Criado, J Maseda, E Navarro, A Escarpa, F Avello. *Induction of anaesthesia with etomidate: haemodynamic study of 36 patients.* Br J Anaesth 1980, **52**, 803.
21. R.M. Grounds, A.J. Twigley, F. Carli, J.G. Whitwam, M. Morgan. *The haemodynamic effects of intravenous induction.* Anaesth., 1985, **40,** 735
22. E.I. Eger. *The pharmacology of isoflurane.* Br. J. Anaesth., 1984, **54,** 71S
23. R.L. Knill, J.L. Clement. *Variable effects of anaesthetics on the ventilatory response to hypoxaemia in man.* Can. Anaesth. Soc. J., 1982, **29,** 93
24. AR Forbes. *Halothane depresses mucocilliary flow in the trachea.* Anesth 1976, **45,** 693.
25. HH Chui, WK Van. *Clinical evaluation of etomidate as an induction agent.* Aneasth Int Care 1978, **6**, 129,.
26. G Rolly, L Versichelen. *Comparison of propofol and thiopentone for induction of anaesthesia in premedicated patients.* Anaesthesia 1985, **40**, 945.
27 SD Choi, BC Spaulding, JB Gross, JL Apfelbaum. *Comparison of the ventilatory effects of etomidate and methohexital.* Anesth 1985, **62**, 442.

Interaction between General Anaesthesia and High Pressure

S. Daniels

WELSH SCHOOL OF PHARMACY, UNIVERSITY OF WALES
CARDIFF, UK

The changes in central nervous system function brought about by general anaesthetics, reduced perception of sensations and unconsciousness are quite different from those arising from exposure to high pressure, increasing excitability. Nevertheless, there is a remarkable connection, the ability of high pressure to reverse general anaesthesia (Lever et al., 1971) and the antagonism of the effects of pressure by general anaesthetics (Miller, 1974).

Pressure reversal of general anaesthesia was first reported by Johnson & Flagler (1950) who showed that tadpoles, anaesthetised with ethanol, had their swimming ability restored by the application of 10MPa hydrostatic pressure. This observation received little attention until it was established that helium acted as a pure transmitter of pressure (Miller et al., 1967) and consequently an equivalent phenomenon could be demonstrated in mammals (Lever et al., 1971). The interaction between general anaesthetics and pressure invites speculation as to the nature of the interaction. The interaction may represent opposing effects at a single site of action or, alternatively, a summation of separate physiological effects. It has also been thought that understanding the interaction will lead to a clearer understanding of the mechanism of action of both general anaesthetics and high pressure.

The interaction between pressure and general anaesthetics was rationalised, initially in terms of a physicochemical interaction within the lipid bilayer of excitable cells in the central nervous system, the Critical Volume Theory (Lever et al., 1971; Miller et al., 1973). However, it is now clear that the anaesthesia is more likely to arise as a result of the action of general anaesthetics on the ligand-gated ion channels responsible for fast synaptic neurotransmission (see Richards; Wann; Lambert; this volume). The questions to be addressed, therefore, are whether pressure has a similar specific action and whether an opposing interaction exists.

Mechanism of Action of Pressure

The effects of pressure are well characterised in both man (Bennett, 1982) and animals (Brauer, 1969) and are referred to, collectively, as high pressure neurological syndrome (HPNS). In man, the signs and symptoms are tremor, dizziness, nausea, psychomotor impairment, EEG changes and occasional myoclonus. In animals, tremor is observed, beginning between 3MPa and 4MPa at the head and forequarters, and becoming increasingly severe with increasing pressure, until frank convulsions occur between 8MPa and 9MPa; convulsions are followed by respiratory depression and death, at still higher pressure (Bowser-Riley et al., 1984). In both man and animals, the signs and symptoms of HPNS appear at lower pressures and are more severe as the rate of compression is increased but they also remit with time if the pressure is held constant.

Pharmacology of pressure effects

Treatment with resperpine reduces the pressure at which mice exhibit HPNS, by approximately 40%, and also abolishes the dependence on compression rate (Brauer et al., 1978). This suggests a possible role for the biogenic amine group of neurotransmitters. However, investigations using more selective agents failed to establish a definitive role (Koblin et al., 1980; Daniels et al., 1981). Experiments using selective neurotoxins indicated that cortical noradrenergic-mediated inhibition did act to modulate the effects of pressure but was not involved in the primary genesis of the effects (Bowser-Riley, 1984).

It appears unlikely that cholinergic mechanisms are involved in the aetiology of HPNS because neither atropine nor mecamylamine have any effect on the expression of HPNS (Wardley-Smith et al., 1984).

A role for GABAergic processes was suggested by experiments showing that drugs that potentiate the action of GABA (aminooxyacetic acid, 2,4-diamino-butyric acid, sodium valproate) postpone the onset of HPNS (Bichard & Little, 1984). The benzodiazepines (clonazepam, diazepam) also provide some protection against the effects of pressure, although at sedative doses (Halsey & Wardley-Smith, 1981). The protective effect of flurazepam is abolished by the selective benzodiazepine antagonist RO15-1788 (Bichard & Little, 1982), indicating that the protective action arises via a potentiation of GABAergic inhibition.

However, neither muscimol, a GABA agonist at $GABA_A$ receptors, nor baclofen, a GABA agonist at $GABA_B$ receptors, have any effect on the development of HPNS (Bowser-Riley, 1984). An alternative to testing the effect of GABA agonists is to test the additivity of action of GABA antagonists with pressure. Bicuculline and picrotoxin both show very marked divergence from the additivity of action expected if pressure was exerting its effect through some action on the $GABA_A$ receptor (Bowser-Riley et al., 1988a; 1988b).

Finally, pressure has no effect on the response of the rat superior cervical ganglion to exogenous GABA nor does pressure reverse the potentiation of the GABA response produced by either ketamine or pentobarbitone (Bichard & Little, 1984). Pressure also has no effect on either spontaneous or electrically evoked release of GABA from frog hemisected spinal cord, unless the cord was electrically stimulated prior to pressurisation in which case an increase in GABA release was observed (Bichard & Little, 1984).

Taken together these results suggest that the action of pressure does not arise through an effect on the $GABA_A$ or $GABA_B$ receptors or through an effect on GABA release.

Glycine is now recognised as one of the principal inhibitory neurotransmitters, particularly in sub-cortical structures (Kuhse et al., 1995). A possible link between glycine mediated inhibitory neurotransmission and the mechanism of action of pressure arose from studies into the interaction between pressure and a group of propandiol-based centrally acting muscle relaxants (Bowser-Riley, 1984). The aromatic compounds (mephenesin & methocarbamol) are both effective anticonvulsants against strychnine and pressure but not against bicuculline or picrotoxin. In contrast, the aliphatic compounds (meprobamate & carisoprodol), that are much more potent centrally acting muscle relaxants, are effective anticonvulsants against bicuculline and picrotoxin but not against strychnine or pressure.

Strict additivity of action, in eliciting seizures, between strychnine and pressure is observed (Bowser-Riley et al., 1988a). Furthermore structure-activity relationship investigations revealed that the efficacy of action against pressure was mirrored by the anticonvulsant efficacy against strychnine (Bowser-Riley et al., 1989).

Finally, a separate group of compounds, related to benzimidazole, that are also centrally acting muscle relaxants and anticonvulsants against strychnine but not against bicuculline or picrotoxin, are also effective against pressure. The structure-activity relationship of this group of compounds against pressure matched that against strychnine and their potency against pressure (or strychnine) was unrelated to their muscle relaxant potency (Bowser-Riley et al., 1988c).

These findings provide a compelling body of evidence in support of a role for glycine mediated inhibitory neurotransmission in the aetiology of HPNS.

A role for glutamate mediated excitatory neurotransmission was postulated following experiments that showed effective protection against pressure by the NMDA antagonist 2-amino-7-phosphonoheptanoic acid (Meldrum et al., 1983). Generally, competitive antagonists at the NMDA-type glutamate receptor provide protection against both tremor and seizure phases of HPNS, although there is considerable variation in potency (Wardley-Smith & Meldrum, 1984; Pearce et al., 1991; 1993; Wardley-smith et al., 1987). Non-competitive NMDA antagonists exhibit widely different effects (Angel et al., 1984; Wardley-Smith & Wann, 1989; Pearce et al., 1990).

The action of pressure at non-NMDA type glutamate receptors is less clear. The kainate/AMPA selective antagonist γ-D-glutamylaminomethylsulphonic acid is effective in the rat against the tremor phase of HPNS but not against the seizures (Wardley-smith et al., 1987) whereas LY 293558 is effective against both phases (Pearce et al., 1994). The non-competitive AMPA antagonist L-glutamic acid diethyl ester (GDEE) has no effect on HPNS in the rat (Wardley-Smith & Meldrum, 1984). However, in the baboon the non-competitive AMPA antagonist 1-(4-aminophenyl)-4-methyl-7,8-methylene-dioxy-5H-2,3-benzodiazepine is effective against the tremor phase (Pearce et al., 1994).

NMDA acts additively with pressure in mice suggesting, by analogy with the additivity observed between pressure and strychnine, that pressure may have a direct effect on glutamate receptors as well as on glycine receptors (unpublished results).

Effects of pressure on isolated postsynaptic receptors

Fagni and colleagues (1987) showed that the postsynaptic excitability of hippocampal CA1 cells was increased at pressure even though afferent input was reduced demonstrating that neurones do exhibit excitability at pressure. Intracellular recordings at pressure from CA_1 cells show bursts of transient depolarisations accompanied by repetitive action potentials and increased excitability (Wann & Southan, 1992).

The action of pressure on the postsynaptic receptors for glycine, GABA and glutamate (both NMDA-type and kainate-type) has been investigated by isolating mammalian receptors using heterologous expression in *Xenopus* oocytes (Daniels et al., 1991; Shelton et al., 1993; 1996; Roberts et al., 1996; Williams et al., 1996). The effects of pressure on these ionotropic receptors show an interesting symmetry. For the receptors mediating excitatory transmission, NMDA-sensitive glutamate receptors show a marked potentiation (128%) at 10MPa whilst kainate-sensitive receptors are relatively unaffected, showing little potentiation of the maximum response (<14%) and no change in the EC_{50}. For the receptors mediating inhibitory transmission, glycine receptors show no change in the maximum response but a significant increase in the EC_{50} (60%) at 10MPa whereas pressure has no effect on the $GABA_A$ receptor. The effects of pressure on both the NMDA and glycine receptor become progressively greater at higher pressures.

In the case of the glycine receptor, a classical interpretation of the dose-response data would be that pressure is acting, pharmacologically, as a simple competitive antagonist. This action could be manifest either as a reduction in the efficacy of glycine binding or in the transduction of binding to channel opening. A direct effect on channel opening or lifetime seems unlikely since the closely related $GABA_A$ receptor is unaffected. The apparent volume change in glycine binding was estimated as 110 $cm^3 . mol^{-1}$, which is very large for simple chemical equilibria and indicates that considerable conformational changes occur when glycine binds. The effect of pressure may therefore be envisaged as opposing the conformational change that follows glycine binding thereby reducing the effective inhibitory transmission available.

The picture with the NMDA receptor is not as simple. In this case the apparent volume change associated with glycine binding must be negative so that the interaction would be augmented by pressure. Pressure must favour the existence of the glycine-bound form of the receptor. In both cases a more detailed understanding of the effect of pressure on glycine binding at these two receptors must await experiments on single channels.

In summary the effects of pressure on the isolated postsynaptic receptors in entirely compatible with the effects of pressure seen *in vivo* and the observed pharmacology.

The Interaction of Pressure and General Anaesthetics at Postsynaptic Receptors

The action of sub-maximal concentrations of GABA at the $GABA_A$ receptor is potentiated by barbiturates (Lambert, this volume). Pressure has no effect on this potentiation as shown in Figure 1A. A potentiation by pentobarbitone (100μM) of approximately 9 fold of the current elicited by application of 5μM GABA ($<ED_{10}$) to oocytes injected with mRNA extracted from rat whole brain was unaffected by the application of 10 MPa pressure (Shelton et al., 1996). Similarly, it has been found that an approximately 6 fold potentiation by pentobarbitone (600μM) of the response to the application of 75μM glycine ($<ED_5$) to oocytes injected with mRNA encoding the human α1 glycine subunit was unaffected by pressures up to 10 MPa (Figure 1B; unpublished results).

However, potentiation by nitrous oxide (1 MPa) of the response to the application of 75μM glycine to oocytes injected with mRNA encoding the human α1 glycine subunit was found to be reversed, in a dose dependent manner, by the application of pressure up to 15 MPa (Figure 1C; unpublished results). It is important to note that this "pressure reversal" was evident after allowing for the fact that pressure attenuates the response of the homomeric α1 glycine receptor to applied glycine.

Summary

The demonstration that pressure and a simple gaseous anaesthetic can exhibit pressure reversal at an isolated postsynaptic receptor provides support for the observation that whilst pressure reversal occurs in tadpoles it does not in shrimps (Smith et al., 1984). It was postulated that shrimps did not exhibit pressure reversal because they do not utilise glycine as an inhibitory neurotransmitter. If the glycine receptor is the only receptor at which pressure reversal occurs then this would be a satisfactory explanation of those early results. It remains to be seen whether pressure reversal occurs at the pressure-sensitive NMDA-type glutamate receptor.

It is tempting to speculate that the observation that pressure reversal can be observed for nitrous oxide but not for pentobarbitone suggests that they have quite distinct mechanisms of action. Further work is in progress to assess whether the simple gaseous anaesthetics differ consistently from intravenous and volatile anaesthetics.

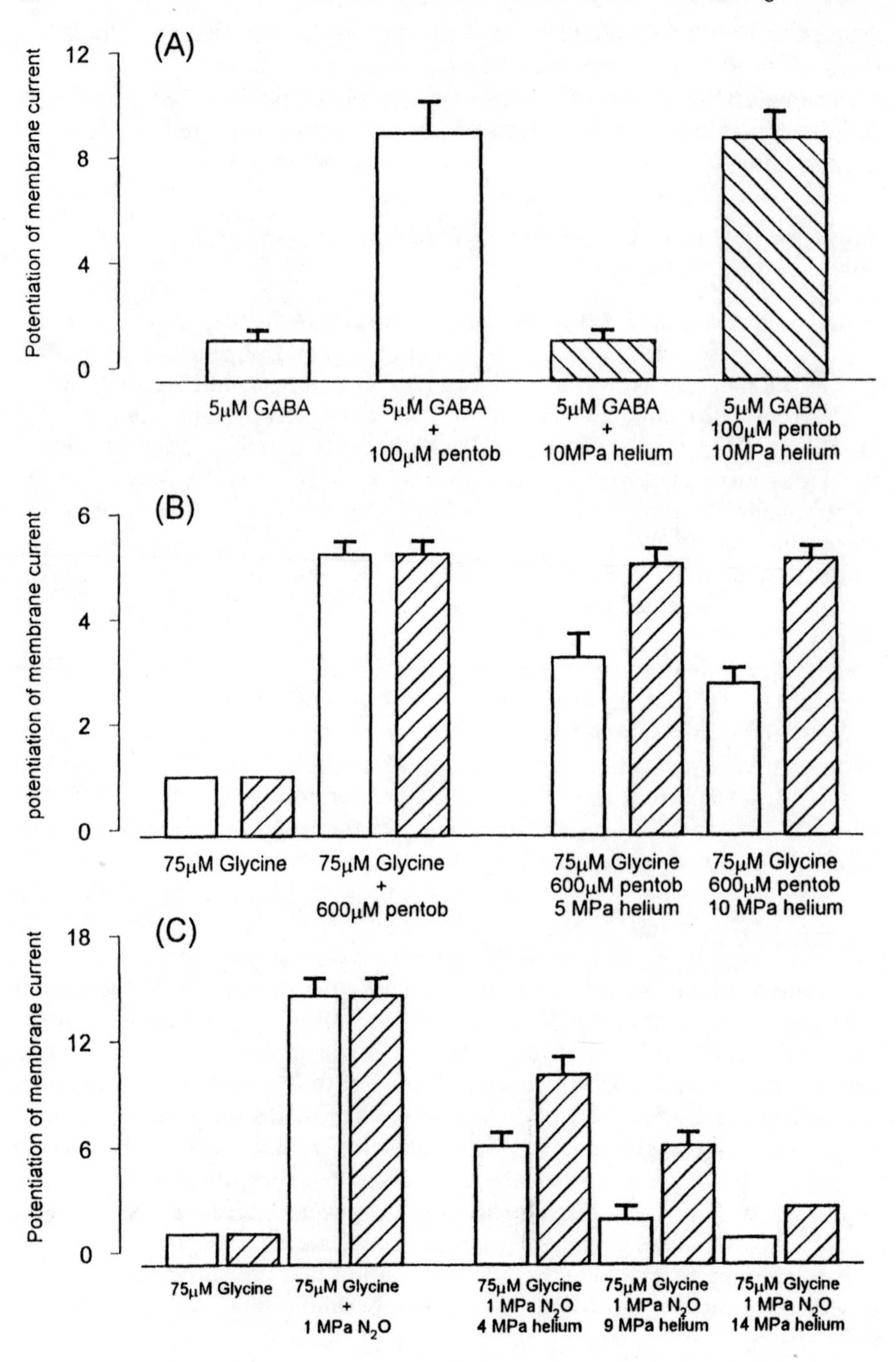
(A)
Potentiation of membrane current
12
8
4
0
5μM GABA
5μM GABA + 100μM pentob
5μM GABA + 10MPa helium
5μM GABA 100μM pentob 10MPa helium
(B)
potentiation of membrane current
6
4
2
0
75μM Glycine
75μM Glycine + 600μM pentob
75μM Glycine 600μM pentob 5 MPa helium
75μM Glycine 600μM pentob 10 MPa helium
(C)
Potentiation of membrane current
18
12
6
0
75μM Glycine
75μM Glycine + 1 MPa N2O
75μM Glycine 1 MPa N2O 4 MPa helium
75μM Glycine 1 MPa N2O 9 MPa helium
75μM Glycine 1 MPa N2O 14 MPa helium

References

Angel, A., Halsey, M. J., Little, H. J., Meldrum, H. J., Ross, J. A. S., Rostain, J. -C. & Wardley-Smith, B., Specific effects of drugs at pressure: animal investigations., *Philosophical Transactions of the Royal Society of London*, B304, 85, 1984.

Bennett, P. B., The high pressure nervous syndrome in man., in *The physiology and medicine of diving*, Bennett, P. B. & Elliott, D. H., Eds., Balliere TIndall, London, 3rd edition, 1982, 262.

Bichard, A. R. & Little, H. J. The benzodiazepine antagonist, RO 15-1788 prevents the effects of flurazepam on the high pressure neurological syndrome. *Neuropharmacology*, 21, 877, 1982.

Bichard, A. R. & Little, H. J. γ-aminobutyric acid transmission and the high pressure neurological syndrome., in *Proceedings of the VIII symposium on underwater physiology*, Bachrach, A. J. & Matzen, M. M. Eds., Undersea Medical Society Inc., Bethesda, 1984, 545.

Bowser-Riley, F., Mechanistic studies on the high pressure neurological syndrome., *Philosophical Transactions of the Royal Society of London*, B304, 31, 1984.

Bowser-Riley, F., Lashbrook, N. M., Paton, W. D. M. & Smith, E. B. An evaluation of mephenesin and related drugs in controlling the effects of high pressure on animals. In *Proceedings of the VIII symposium on underwater physiology*, Bachrach, A. J. & Matzen, M. M., Eds., Undersea Medical Society Inc., Bethesda, 1984, 557.

Figure 1 *(A) Potentiation by pentobarbitone (100μM) of the current elicited from* Xenopus *oocytes expressing $GABA_A$ receptors (rat brain mRNA) following bath application of 5μM GABA (open bars). Oocytes were perfused with standard Frog Ringer (120mM NaCl, 2mM KCl, $CaCl_2.2H_2O$, 10mM, HEPES, pH7.4) and voltage clamped at −60mV. The current recorded following application of 5μM GABA shows no effect of 10 MPa helium pressure (hatched bar). Finally, the pentobarbitone-potentiated GABA response is also unaffected by 10 MPa helium pressure (hatched bar) Error bars denote standard error of the mean. (B) Potentiation by pentobarbitone (600μM) of the current elicited from* Xenopus *oocytes expressing glycine receptors (human α1 cRNA) following bath application of 75μM glycine and the effect of applying 5 MPa and 10 MPa pressure with helium (open bars). Oocytes were perfused with standard Frog Ringer and voltage clamped at −40mV. However, the glycine receptor is sensitive to pressure (Roberts et al., 1996) and the observed currents must therefore be corrected for this pressure sensitivity. The glycine current would be attenuated by a factor of 0.6 and 0.41 at 5 MPa and 10 MPa, respectively. The corrected membrane currents (hatched bars) show that there was no pressure reversal of the potentiation caused by pentobarbitone. Error bars denote standard error of the mean. (C) Potentiation by nitrous oxide (1 MPa) of the current elicited from* Xenopus *oocytes expressing glycine receptors (human α1 cRNA) following bath application of 75μM glycine and the effect of applying 5 MPa, 10 MPa and 15 MPa pressure with helium (open bars). Oocytes were perfused with standard Frog Ringer and voltage clamped at −40mV. The glycine current would be attenuated by a factor of 0.6, 0.41 and 0.23 at 5 MPa, 10 MPa and 15 MPa, respectively. The corrected membrane currents (hatched bars) show that there was genuine pressure reversal of the potentiation caused by pentobarbitone amounting to 31%, 59% and 83% at 5 MPa, 10 MPa and 15 MPa, respectively. Error bars denote standard error of the mean.*

Bowser-Riley, F., Daniels, S. & Smith, E. B. Investigations into the origin of the high pressure neurological syndrome: The interaction between pressure, strychnine and 1,2-propandiols in the mouse. *British Journal of Pharmacology*, 94, 1069, 1988a.

Bowser-Riley, F., Daniels, S., Hill, W. A. G., Learner, T. S. & Smith, E. B. The additive effects of pressure and chemical convulsants in mice. *Journal of Physiology*, 409, 36P, 1988b.

Bowser-Riley, F., Daniels, S. & Smith, E. B. The effect of benzazole-related centrally acting muscle relaxants on HPNS. *Undersea Biomedical Research*, 15, 331, 1988c.

Bowser-Riley, F., Daniels, S., Hill, W. A. G. & Smith, E. B. An evaluation of the structure-activity relationships of a series of analogues of mephenesin and strychnine on the responses to pressure in mice. *British Journal of Pharmacology*, 96, 789, 1989.

Brauer, R. W., The high pressure nervous syndrome: animals., in *The physiology and medicine of diving*, Bennett, P. B. & Elliott, D. H., Eds., Balliere Tindall, London, 1st edition, 1969, 231.

Brauer, R. W., Beaver, R. W. & Sheehan, M. E., Role of monoamine neurotransmitters in the compression rate dependence of HPNS convulsions., in *Proceedings of Sixth Symposium on Underwater Physiology*, Shilling, C. S. & Beckett, M. W., Eds., FASEB, Bethesda, 1978, 49.

Daniels, S., Green, A. R., Koblin, D. D., Lister, R. G., Little, H. J., Paton, W. D. M., Bowser-Riley, F., Shaw, S. G. & Smith, E. B. Phamacological investigation of the high pressure neurological syndrome: Brain monoamine concentrations., in *Proceedings of the VIIth Symposium on Underwater Physiology*, Bachrach, A. J. & Matzen, M. M. Eds., Undersea Medical Society Inc., Bethesda, 1981, 329.

Daniels, S., Zhao, D. M., Inman, N., Price, D. J., Shelton, C. J. & Smith, E. B., Effects of general anaesthetics and pressure on mammalian excitatory receptors expressed in Xenopus oocytes. In *Molecular and Cellular Mechanisms of Alcohol and Anesthetics* ed. ROTH, S.H. & MILLER, K.W. Annals of the New York Academy of Sciences, 625, 108, 1991.

Fagni, L., Zinebi, F. & Hugon, M., Evoked-potential changes in rat hippocampal slices under helium pressure., *Experimental Brain Research.*, 65, 513, 1987.

Halsey, M. J. & Wardley-Smith, B., The high pressure neurological syndrome: do anticonvulsants prevent it?, *British Journal of Pharmacology*, 72, 502, 1981.

Johnson, F. H. & Flagler, E. A. Hydrostatic pressure reversal of narcosis in tadpoles, *Science*, 112, 91, 1950.

Koblin, D. D., Little, H. J., Green, A. R., Daniels, S., Smith, E. B. & Paton, W. D. M. Brain monoamines and the high pressure neurological syndrome. *Neuropharmacology*, 19, 1031, 1980.

Kuhse, J., Betz, H. & Kirsch, J. The inhibitory glycine receptor: architecture, synaptic localisation and molecular pathology of a postsynaptic ion-channel complex, *Current Opinion in Neurobiology*, 5, 318, 1995.

Lever, M. J., Miller, K. W., Paton, W. D. M. & Smith, E. B. Pressure reversal of anaesthesia. *Nature*, 231, 368, 1971.

Meldrum, B., Wardley-Smith, B., Halsey, M. J. & Rostain, J.-C., 2-amino-phosphono-heptanoic acid protects against the high pressure neurological syndrome, *European Journal of Pharmacology*, 87, 501, 1983.

Miller, K.W., Paton, W.D.M., Streett, W.B. & Smith, E.B. Animals at very high pressures of helium and neon. *Science*, 157 , 97-98, 1967.

Miller,K. W., Paton, W. D. M., Smith, R. A. & Smith, E. B., The pressure reversal of general anaesthesia and the critical volume hypothesis. *Molecular Pharmacology*, 9, 131 , 1973.

Miller, K. W., Inert gas narcosis, the high pressure neurological syndrome and the critical volume hypothesis, *Science*, 185, 867, 1974.

Pearce, P. C., Dore, C. J., Halsey, M. J., Luff, N. P., Maclean, C. J. & Meldrum B. S. The effects of MK801 on the high pressure neurological syndrome in the baboon. *Neuropharmacology.*, 29, 931, 1990.

Pearce, P. C., Halsey, M. J., Maclean, C. J., Ward, E. M., Webster, M. T., Luff, N. P., Pearson, J., Charlett, A. & Meldrum, B. S. The effects of the competitive NMDA receptor antagonist CCP on the high pressure neurological syndrome in a primate model. *Neuropharmacology*, 30, 787, 1991.

Pearce, P. C., Halsey, M. J., Maclean, C. J., Ward, E. M., Pearson, J., Henley, M. & Meldrum, B. S. The orally active NMDA receptor antagonist CGP 39551 ameliorates the high pressure neurological syndrome in *Papio anubis*. *Brain Research*, 622, 177, 1993.

Pearce, P. C., Maclean, C. J., Shergill, H. K., Ward, E. M., Halsey, M. J., Tindley, G., Pearson, J. & Meldrum, B. S. Protection from high pressure induced hyperexcitability by the AMPA Kainate receptor antagonists GYKI-52466 and LY-293558. *Neuropharmacology.*, 33, 605, 1994.

Roberts, R. J., Shelton, C. J., Daniels, S. & Smith, E. B., Glycine activation of human homomeric $\alpha 1$ glycine receptors is sensitive to pressure in the range of the high pressure nervous syndrome. *Neuroscience Letters*, 208, 125, 1996.

Shelton, C. J., Doyle, M. G., Price, D. J., Daniels, S. & Smith, E. B., The effect of high pressure on glycine and kainate sensitive receptor channnels expressed in *Xenopus* oocytes. *Proceedings of the Royal Society of London Series B*, 254, 131, 1993.

Shelton, C. J., Daniels, S. & Smith, E. B., Rat brain $GABA_A$ receptors expressed in *Xenopus* oocytes are insensitive to high pressure. *Pharmacology Communications*, 7, 215, 1996.

Smith, E.B., Bowser-Riley, F., Daniels, S., Dunbar, I.T., Harrison, C.B. & Paton, W.D.M. Species variation and the mechanism of pressure-anaesthetic interactions. *Nature*, 311, 56-57, 1984.

Wann, K. T. & Southan, A. P., The action of anaesthetics and high pressure on neuronal discharge patterns., *General Pharmacology.*, 23, 993, 1992.

Wardley-Smith, B. & Meldrum, B., Effect of excitatory amino acid antagonists on the high pressure neurological syndrome in rats. *European Journal of Pharmacology*, 105, 351, 1984.

Wardley-Smith, B., Angel, A., Halsey, M. J. & Rostain, J.-C., Neurochemical basis for the high pressure neurological syndrome: are cholinergic mechanisms involved? in *Proceedings of the VIIIth Symposium on Underwater Physiology*, Bachrach, A. J. & Matzen, M. M. Eds., Undersea Medical Society Inc., Bethesda, 1984, 621.

Wardley-Smith, B., Meldrum, B. S. & Halsey, M. J. The effect of two novel dipeptide antagonists of excitatory amino acid neurotransmission on the high pressure neurological syndrome in the rat. *European Journal of Pharmacology*, 138, 417, 1987.

Wardley-Smith, B. & Wann, K. T. The effect of non-competitive NMDA receptor antagonists on rats exposed to hyperbaric pressure. *European Journal of Pharmacology.*, 165, 107, 1989.

Williams, N., Roberts, R. J. & Daniels, S., A study into pressure sensitivity of ionotropic receptors. *Progress in Biophysics and Molecular Biology*, 65 (Suppl. 1), 113, 1996.

A Genetic Approach to Understanding Anesthesia

Shantadurga Rajaram, Bernhard Kayser, Phil G. Morgan, and Margaret Sedensky

DEPARTMENT OF ANESTHESIOLOGY & GENETICS, CASE WESTERN RESERVE UNIVERSITY, CLEVELAND, USA

Introduction

Upon first examination it appears a remarkable curiosity that the mechanism of action of volatile anesthetics remains unknown despite 150 years of clinical use. However, upon closer scrutiny, it is clear that reversible loss of consciousness in humans is quite a complex phenomenon. The human brain, the presumed seat of all consciousness, contains roughly 10^{10} neurons, each with 10^3 to 10^4 synapses; the brain appears capable of encoding a nearly incomprehensible number of bits of information.[1] Although a great deal of progress is being made in many areas of the neurosciences, the brain is still a relative black box with respect to human behavior. However, the complexity of the target organ is not the only daunting aspect of understanding how volatile anesthetics work. Volatile anesthetics themselves encompass an array of molecules that seems to defy categorization. A noble gas like xenon as well as a complex halogenated ether like isoflurane can each produce general anesthesia in humans.[2,3] This lack of any clear structure/function relationship makes the general adherence of all volatile anesthetics to the Meyer-Overton rule one of the more tenaciously-held dogmas of anesthetic research.[4,5] But their very property of lipid solubility makes their effects quite ubiquitous; volatile anesthetics can, to some degree, perturb nearly any system one cares to measure. Interpreting these perturbations and relating them in some way to loss of consciousness, a whole animal phenomenon, is often a most difficult task.

For the above reasons we have embarked upon a very simple-minded approach to investigate how the volatile anesthetics work. We would like to understand anesthetic action at the molecular level. To that end we are trying to identify an anesthetic site of action in a whole animal model. We are using genetics to tackle this question in a very simple metazoan, *C. elegans*. Genetics, in general, may be a tool without rival in understanding how volatile anesthetics work. *C. elegans*, in particular, offers many features that make it uniquely useful for exploitation.

Use of Genetics in Several Models

Genetics can be a valuable tool in understanding how volatile anesthetics work because it does not presuppose the answer to the question. In general, researchers in genetics have sought a heritable change in a behavior upon exposure to a volatile agent. Such a change in behavior can ultimately be traced back to a change in the organism's genetic code, its DNA. DNA is the molecule that is responsible for the synthesis of all of our cellular components. It is a double stranded helix consisting of a sugar and phosphate backbone, with the two strands of the helix held together by hydrogen bonding of purines and pyrimidines across the interior of this long molecule. The specific order of these interior bases is the code for synthesis of the specific gene product. Although there are only four bases in DNA, (adenine, guanine, cytosine, and thymine) their order can be specified along great lengths. If genes are thought of as sentences written with a four letter alphabet, sentences can go on for hundreds or thousands of words. These words are in the form of 3 base sets, each of which can specify a unique amino acid when the protein product of the gene is synthesized (For a review see reference 6). A change in the order of just one letter in a word of a sentence can make a significant change in its meaning. For example, when GAG (guanine/adenine/guanine) is changed to GTG (thymine replaces adenine) in the genetic code, valine is inserted into the gene product instead of glutamic acid. If this happens in the sixth amino acid of the β-chain of the hemoglobin molecule, the result is HgS, the basis of sickle cell disease.[7] Knowing the order of the bases within a gene is called sequencing, and can reveal much about a gene product's predicted function.

DNA is the permanent blueprint for all of an organism's individual molecular constituents, including, of course, the inherited constituents that interact with volatile anesthetics. A genetic change in DNA, i.e. a mutation, can lead to the identification of the structure and function of the product encoded by that piece of DNA. A mutation which changes responses to volatile anesthetics can lead to a discrete change in a molecular component of the anesthetic site of action, regardless of the particular chemical nature of that site. Analysis of this change will directly relate to mechanisms of action of volatile anesthetics. Incidental or non-specific effects of volatile anesthetics are thus avoided, as are presuppositions about the nature of the molecular target.

Yeast

A paragon of eukaryotic genetics is the yeast *S. cerevisiae*. Keil *et al*[8] have been able to isolate mutations in this organism that can grow in 12% isoflurane (normally yeasts are inhibited in growth at this concentration). They have shown that a single mutation in a gene called ZZZ4 causes this altered response. This gene's product looks similar to phospholipase A_2 activating protein, a molecule implicated in signal transduction in other systems. They are now in the process of further analyzing the interaction of this protein with isoflurane, as well as the effects of other genes they have isolated which affect anesthetic sensitivity.

Drosophila

The simplicity of yeast makes it an ideal model system in which to understand very fundamental interactions of volatile anesthetics with subcellular targets; changes in these targets change an organism's normal function in volatile anesthetics. However, a very time-tested genetic model, the fruit fly *Drosophila melanogaster*, offers layers of additional complexity as a model system. It possesses a well-defined nervous system and complex behaviors, including a long-recognized reversible paralysis in response to volatile anesthetics. Over 15 years ago Gamo *et al*[9] studied several spontaneous mutations that caused a strain of flies to be resistant to diethyl ether. These original mutations were genetically mapped to specific chromosomes, but the sorting out of an individual gene's contribution to the animal's response to volatile anesthetics was at that time quite difficult. She and her co-workers have since mutated animals to produce clear single gene mutations that cause resistance to diethyl ether.[10] Three of these genes are currently under investigation. The product that they make is found primarily within the central nervous system of the animal; the final identification of these products is currently under way.

Nash and co-workers have isolated single gene mutations in *Drosophila* that confer resistance to halothane.[11] They performed a sort of column fractionation of mutagenized animals in an 'inebriometer', a column of baffles through which anesthetics are run. Animals that remained in the device longer than normal, non-mutated flies were collected and studied further. Four mutations that confer halothane resistance have been genetically mapped, i.e., localized to specific areas of the animals' total genetic material. Interestingly, these animals are only changed in response to specific anesthetics.[12] Nash and others have also studied mutations that are known to affect *Drosophila* ion channels.[13,14] In Nash's assay, mutations in any channel produced a change in anesthetic sensitivity to at least one of the three gasses tested. Effects were strikingly non-uniform; some mutations rendered the fly sensitive to halothane but not to trichloroethylene, while others produced hypersensitivity to trichloroethylene but not halothane. Whether ion channels are direct targets of volatile anesthetics is not known. Decapitating the mutant flies could make both resistant and sensitive strains more like controls.

Mammals

In an effort to use an experimental system as closely related to the ourselves as possible, rodents have been used to study the genetic control of the general anesthesia. A great deal of progress has occurred in alcohol research in some of these models. (For a Review see 15.) However, in 1980, Koblin *et al* had already published the results of selective breeding of mice for differential sensitivity to nitrous oxide.[16] They were able to maintain strains of mice that differed by over 50% in sensitivity to this gas. Mice that have been bred for different sensitivity to ethanol also have differing sensitivities to certain gaseous anesthetics. For example LS (Long Sleep) and SS (Short Sleep) mice differ in sensitivity (as assayed by a variety of behaviors) to enflurane and isoflurane, but not to halothane.[17,18,19] The difference in ethanol sensitivity between LS and SS mice is thought to be related to function of the $GABA_A$ receptor.[20] Likewise, mice bred for differences in sensitivity to diazepam are also different in halothane and enflurane sensitivity.[21,22] Rats bred for a differential hypnotic response to ethanol are, unlike mice, different from each other in response to enflurane, isoflurane, and halothane.[23] In fact, the LAS (Low Alcohol Sensitivity) strain of rats awaken from an intraperitoneal injection of halothane at a brain level of the drug that is nearly twice that of the HAS (High Alcohol Sensitivity) rats. All in all, these studies hold great promise for future genetic dissection of anesthetic response in mammals. However the complexity of the mammalian brain, the interpretation of multiple behavioral assays, and the relatively recent ability to apply molecular genetics to mammalian systems does place some limitations on what can be learned at present.

Use of Genetics in C. elegans

Behavioral studies

Our laboratory has tried to find a tractable model system for the molecular characterization of an anesthetic site of action. We have chosen the nematode *C. elegans* for the unique advantages it offers in this endeavor (For a review of the organism, see reference 24).

C. elegans is a simple, nonparasitic soil nematode that is about a millimeter long. In the laboratory the animals grow on agar plates spread with *E. coli*. The adult consists of a cuticle, a gut, muscles, nerves, and a gonad (Figure 1). Its usual form is that of a self-fertilizing hermaphrodite that lays about 300 eggs which grow to adulthood in approximately $3\frac{1}{2}$ days. Males do exist that can be easily propagated and used for genetic manipulation. The nonmutated hermaphrodite always consists of exactly 959 somatic cells, of which 302 are neurons. The embryonic ancestry of every cell has been traced back to the original zygote, and the entire worm has been serially sectioned such that every neuronal synapse is known. The haploid genetic content of the animal is only about 25 times that of *E. coli*, and nearly all of its DNA has been ordered relative to a map of known genetic mutations. In fact, over 60% of its DNA has been sequenced, which means that the genetic code for many areas of certain chromosomes is completely spelled out. This is an invaluable tool for use of the powerful techniques of molecular genetics.

Although *C. elegans* is simple, it has complex behaviors that are mediated by neurotransmitters like acetylcholine, GABA, serotonin, and dopamine. There is a huge collection of well-studied mutants, which can be shipped easily through the mail. Over 100 mutants exist with presumed derangements in

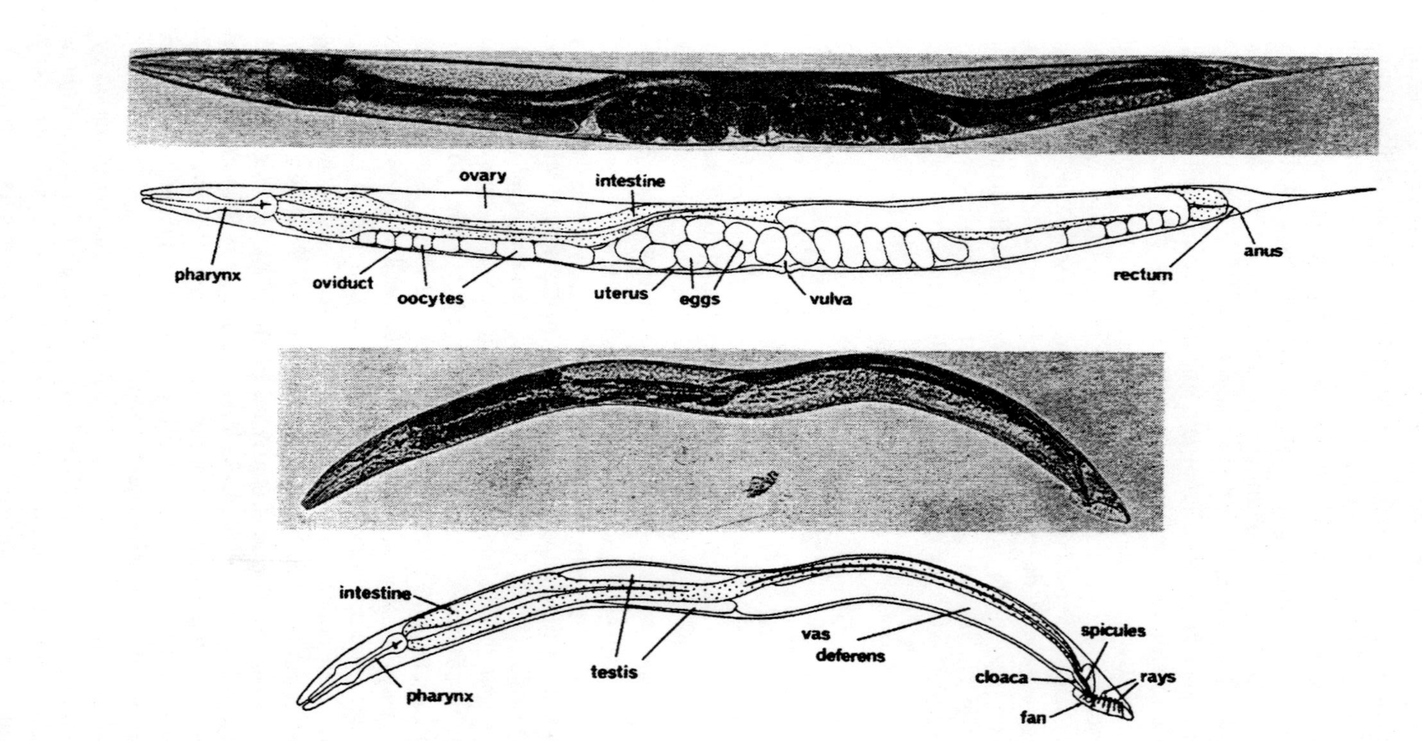

Figure 1 *Photomicrographs of the nematode* C. elegans *hermaphrodite (top) and male (below). Animals are shown in a lateral view. The hermaphrodite is approximately 1 millimeter long. (Reprinted with permission from Sulston JE and Horvitz HR. Dev Biol 56*:110-156, 1977.)

neuromuscular function. A large community of scientists is involved in the study of many different aspects of this animal's development and behavior. The tools to study *C. elegans* at the molecular level are very powerful and readily available.

It initially appeared to us that *C. elegans* was a simple enough model in which to ask questions at the molecular level about how volatile anesthetics work, yet complicated enough that the answers might be relevant to more complex species. The first question was, ' Do worms go to sleep?'

When non-mutated worms, called N2, are placed in a volatile agent, they initially become hyperactive, as if avoiding a noxious stimulus. Their usually sinuous movement across the agar plate becomes progressively uncoordinated, until it ceases altogether. This flaccid paralysis is quickly reversed upon removing the animals from the agent; subsequent feeding, mating, and lifespan appear unaffected[25]. Using reversible paralysis as our endpoint, we exposed N2 to a range of gaseous agents whose O/G partition coefficients varied from 48 to 7230. We found that N2 exactly followed the Meyer-Overton rule: a log/log plot of EC_{50} versus anesthetic concentration yielded a very good straight line fit with a slope of -1[26]. Thus encouraged, we mutagenized worms and looked for those with changes in sensitivity to halothane. The first animal we isolated was a new version, a new allele, of a mutant gene called *unc-79*[27]. The term 'unc' denotes the fact that this animal is uncoordinated (although it generally moves very well), while 79 signifies that it is the 79th such uncoordinated mutant mapped to a chromosome. *unc-79* is extremely sensitive to halothane, with an EC_{50} only 1/3 that of N2. It is a recessive mutation, and appears to be a mutation that causes the animal to entirely lose the normal product of that gene; thus, the normal product of this gene causes worms to have a higher EC_{50} like that of N2. *unc-79* is also very sensitive to chloroform, methoxyflurane, and thiomethoxyflurane, the four most lipid soluble agents that we tested (see Figure 2). However, *unc-79* is slightly resistant to enflurane and flurothyl (hexafluorodiethyleher). It is not very changed in its behavior in isoflurane or fluroxene from N2, and has a mild increase in sensitivity to diethylether.[26] In summary, this single mutation produces a very sharp deviation from the Meyer-Overton rule.

While intent on characterizing *unc-79* at the molecular level, we also needed to isolate other mutations that controlled anesthetic sensitivity in this animal. It is clear that the phenomenon of 'anesthesia' is a complicated process, one that might be disrupted at any one of a number of levels. The entire picture can best be visualized with a family of related or interacting gene products. Therefore, we isolated a class of mutations that suppress the mutation *unc-79*. A suppressor mutation, when added to the *unc-79* mutation, causes the animal to behave normally in gaseous anesthetics. One of the best supressors we have found is the mutant *unc-1*.

unc-1 is an X-linked gene that can restore the behavior of *unc-79* in halothane exactly back to normal; the EC_{50} of a worm carrying both mutations *unc79;unc-1* is identical to that of N2. In fact it restores the extreme sensitivity of *unc-79* to chloroform , methoxyflurane, and thiomethoxyflurane all back to

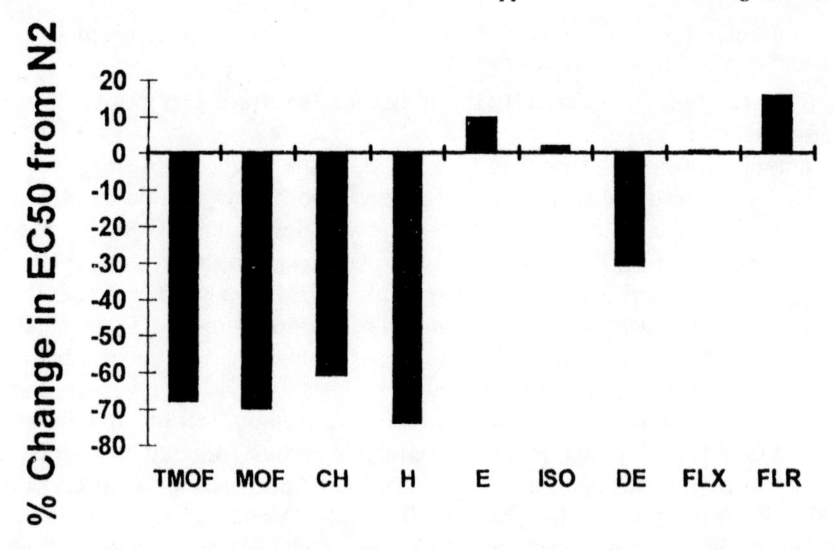

Figure 2 *Percent change in EC_{50}s of* unc-79 *in nine anesthetics compared to the non-mutated worm, N2. Calculated as (*unc-79 *EC_{50} - N2 EC_{50}) /N2 EC_{50}. The response of N2 is the baseline. The mutant animal is most sensitive to the most lipid soluble volatile anesthetics. Abbreviations: TMOF, thiomethoxyflurane; MOF, methoxyflurane; CH, chloroform; H, halothane; E, enflurane; ISO, isoflurane; DE, diethylether, FLX, fluroxene; FLR, flurothyl.*

baseline, as well as returning *unc-79*'s resistance to enflurane and flurothyl back to normal. However, it does not restore the animal's behavior in diethyl ether to normal. *unc-1* itself is normal in sensitivity to all volatile agents tested, except for diethyl ether, to which it is hypersensitive.[28]

These results are consistent with the notion that not all anesthetics work at identical sites of action. The fact that different mutations have completely different effects for specific gasses supports the hypothesis that there are multiple sites of action for volatile anesthetics, even in a very simple animal like *C. elegans*. For example, one can postulate that there are two sites (or neural pathways or domains of a single molecule) that are affected by mutations in *unc-79*. One of them is restored to normal by mutations in *unc-1*, and one is not. However, there is another type of site, one that interacts with isoflurane, that is not affected by *unc-79* or *unc-1* mutations.[28]

By placing *unc-79* and *unc-1* together in one animal and observing its behavior in anesthetics, it is also possible to order them in a genetic pathway. Because the *unc-79;unc-1* animal appears like *unc-1* in its behavior, the *unc-79* gene product must exert its effects somehow through *unc-1*, i.e. *unc-79* cannot show its behavior unless the normal *unc-1* gene product is present.

unc-1 is a very interesting gene in that there are many different sorts of mutations possible within it. In fact, there are 5 classes of *unc-1* mutations. Some are dominant and make the animal coil, while others are recessive mutations and cause the animal to kink. Another class is sensitive to cold, and kinks at 11 °C but not at 20 °C.[29,30] Only those versions of *unc-1* that are

unc-79 ➡ **unc-1** ➡ **fc21** ➡ **Anesthetic Sensitivity**

Figure 3 *A genetic pathway for three genes involved in anesthetic response in* C. elegans. *Construction of mutants bearing multiple mutations orders the genes' relative positions within the pathway.*

kinked in appearance are able to suppress the *unc-79* mutation.[28] This includes the cold-sensitive allele, which only suppresses at 11 °C (Shanta Rajaram, unpublished results). We can eventually correlate the structure of this gene's product to its function, as we discover the exact location of each mutation within the *unc-1* gene (see below). There are also other excellent suppressors of *unc-79*; they are also kinked in appearance, and are undergoing molecular characterization by other laboratories (see below).

Another mutation under intense investigation in our laboratory is *fc21*. It was originally selected as part of a search for mutations that change the response of the animal to gases like isoflurane or enflurane, agents for which we lacked any hypersensitive mutants. f*c21* moves very well in air. However, its EC_{50} in isoflurane is 1.3%, less than 1/5th that of N2 (EC_{50} =7.2%).[31] Unlike *unc-79* or *unc-1* however, it is exquisitely sensitive to every anesthetic gas in which it has been tested. Its EC_{50} in halothane is 1.1%, the same as *unc-79*. Moreover, *fc21* is not suppressed by *unc-1* or any other suppressor. Animals carrying the *fc21* mutation and any other mutation, be it *unc-79* or *unc-1*, behave exactly like *fc21*. Therefore, *fc21* is further downstream than either *unc-1* or *unc-79* in a genetic pathway, perhaps the closest of any of our current mutations to the anesthetic target site itself (Figure 3). *fc21* is also a recessive mutation that appears to be an incomplete loss of function of the protein product. Besides its hypersensitivity to gaseous anesthetics, this mutation also causes the worms to develop a little slower than normal, and to produce fewer eggs.

In addition to these three key genes, we have identified several other genes, with many different versions of particular genes, that can change anesthetic sensitivity in *C. elegans*. One of them for example, *unc-80*, behaves very much like *unc-79*, but maps to a completely different chromosome.[27] Its effects are not additive to *unc-79*, indicating that it is within the same genetic pathway as *unc-79*. Another mutation, *fc34*, is hypersensitive to all anesthetic gases but becomes paralyzed while shrinking in the gas.[31] It also can die with more prolonged exposures to the volatile anesthetics. We have recently isolated suppressors of the suppressor genes. These are mutations that counter the effects of *unc-1* on *unc-79*: a triple mutation *unc-79;unc-1;mutation-X* in an animal will make the animal behave like *unc-79* again.

All of the above studies are directed at determining changes in single genes which affect sensitivity to volatile anesthetics. Crowder and colleagues have adopted a different approach for studying *C. elegans*.[32] They have crossed two different strains of *C. elegans* which have similar sensitivities to volatile anesthetics. These strains are both non-mutated, 'normal' worms, but they are genetically separate by virtue of geographic isolation. They have collected their second generation offspring and noted that some individuals have wide variations from the parental strains in sensitivities to halothane. These worms are then used to start new, hybrid strains in which the new sensitivities breed true. It is important to note that these hybrid strains do not have new mutations; rather they carry a novel mix of gene products from two normal parental strains. In general large stretches of each chromosome are derived from different parental strains. The original combinations present in the parents led to 'normal' sensitivities; the new combinations are altered. The hybrids are potentially genetically different in many ways compared to either individual parent. Using this approach, they have isolated hybrid strains with large increases in resistance to halothane. The new sensitivities to specific anesthetics are mapped to specific locations in the chromosomes. Such a technique is called QTL (Quantitative Trait Loci) mapping and is the same approach used in humans for assigning importance to genes affecting complicated traits such as hypertension. It will be of great interest to combine these two approaches in the study of volatile anesthetics in *C. elegans.*

One other tactic that we have used to genetically dissect mechanisms of anesthetic action is to expose our sundry mutants to stereoisomers of volatile anesthetics. We have exposed key mutant strains to enantiomers of both halothane and isoflurane.[33,34] The nonmutated animal N2 is more sensitive to + than to − halothane, and is also more sensitive to + isoflurane that its − enantiomer. Since *unc-79* is like N2 in its behavior to racemic isoflurane, we expected it to behave like N2 when exposed to stereoisomers of isoflurane. In fact it did, indicating that this gene product is not responsible for the stereospecific behavior of the normal worm to isoflurane. Overall, patterns of stereospecificity to each gas for specific mutants were quite complex and highly individual. For example, the shrinking worm, *fc34*, is much more sensitive to the + than to the − form of halothane, but does not show much stereospecific difference in its responses to enantiomers of isoflurane. In contrast to *fc34*, *fc21* is stereoselective in its response to isoflurane, but not to halothane. These data support the contention that the interactions of volatile anesthetics are NOT non-specific, and probably primarily with a protein target.

Our goal in accumulating a family of mutations that represent a spectrum of behaviors is to find all the genes that can control anesthetic response in *C. elegans*. It is clear that it is a complicated pathway even in this simple animal, and that ultimately many molecules will interact to produce the phenomena we observe. A pathway of mutations is necessary to make decisions as to which mutations are important to characterize at the molecular level. Secondly, in order to interpret the molecular data that we do accumulate, it is indispensable to correlate gene products with observed behaviors.

Molecular characterization of genes controlling anesthetic response in *C. elegans*

We have proceeded with a molecular analysis of all three genes discussed above, that is *unc-1*, *fc21*, and *unc-79*. The ability to clone genes in *C. elegans* and analyze their expression within the animal is a key advantage of this model.

unc-1

Mutations in *unc-1* causes the hypersensitive mutant *unc-79* to behave normally in halothane and several other gases. The *unc-1* gene product is necessary for this mutation to have an effect. In order to characterize this gene we had to first localize it to a very small region of the X chromosome on which it resides. Through standard genetic mapping we eventually narrowed it down to a region that was only about 1/100 of the entire chromosome. Because the entire DNA content of *C. elegans* is so well studied, we were able to correlate this genetic area to actual physical pieces of DNA that have been ordered relative to each other and to this region of the X chromosome. These pieces of DNA have already been cloned, that is physically isolated from all of the other worm DNA and carried in a foreign host, like a virus within a bacterium. To finally pinpoint the gene, we used a technique in *C. elegans* called mutant rescue.[35]

In the *C. elegans* hermaphrodite the gonad exists as a tube which carries immature oocytes within a syncytium of ooplasm. As the oocytes mature they compartmentalize, completing the process just before fertilization. It is possible to microinject within the ooplasm pieces of DNA. These pieces are taken up by the oocytes, and can be expressed in the offspring of the injected parent. In mutant rescue then, one microinjects into a mutant animal DNA that is hypothesized to carry a normal copy of the gene the mutant is lacking. If the hypothesis is correct, offspring of a mutant animal will be normal for the abnormal characteristic the parent displayed. In the case of *unc-1*, it is very easy to see normal moving animals as offspring of a parent that is normally very kinked in its motion. Therefore, once we had the potential region of *unc-1* restricted to just a few pieces of already cloned DNA, we microinjected them in smaller and smaller subsets of DNA, until we identified the smallest rescuing piece. Because the DNA in this region is already sequenced, i.e., the order of its constituent bases is known, we immediately knew that this small piece still contained several possible genes. In fact, in the region in which we were most interested, two genes overlapped on the same piece of the double-stranded DNA molecule. One predicted gene was encoded on one strand of DNA and read in a certain direction, while the other was read in the opposite direction on the opposite strand. One gene codes for a protein called stomatin, while the other codes for a product called neurocalcin.

Although computer-generated predictions of how regional DNA sequences make sense as a gene are invaluable, they are only starting points for confirming a gene's location. In our case, we were able to identify the intermediate molecules that DNA uses to generate its protein product from the starting material, chromosomal DNA. These molecules, called messenger RNAs, contain only those pieces of DNA that actually get made into protein. Their base sequence confirms or modifies the computer's prediction of exactly where a gene starts or stops. We were able to show that the two genes in the region do not make overlapping messages, and that the very beginning of the neurocalcin gene was completely lacking in the smallest piece of DNA that strongly rescued the *unc-1* mutant.

At this point the usefulness of having many mutations in a specific gene comes into play. We were able to sequence our collection of mutations in *unc-1*, and compare the order of bases in the mutants to that of the normal animal's genes for this region. Along with single base changes in the various classes of *unc-1* mutants, we have generated a small deletion of DNA within the region. All of the mutations we have sequenced to date have significant mutations in the stomatin gene. The protein product of *unc-1* contains 285 amino acids, with a molecular weight of about 31 kD.

Another weaker suppressor of *unc-79*, *unc-24* has also been cloned.[36] Part of its sequence is very similar to stomatin. Mutations in two other genes, *unc-7* and *unc-9* are excellent suppressors of *unc-79*.[28] They have each been cloned. Their DNA sequences are related, and predict a transmembrane protein with an as yet unknown function.[37, 38]

In humans stomatin is an integral membrane protein that is thought to regulate cation conductance in red blood cells. (For a review see reference 39.) An autosomal dominant hemolytic anemia, hereditary stomatocytosis, results when stomatin is lost in red blood cells. Red blood cells swell and lyse, presumably due to changes in membrane conductance. Patients with the disease show a reversal in the Na^+/K^+ ratio within their red blood cells.[39,40] In *C. elegans* a stomatin-like protein is required for the animals to be sensitive to light touch.[41] It is thought to link a mechanosensory channel to the microtubular cytoskeleton of touch receptor neurons, probably at its carboxy-terminal end, which is within the cell cytoplasm. It is this region that is also postulated to link human stomatin to the red cell's cytoskeleton. The protein is thought to support, activate and/or regulate an associated channel in a ball and chain model of receptor regulation (Figure 4).

All of our mutations sequenced to date fall within the large intracytoplasmic domain of the stomatin, which is thought to be its regulatory domain. This includes *n494*, the version of *unc-1* that does not suppress *unc-79's* halothane sensitivity. Comparing those changes that suppress *unc-79*'s halothane sensitivity to those that do not will give us some further clues as to the mechanism of *unc-1*'s control of anesthetic response. In addition to characterizing and interpreting the nature of the mutations in various *unc-1* types of animals, we are studying the localization of the protein product within the animal. Is the protein primarily in neurons? Is it present at all stages of development? What happens to it when the animal is also mutant for *fc21*? These and questions like them are currently under investigation.

fc21

In the case of *fc21*, the mutation that causes animals to be hypersensitive to all gaseous anesthetics tested, we have pursued lines of investigation similar to those described for *unc-1*. After fine genetic mapping, we localized the gene to a 4.6 kb (1 kb = 1000 base pairs) piece of DNA using the technique of mutant

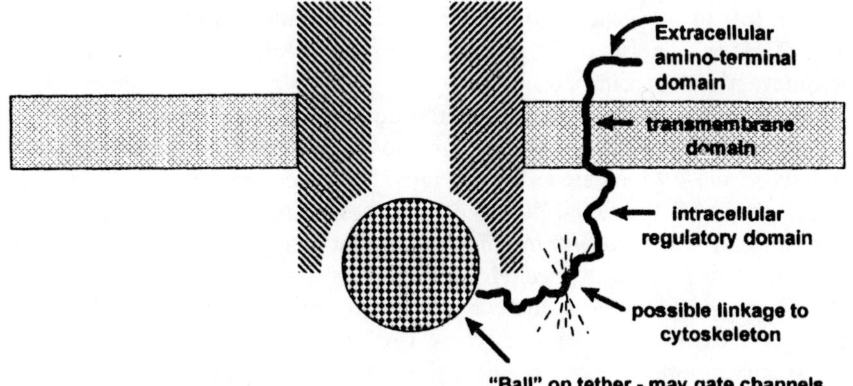

Figure 4 *A possible model of stomatin as a 'ball and chain' molecule that regulates membrane channels. The large cytoplasmic domain may act as a 'plug' that is tethered to the membrane. Here it is drawn as if it blocks a theoretical transmembrane channel.*

rescue. The smallest rescuing fragment contained only 1 complete gene, the 49kD-type subunit of NADH:ubiquinone oxidoreductase. This protein is part of a huge enzyme complex embedded in the mitochondrial membrane.[42] It is the first complex in the electron transport chain, for it removes electrons from NADH and passes them to ubiquinone.

Based on the length of the messenger RNA we have found, the protein of the *C. elegans* molecule contains 482 amino acids, a length that is similar to that found in other organisms. Sequencing the DNA of the gene in our mutant animal revealed a single base change that substitutes arginine for lysine in a position of the protein that is otherwise absolutely conserved in 12 species ranging from bacteria to fungi to plants to invertebrates to cows.[43] Thus, the mutation alters an essential position in the protein.

In other species it is estimated that there are over 20 protein molecules within the NADH:ubiquinone reductase complex. Exactly how this single base change causes such a profound alteration in anesthetic response is under intense study. It is known that mitochondrial function alters many aspects of cellular physiology, however. These include energy generation, calcium homeostasis, and physiologic state of the cell membrane. It is interesting to speculate that one of these types of changes affects sensitivity to volatile anesthetics. We are generating more versions of mutations in this gene, and by studying the pattern of protein expression in the animal, we hope learn more of this protein's function.

unc-79

unc-79 was the first mutation we found that changed the response of *C. elegans* to volatile anesthetics, yet has remained the most problematic. In a genetic sense, it is the most upstream in our genetic pathway controlling anesthetic sensitivity, and perhaps the farthest removed from an anesthetic target site.

However, loss of this gene's product makes the animals very sensitive to only a subset of gases; analysis of its product may give us clues as to how different gases interact with specific sites.

Using genetic mapping followed by mutant rescue, we were able to localize *unc-79* to a small region of the third chromosome. The smallest piece of DNA that rescued *unc-79* indicated that the *unc-79* gene coded for a single protein from a region of DNA called E03A3.6 , based on computer prediction of genes in the area. It was also within this region that many of our genes showed changes in their DNA sequence. However, it appears that the messenger RNA from this region actualiy includes some of this gene and a predicted neighboring gene. Sequence analysis of these two adjacent stretches of DNA that seem to encode one product predicts that the *unc-79* gene is most like a C-lectin type receptor. These preliminary data are quite exciting to us, and once confirmed will afford us many possibilities to correlate this protein's function to a potential interaction with stomatin.

Summary

The intent of this brief exposure to some genetic studies currently underway is to illustrate the potential power of molecular genetics to unravel the mystery of how volatile anesthetics work. Even in such a simple model as ours gaseous anesthetics seem to interact with an array of molecules in a complex manner. However, if headway can be made in this simple animal, these genes may be used as probes in more complicated animals like flies or mammals. It is important to remember that virtually all of our basic knowledge about the genetic code and the synthesis of proteins came from studies in very simple models like viruses and prokaryotic cells. With only minor variations these fundamental processes are the same in humans and in *E. coli*. The phenomenon of 'anesthesia' produced by volatile anesthetics appears to be a highly conserved process shared by many phyla. The adherence of so many different species to the Meyer-Overton rule speaks to this point. Many genes which specify basic neuronal function, including families of genes coding for membrane receptors, are very similar in their DNA content in different species, attesting to phylogenetic conservation of basic neuronal constituents.[44,45,46] Ultimately a behavioral response in a simple animal like *C. elegans* may be used to search for a ancient gene still operative in human behavior.

To date, no studies have given a molecular picture of an anesthetic site of action. The field of molecular genetics is now providing us with the tools to begin work. In the near future a very old and fundamental question may be answered by these new and powerful techniques.

Acknowledgments

This work was supported in part by NIH grants GM-45402, GM41385, and AA09144. In addition, MMS was supported by the B. B. Sankey Award from the International Anesthesia Research Society and PGM was the recipient of a Research Starter Grant from the American Society of Anesthesiology.

References

1. Sagan, C: The Dragons of Eden: Speculations on the Evolution of Human Intelligence. Ballantine Books, NY. Chapter 2. Copyright 1977.
2. Cullen SC, Gross EG: Anesthetic properties of xenon and krypton. *Science. 113*:580, 1951.
3. Cullen SC, Eger EI, Cullen BF, Gregory P: Observations on the anesthetic effect of the combination of xenon and halothane. *Anesthesiology. 31*:305, 1969.
4. Meyer HH: Zur Theorie der Alkoholnarkose. I. Mitt. Welche Eigenschaft der Anasthetika bedingt ihre Narkotische Wirkung? *Arch Exper Pathol Pharmak. 42*:109-119, 1899.
5. Overton E: Studien uber die Narkose. Jena Verlag Von Gustav Fisher, 1901.
6. Watson JD, Hopkins NH, Roberts JW, *et al*: *The Molecular Biology of the Gene.* Menlo Park, California, Benjamin/Cummings Publishing Co., 1987.
7. Ingram VM: Gene mutations in human hemoglobin: The chemical difference between normal and sickle cell hemoglobin. *Nature. 180*:326-328, 1957.
8. Keil RL, Wolfe D, Reiner T, Peterson CJ, Riley JL: Molecular genetic analysis of volatile anesthetic action. *Mol Cell Biol. 16*:3446-3453, 1996.
9. Gamo S, Ogaki M, Nakashima-Tanaka E: Strain differences in minimum anesthetic concentrations in *Drosophila melanogaster*. *Anesthesiology. 54*: 289-293, 1981.
10. Gamo S, Taniguchi F, Morioka K, Tanaka Y, Michinomae M, Inoue Y: Resistant mutants to diethylether anesthesia in *Drosophila melanogaster*. *Prog Anesth Mech. 2*:11-20, 1994.
11. Krishnan KS, Nash HA: A genetic study of the anesthetic response: Mutants of *Drosophila melanogaster* altered in sensitivity to halothane. *Proc Natl Acad Sci. 87*:8632-8636, 1990.
12. Campbell DB, Nash H: Use of *Drosophila* mutants to distinguish among volatile general anesthetics. *Proc Natl Acad Sci. 91*:2135-2139, 1994.
13. Tinklenberg JA, Segal IS, Tianzhi G, and Maze M: Analysis of anesthetic action on the potassium channels of the Shaker mutant of *Drosophila. Annals N Y Acad Sci.* 625:532-539, 1991.
14. Leibovitch BA, Campbell DB, Krishnan KS, Nash HA: Mutations that affect ion channels change the sensitivity of *Drosophila* melanogaster to volatile anesthetics. *J. Neurogen. 10*:1-13, 1995.
15. Crabbe J, Belknap J, Buck K: Genetic animal models of alcohol and drug abuse. *Science. 264*:1715-1723, 1994.
16. Koblin DD, Dong DE, Deady JE, *et al*: Selective breeding alters murine resistance to nitrous oxide without alteration in synaptic membrane lipid composition. *Anesthesiology. 52*:401-407, 1980.
17. Baker R, Melchior C, Deitrich R: The effect of halothane on mice selectively bred for differential sensitivity to alcohol. *Pharmacol Biochem Behav. 12*:691-695, 1980.
18. Koblin DD, Deady JE: Anaesthetic requirement in mice selectively bred for differences in ethanol sensitivity. *Brit J Anaesth. 53*:5-10, 1981.
19. Simpson VJ, Baker RC, Timothy BS: Isoflurane but not halothane demonstrates diffential sleep time in long sleep and short sleep mice. *Anesthesiology. 79*(3A):A387, 1993.

20. Wafford KA, Burnett DM, Dunwiddie TV, Harris, RA: Genetic Differences in the ethanol sensitivity of $GABA_A$ receptors expressed in Xenopus oocytes. *Science. 249*:291-293, 1990.
21. McCrae A, Gallaher E, Winter P, Firestone L: Volatile anesthetic requirements differ in mice selectively bred for sensitivity or resistance to diazepam: implications for the site of anesthesia. *Anesth Analg. 76*:1313-1317, 1993.
22. Quinlan J, Jin K, Gallaher E, McCrae A Firestone L: Halothane sensitivity in replicate mouse lines selected for diazepam sensitivity or resistance. *Anesth Analg. 79*:927-932, 1994.
23. Deitrich RA, Draski LJ, Baker RC: Effect of pentobarbital and gaseous anesthetics on rats selectively bred for ethanol sensitivity. *Pharmacol Biochem Behav. 47*(3):721-725, 1994.
24. *C. elegans* II edited by Riddle, DL, Blumenthal, T, Meyer, BJ, and Priess JR. Cold Spring Harbor Laboratory Press. 1997.
25. Morgan PG, Cascorbi HF: Effect of anesthetics and a convulsant on normal and mutant *Caenorhabditis elegans. Anesthesiology. 62*:738-744, 1985.
26. Morgan PG, Sedensky MM, Meneely PM, et al: The effect of two genes on anesthetic response in the nematode *Caenorhabditis elegans. Anesthesiology. 69*:246-251,1988.
27. Sedensky MM, Meneely PM: Genetic analysis of halothane sensitivity in *C. elegans. Science. 236*:952-954, 1987.
28. Morgan PG, Sedensky MM, Meneely PM: Multiple sites of action of volatile anesthetics in
29. *C. elegans. Proc Natl Acad Sci. 87*:2965-2969. 1990.
30. Park EC, Horvitz HR: Mutations with dominant effects on the behavior and morphology of the nematode *Caenorhabditis elegans. Genetics.113*:821-852, 1986.
31. Hecht RM, Norman TV, Jones W: A novel set of uncoordinated mutants in *C. elegans* uncovered by cold sensitive mutations. *Genome. 39*:459-464. 1996.
32. Morgan PG, Sedensky MM: Mutations conferring new patterns of sensitivity to volatile anesthetics in *C. elegans. Anesthesiology. 81*(4):888-898, 1994.
33. Crowder M: *Proc Natl Acad Sci.* In press.
34. Sedensky MM, Cascorbi HF, Meinwald J, Radford P, Morgan PG: Genetic differnces affecting the potency of stereoisomers of halothane. *Prog Anesth Mech. 3*:20-225, 1995.
35. Morgan PG, Usiak M, Sedensky MM: Genetic differences affecting the potency of stereoisomers of isoflurane. *Anesthesiology. 85*(2): 385-392, 1996.
36. Fire A: Integrative transformation of *C. elegans. The EMBO Journal. 5*(10):2673-2680, 1986.
37. Barnes T, Jin Y, Howitz HR, Ruskin G, Hekimi S: *J. Neurochem. 67*(1):46-57, 1996.
38. Starich TA, Herman RK, Shaw JE: Molecular and genetic analysis of *unc-7*, a *C. elegans* gene required for locomotion. *Genetics. 133*:527-41, 1993.
39. Barnes T: Personal communication
40. Stewart GW, Argent AC, Dash BCJ: Stomatin: a putative cation transport regulator in red cell membrane. *Biochim Biophys Acta. 1225*:15-25, 1993.
41. Stewart, GW, Hepworth GW, Jones BE, Keen JN, Dash BCJ, Argent AC, Casimir, CM: Isolation of cDNA coding for a ubiquitous membrane protein deficient in high NA^+, low K^+, stomatocytic erythrocytes. *Blood. 79*:1593-1601, 1992.

42. Huang M, Gu G, Ferguson E, Chalfie M: A stomatin-like protein necessary for mechanosensation in *C. elegans*. *Nature*. *378*:292-295, 1995.
43. Walker JE: The NADH: ubiquinone oxidoreductase (complex I) of respiratory chains. *Quart Rev Biophys*. *5*(3):253-324, 1992.
44. Fearnley IM, Walker JE: Conservation of sequences of subunits of mitochondrial complex I and their relationships with other proteins. *Biochem Biophys Acta*. *1140*:105-134, 1992.
45. Smith GB, Olsen RW: Functional Domains of $GABA_A$ receptors. *Trends Pharm Sci*. *16*(5): 162-168, 1995.
46. Franks NP, Lieb WR: An anesthetic-sensitive superfamily of neurotransmitter-gated ion channels. *J Clin Anesth*. *8*:3S-8S, 1996.
47. Catterall WA: Molecular properties of a superfamily of plasma-membrane cation channels. *Curr Opin Cell Bio*. *6*(4):607-615; 1994.

Closing Plenary Lecture
Do We Need New Anaesthetic Drugs?

R. M. Jones

IMPERIAL COLLEGE SCHOOL OF MEDICINE, ST MARY'S HOSPITAL, LONDON, UK

General anaesthesia for patients undergoing surgery is now an extremely safe process and anaesthesia is the cause of death in less than 1 in 100,000 patients (1). Indeed, fatal complications of anaesthesia are now so rare that studying their causes is a time consuming and expensive process. In these circumstances it is clearly appropriate to ask whether resources should be invested in developing new anaesthetic drugs.

In affirming the need for new anaesthetic drugs an analogy with air travel is appropriate. Air travel has been a safe form of transport for some decades but newer aircraft are constantly being developed and introduced on the world airlines. Today safety is taken for granted and the decision concerning whether to fly with one airline or another is taken on grounds of not which is the most likely to reach the destination but which is the most convenient, comfortable and offers best over all value for money. It is the same for an anaesthetic. Patient comfort, overall convenience and cost effectiveness are the major influences in choosing any given anaesthetic regime for an individual patient. Thus, the emphasis is increasingly on quality of care and in this respect there remains an urgent need for newer drugs which provide for a better standard of operative and peri-operative experience for the patient. The patient need no longer worry about surviving the anaesthetic and the focus is now on ensuring that they waken quickly, in no pain and not feeling sick.

In the past drugs such as diethyl ether provided lack of awareness (hypnosis), muscle relaxation and probably some degree of analgesia as well. In contemporary anaesthetic practice these individual elements – hypnosis, muscle relaxation, analgesia – are provided by different categories of drug not simply a single agent; this approach has been termed *balanced anaesthesia* care, In no single category of drug is a theoretically ideal agent currently available. It is probably true to state that when the molecular site of action of the category of drug in question has been well described the choice of drugs available more nearly approach the theoretical ideal. However, if the basis of action is less well understood the choice is between drugs falling (sometimes far) short of the ideal.

Thus, the molecular site of action of muscle relaxant drugs is known to be on the alpha subunits of the nicotinic acetylcholine receptor, the structure and three dimensional shape of which is well categorised. Not only are there a variety of drugs available which interact with this receptor in a very specific manner (and with no others) but reversal agents are available to antagonise their action at the end of surgery in order to reverse residual (unwanted) paralysis. Drugs such as cisatracurium are of intermediate duration of action and therefore easily reversible, but have few if any side effects if administered in normal clinical doses. This indicates their specificity of action at the nicotinic receptor. Indeed, cistracurium and its predecessor, atracurium, are examples of designer drugs which were developed specifically to fit the appropriate site on the acetycholine receptor and also to undergo non-enzymatic, organ independent elimination by Hofmann degradation. Their duration of action is therefore independent of liver and renal function, making them safe and predictable to use in a variety of pathophysiological circumstances.

However, the molecular basis underlying hypnosis and analgesia is much less well understood than that of the (physiologically) simpler process of neuromuscular transmission. Within the categories of hypnotic and analgesic drugs there remains much further work needed on the physiology of hypnosis and analgesia before major therapeutic advances are likely to occur. Hypnotics, be they injected such as propofol or inhaled such as isoflurane, have relatively low therapeutic indices (<3 or 4) and a host of side effects related to depression of the cardiorespiratory systems. These deleterious properties imply that the drugs are targetting a variety of receptors and ion channels some of which are responsible for the unwanted side effects (e.g. calcium channels in the heart). Similarly, potent analgesic drugs all have deleterious side effects including respiratory depression and nausea and vomiting. Clearly, the currently available potent analgesics are acting at receptor types or subtypes in a non-discriminatory manner. Although there have been great advances in our understanding of the nature of the opioid receptor with the potential for sub-types of individual receptors, at present an agonist mediating analgesia but lacking respiratory depression has not been developed (Respiratory depression is a particularly serious side effect in the postoperative period when patients are often not as well monitored as they are in the operating theatre or recovery room.).

As an example of some of the general principles concerning drug development it is instructive to review the development of inhaled anaesthetics from the early first generation drugs such as ether to those that have recently entered clinical practice such as sevoflurane and desflurane.

Development of Third Generation Inhaled Anaesthetics

Until the 1930's all new inhaled anaesthetics had been introduced by chance observations. This is true of the 19th century drugs, such as diethyl ether and nitrous oxide, as well as drugs introduced in the early 20th century such as ethylene (the botanists Crocker and Knight observed in 1908 that ethylene would stop carnation buds from opening and it has subsequently been used to ripen fruit as well as stopping patients eyes from opening during surgery!).

The 1930's was a decade of expanding understanding in structure activity relationships of drugs. The first anaesthetic to be introduced with some understanding of structure activity principles was divinyl ether when in 1930 Leake and Chen in the USA discovered its anaesthetic properties when trying to combine the advantages of diethyl ether and ethylene (2). Soon after this, again in America, Booth and Bixby wrote a paper in which the following prophetic words are included "*A survey of the properties of 166 known gases suggests that the best possibility of finding a new non-combustible anaesthetic lays in the field of organic fluoride compounds. Fluorine substitution for other halogens lowers the boiling point, increases stability and generally decreases toxicity*". (3)

Therefore, it was known for about 20 years before fluorinated anaesthetics were first introduced in the 1950's that they would more nearly approach a theoretical ideal, and be non-explosive – an important consideration with the increasing use of surgical diathermy. However, fluorine is the most electronegative of all the elements and in the 1930's and 40's a sufficient variety of compounds could not be synthesized and no fluorine containing anaesthetic was developed. Here is a parallel with the current situation with respect to potent analgesics. There is a good notion of what needs to be done but current molecular biology techniques are such that the goal remains elusive. It was two military goals during world war 2 that resulted in major advances in fluorine chemistry thus laying the ground for modern day anaesthesia. Hydrogen fluoride is necessary for the production of high octane aviation fuel and uranium hexafluoride in vital for the production of uranium 235. Thus, after the war fluorine chemistry was sufficiently advanced for a variety of compounds to be synthesized and fluoroxene and halothane were introduced into clinical practice in the 1950's. The latter represented a significant pharmacological advance compared with the first generation of non-fluorinated anaesthetics because it was non-flammable and (relatively) non-toxic. It was the first clinically useful second generation inhaled anaesthetic and was introduced with an understanding of structure activity relationships coupled with advances in fluorine chemistry. Thus the means to accomplish what was already theortically appreciated came from advances not remotely associated with with the goal in question. The principle that major advances may derive from unexpected developments remains true today and because of this foretelling the future is a hazardous excercise. (The future is an opaque mirror. Anyone who tries to look into it sees nothing but the dim outline of an old and worried face (4).)

Further advances have taken place and the third generation agents, desflurane and sevoflurane, are halogenated entirely with fluorine and have very low solubilities. They are thus are characterised by rapid uptake and elimination, properties generally perceived as being advantageous epecially with the increase in day case surgery.

However, with the possible exception of xenon, these two agents probably represent the maximum development that can take place in the absence of a clearer understanding of precisely how and where inhaled anaesthetics act within the CNS and what, at cellular level, is mediating their unwanted side effects.

Therefore, the major principles governing the search for new anaesthetics are: (i) Any new drug must have as safe a profile as possible. Anaesthesia in the 1990's is very safe and any new drug or technique must not detract from this safety. (ii) Major advances in terms of hypnotics – injected or inhaled – as well as analgesics – require further clarification of the molecular physiology of hypnosis and of acute and chronic pain.

Although this symposium has summarised the current advances taking place in anaesthesia it is important that the concept of quality of care in not lost in the minutiae of molecular biology. Quality, in a clinical context, is difficult to define and what the clinician thinks is important may not be what most concerns the patient. A good example of this emerged from a study carried out by Orkin in America recently (5). He asked patients to categorise, in order, the things they were most concerned about in the postoperative period. I suspect that the majority of clinicians would anticipate patients to be most concerned about postoperative pain. However, less than 10% of Orkin's patients mentioned this whereas about 75% were most concerned about postoperative sickness. It is clear that as in all branches of medicine we must learn again the value of listening to our patients in order to most appropriately invest resources in developing new drugs.

References

1. Campling E A, Devlin H B, Hoile R W, Lunn J N. National Confidential Enquiry into Periopertive Deaths 1992/1993.
2. Leake C D, Chen M Y. Anesth Analg 1931; 10: 1.
3. Booth H S, Bixby E M. J Ind Eng Chem 1932; 24: 637.
4. Bishop J. New Y J Amer 1959; 14 March.
5. Orkin F K. Anesth Analg 1992; 74: S255

Subject Index

Printed in the United Kingdom
by Lightning Source UK Ltd.
111838UKS00001B/76